Knowing Science

Knowing Science

Alexander Bird

OXFORD
UNIVERSITY PRESS

OXFORD
UNIVERSITY PRESS

Great Clarendon Street, Oxford, OX2 6DP,
United Kingdom

Oxford University Press is a department of the University of Oxford.
It furthers the University's objective of excellence in research, scholarship,
and education by publishing worldwide. Oxford is a registered trade mark of
Oxford University Press in the UK and in certain other countries

First Edition published in 2022

Published in the United States of America by Oxford University Press
198 Madison Avenue, New York, NY 10016, United States of America

British Library Cataloguing in Publication Data
Data available

Library of Congress Control Number: 2022942402

ISBN 978–0–19–960665–8

10.1093/oso/9780199606658.001.0001

Printed and bound by
CPI Group (UK) Ltd, Croydon, CR0 4YY

For Sebastian, Benedict, and Ishbel,
and for Stephanie, Arthur, Ottilie, and Toby

PREFACE

The central messages of this book can be summed up as saying, regarding philosophy of science:

knowledge good, empiricism bad.

Empiricism has long interested me. My first academic job lasted for ten years at the University of Edinburgh, with an office in the David Hume Tower.[1] The history of empiricism was present to those of us who walked the same streets as Hume and his contemporaries. I lived just off Princes Street at one end, Hume's house was just off Princes Street at the other end. My cycle route to work took me past the Mercat Cross (market cross) in Parliament Square, standing where, in the late eighteenth century George III's chemist John Amyatt famously remarked, one could 'in a few minutes, take fifty men of genius and learning by the hand'.

For a philosopher of science, empiricism was also orthodoxy. It was science's foundation in experience that differentiated it from other areas of intellectual activity and which explained its remarkable success. The task for a philosopher of science, if she is a realist, is to show exactly how the great edifice of scientific knowledge, knowledge of a world of imperceptibly small and large entities, of unobservable processes, of long past events and changes to come, is built on such foundations. Not all empiricists felt this could be done—*knowledge* of these things cannot be achieved when our evidential resources are limited to our experiences. In which case the task of the philosopher of science would be to show how science is still an important intellectual achievement, even if it is not quite what one might think it to be. (Some empiricists try to square the circle by adopting an unorthodox metaphysics. Experience can give us knowledge of the world in its entirety—if what the world is made of is experience itself.)

So I had both historical and contemporary reasons to be interested in empiricism. The first large, international conference I organised (with Richard Foggo, now a senior mandarin in the Scottish civil service) was thus entitled *The Legacy of Empiricism*, taking place in September 1996. (This conference was held in honour of George Davie, the historian of Scottish philosophy.) After this conference my thinking about empiricism began to change. A given for me was that the standard picture of science as capable of delivering knowledge (at least sometimes), is right. I had taken David Armstrong's lesson to heart, that if a fancy philosophical argument tells us that X is true, but you have every (non-philosophical) reason to think that X is false, then

[1] The David Hume Tower has since been re-named. I had always felt that having failed to give Hume the chair of moral philosophy, the University of Edinburgh was adding insult to injury by naming a nondescript office block after the great enlightenment thinker and one of the first people to live in the neo-classical splendour of Edinburgh's New Town—especially considering that the university demolished some fine eighteenth-century houses to build that office block. So perhaps Hume would not be too disappointed by the re-naming.

you have good reason to doubt that the philosophical argument is sound. I began to agree with the above-mentioned empiricists who argued that a realist picture of science cannot be supported by empiricism. Yet I could not agree with empiricist anti-realism, let alone with a non-standard metaphysics. It began to dawn on me that empiricism itself was the problem.

At that conference I listened to Timothy Williamson's new paper 'Knowledge as evidence' in which he expounded his now famous view that a subject's evidence is precisely what they know, E=K. (This paper was published the following year as Williamson (1997). Tim was then the recently arrived Professor of Logic and Metaphysics at Edinburgh.) Although he did not particularly present it as such, it struck me that this equation has profoundly anti-empiricist implications. Empiricism and the orthodoxy in the philosophy of science took it that evidence claims must have a very special kind of origin—they must come directly from sense-experience. Consequently they must have a particular kind of content—they can only be about things that can be experienced. And it was this restriction that was the source of empiricism's anti-realist implications. E=K, however, places no such restrictions on what can be evidence—anything that can be known can be evidence. Furthermore, if one is an epistemological externalist, anything can be known with which one can be in the right kind of external connection (e.g. a reliable belief-forming connection).

So the view I found myself developing was one that rejected empiricism and embraced epistemological externalism, and in particular Williamson's knowledge-first approach. This book is the product of that thinking. Some parts of the book, such as the first drafts of 'What is scientific progress?' (Bird 2007*d*) which has become Chapter 3, date back to my time at Edinburgh and owe a great deal to the knowledge-first programme that I was fortunate enough to see developing at first hand. Much of the remainder was developed while I was at the University of Bristol, where I benefitted from many exchanges with James Ladyman and Samir Okasha, and subsequently at King's College London, again learning from my colleagues, Clayton Littlejohn, Julien Dutant, and David Papineau.

Many people have commented on various parts of the book or provided useful advice, information, or discussion. I am very grateful for their generosity. We all have so little time to spare in our working lives, and so I am conscious of the value of the effort put into helping me with this work as well as the value of the comments themselves. In particular I would like to thank Chris Allton, Kabir Bakshi, Helen Beebee, Bill Brewer, Alex Broadbent, Armando Cintora, Julien Dutant, Ludwig Fahrbach, Robin Hendry, Alison Hills, Milena Ivanova, James Ladyman, Tim Lewens, Clayton Littlejohn, Kevin McCain, Hugh Mellor, Samir Okasha, David Papineau, Richard Pettigrew, Stathis Psillos, Mona Simion, Deborah Tollefsen, Peter Vickers, Ralph Wedgwood, Alan Weir, Oscar Westerblad, and Timothy Williamson. Especial thanks are due Ted Poston, who wrote extensive comments on every chapter that have helped improve the book immeasurably, to Roy Sorensen, who suggested its title, and to Ned Campbell, who helped me find the cover artwork. I am grateful to my editors, Henry Clarke and Peter Momtchiloff, for their support, encouragement, and patience.

Peter Lipton and his work have been a key influence on this book. It has been said (Salomon 1945: 596) that the work of Max Weber can be understood as an extended debate with the ghost of Karl Marx. I have often felt that my work on this book has been a debate with Peter Lipton and his memory. (This is an analogy that flatters me and flatters Marx.) I have often wondered what Peter would have said in response to what I have written here.

My thanks are due also to the Warden and Fellows of All Souls College, Oxford, where I was a visiting fellow while writing Chapter 8. I am grateful to the Arts and Humanities Research Council for its support for this project (grant no. AH/I004432/1).

(A typographical note. It is standard to italicize non-English words and phrases. English, however, benefits greatly from welcoming foreign words and making them part of the language. At which point, then, do they no longer need to be marked out by being italicized? I regard some common Latin terms and phrases that are essential parts of philosophical writing to be English terms and phrases too, including what would have been '*a priori*', '*ad hoc*', and '*prima facie*'. On the other hand, 'Does anyone have a priori knowledge?' looks a little odd, and so I have chosen to write 'apriori' rather than 'a priori'. Furthermore, not being overly disposed to radical change, I have not dropped italicization for all Latin words and phrases, even some fairly common ones. Perhaps both the departure from typographical orthodoxy and the failure to implement it completely will irritate some. I hope they will find compensation in the content.)

St John's College
Cambridge
January 2022

CONTENTS

1

INTRODUCTION: SCIENCE AND THE PURSUIT OF KNOWLEDGE

I introduce the two central themes of this book: the importance of knowledge in science and philosophy of science, and the thoroughgoing rejection of empiricism. These themes are illustrated with some brief remarks on the history of science and then through three more detailed examples of modern science from palaeontology, meteorology, and particle physics. I conclude with a synopsis of the remainder of the book.

1.1 Science, knowledge, and empiricism

1.1.1 *Science and the desire to know*

Aristotle tells us:

> All people by nature desire to know.

More recently, Stephen Hawking echoed Aristotle when commenting on the New Horizons probe's flyby of Pluto. 'We explore,' said Hawking, 'because we are human and we long to know.'[1] Science is the principal institutional embodiment of this desire.[2,3]

This introduces the first of two key themes of this book—that in order to understand the epistemology of science we need to pay greater attention to the concept of knowledge. In the next chapter I argue that the aim of science is the accumulation of knowledge. I expand on the central role of knowledge throughout this book.

As an individual, Aristotle's thirst for knowledge inspired him to make remarkable contributions to science, above all in biology. Aristotle was the first to describe the life-cycle of the cicada. He studied the sex-lives of snakes. He noted that dolphins and

[1] Quoted in the *Guardian*, 14 July 2015.

[2] We can desire to know all sorts of things, and we might institutionalize these desires in different ways. Wisden is an institution aiming at delivering knowledge of cricket statistics to fans, but Wisden is not a form of science. So the kind of knowledge that is in question needs to be circumscribed. Hoyningen-Huene (2013) offers an account of scientific knowledge in terms of systematicity that is relevant here. That account understands 'science' in a broad way that includes the humanities and other disciplines that aim at systematic knowledge, rather as 'Wissenschaft' is understood in German. If we want to restrict 'science' to the natural sciences, then further restrictions will need to be made. For present purposes I am more interested in what science has in common with other knowledge-producing institutions, and less in what distinguishes science from them.

[3] Is science essentially social, as this claim implies? Can't it be conducted in a solitary way? I do not think so. Yes, a scientist might carry out solitary fieldwork but this will be pursued against a background of socially produced and acquired knowledge. Perhaps someone without any scientific background could engage in an activity that looks much like science. We might call that activity 'scientific' to mark that fact. But they would be engaging in proto-science at most, not science itself.

Knowing Science. Alexander Bird, Oxford University Press. © Alexander Bird 2022.
DOI: 10.1093/oso/9780199606658.003.0001

whales breathed air, gave birth to live young and suckled them like land mammals, and so he categorized them apart from fish.

Aristotle also recognized the importance of collaboration in the production of knowledge. His pupil, Alexander the Great, arranged for plants and animals and reports to be sent back to Aristotle from the remote places to which he and his armies travelled. Aristotle founded a school, the Lyceum, whose pattern we can recognize today: researchers convey their discoveries to their peers and students through writing and lecturing. The Lyceum had a library, a repository of the knowledge accumulated by Aristotle's predecessors and contemporaries, both from Athens, from elsewhere in the Greek world, and beyond. Theophrastus, Aristotle's successor as *scholarch* of the Lyceum continued his work. It was Theophrastus who introduced Aristotle to biology, which together they studied from Theophrastus's home on Lesbos. Theophrastus added to and systematized Aristotle's discoveries in zoology, while he himself instituted a parallel study of botany. At the Lyceum, he expanded the library and the size of the student body.[4] Around the same time, Alexander's most trusted general and long-time friend, Ptolemy Soter, having become ruler of Egypt, founded the library and 'museum' at Alexandria, the museum being a research institute that brought together scholars from all parts of the Greek-speaking world.

The Greek model of research and teaching was preserved and developed, first in the courts and madrassas of the Umayyad and Abbasid caliphates, during the last third of the first millennium, and thereafter in the universities and monasteries of western Europe. Scholars disagree about the extent to which medieval Islamic and Christian science was dominated by the interpretation and transmission of existing knowledge relative to extending it with new discoveries. While there were many important additions to knowledge in this period, a heavy emphasis on transmission brought with it the danger that errors are preserved with few challenges and thereby become firmly established. Aristotle held that the individuals of many species are created by spontaneous generation. Failing to find the sexual organs of eels, he believed that they originated in river mud. Observing that maggots developed within the rotting flesh of dead animals, he believed that flies, as well as other insects, were also spontaneously generated. This opinion held sway for over 1,500 years.

Seeking to add to existing beliefs is an important mechanism for testing them. Why? For two reasons. First, a natural place to start in trying to extend someone else's discoveries is to repeat them. Did they miss something? If we do things slightly differently, what do we then discover? If I master the technique in the (supposedly) established case, what could I discover by applying it to a new case? By repeating or varying an experiment or by applying it to new conditions, prior claims about its outcomes and what it shows are put to the test. Secondly, old discoveries will be used as auxiliary hypotheses in the attempt to make new discoveries. If those attempts fail, that may lead us to question those auxiliaries. Consequently, without the desire not only to know but also to add to what is already known, it becomes difficult to discern the difference between genuine knowledge and those beliefs that fail to be

[4] For details of the life and work of Theophrastus, see Ierodiakonou (2016).

knowledge (because erroneous or because supported by insufficient evidence or by mistaken reasoning). That it seeks to progress is key to the success of science.

'What Des-Cartes did was a good step. You have added much several ways, & especially in taking the colours of thin plates into philosophical consideration. If I have seen further it is by standing on the shoulders of Giants.' Newton's famous lines come in a letter to Hooke concerning research into the nature of light, carried out by Newton himself as well as by Hooke and by Descartes. Those words have come to represent the idea of scientific progress, the idea that science is a collective enterprise in which scientists add to the edifice upon which their colleagues and predecessors have been labouring. While the metaphor is an old one, it is only in the early modern period that thinkers came to view history and culture—science in particular—in terms of progress. Descartes (1960: 85) himself, in the *Discourse on Method*, having remarked on the state of knowledge in medicine—that almost nothing is known compared to what remains to be known—invites 'men of good will and wisdom to try to go further by communicating to the public all they learn. Thus, with the last ones beginning where their predecessors had stopped, and a chain being formed of many individual lives and efforts, we should go forward all together much further than each one would be able to do by himself.' Descartes's desire found its expression in the scientific societies of the time, such as the Académie des Sciences (1666) and the Royal Society (1660), and in learned journals, such as the *Journal des Sçavants* and the *Philosophical Transactions of the Royal Society*. Late Renaissance artisans published knowledge of their crafts in the spirit of contributing incrementally to the further growth of the public good or understanding (Zilsel 1945). But it is in the work of Francis Bacon that progress as an ideal for science is first promoted as such. In *The Advancement of Learning* and *New Organon*, Bacon lays down the growth of knowledge as a collective goal for scientists, knowledge which would lead to social improvement also. And in his *New Atlantis* (1627) Bacon articulates a vision of a society centred upon Salomon's House, a state-supported college of scientists working on cooperative projects, and the model for the learned academies founded later in the seventeenth century.

1.1.2 *Empiricism*

Aristotle's claim about the desire to know comes from the following passage in his *Metaphysics*:

> All men [people] by nature desire to know. An indication of this is the delight we take in our senses; for even apart from their usefulness they are loved for themselves; and above all others the sense of sight. For not only with a view to action, but even when we are not going to do anything, we prefer sight to almost everything else. The reason is that this, most of all the senses, makes us know and brings to light many differences between things. (Translation from Lear 1988: 1).

Aristotle's evidence for what he reports in his biological works comes from the senses, primarily sight. As a zoologist he was a careful and acute observer. He made important contributions to comparative anatomy as a result of dissections on a wide range of animals. Ptolemy encouraged anatomy in Alexandria, permitting human dissections (which was *very* unusual in the ancient world). As a result Herophilos was able to describe in detail the various components of the eye, the nervous system, and the difference between the veins and the arteries, among a wide range of other anatomical phenomena. Galen built on Herophilos's work, adding to it by making a vast number of his own observations, particularly from the dissection of monkeys.

This process corrected many prior errors. Having dissected hens' eggs to discern the order of the development of the organs in an embryo, Aristotle thought that the heart was the first and was therefore the most important organ, and so the seat of thought and emotion. In fact what he was observing was instead the spinal cord, and it was Herophilos who demonstrated that the brain was responsible for intellectual and emotional activity. The process also introduced new errors: Galen mistakenly said that the human mandible has two bones and that the sternum has seven parts, the result of making observations on dogs and apes not on humans. These errors became ossified as a result of the medievals' failure to attempt to corroborate or to develop Galen's work. Error was transmitted alongside knowledge when learning was received primarily through books and lectures rather than acquired through personal experience. Not until Andreas Vesalius, determined to carry out his own dissections, cut open a human cadaver in 1641 to see for himself, was it discovered that the human mandible is a single bone and that the sternum has only three parts. Aristotle's belief in the spontaneous generation of insects was finally refuted only by Francesco Redi's experimental observations published in 1668.[5]

When the errors of Aristotle and Galen among many other falsehoods were discovered and corrected by employing the senses, it was natural for some philosophers to endow sense-perception with an essential role in science, and indeed in thought and knowledge more generally. It is tempting to see the empiricism of Hobbes and Locke as the unofficial philosophy of the scientific revolution and by extension of science in general. Empiricism in its simplest form holds that sense-experience is (i) the origin or foundation of all knowledge, and (ii) the origin or foundation of all concepts. In the hands of Mill, and then allied to the positivism of Comte, empiricism developed as a philosophy of science in particular, reaching its most sophisticated forms in the logical empiricism/positivism of the early twentieth century (I articulate some of these developments in a little more detail in Chapter 6.) While the most extreme forms of empiricism have long been abandoned by philosophers of science, a moderate but nonetheless significant strand of empiricism remains as a standard component of many philosophies of science: observation is the foundation of scientific knowledge—evidence is observational in nature—and observation is primarily a matter of sense-perception.

This gives us the second key theme of this book: empiricism. Or, more precisely, the rejection of empiricism. Not all evidence is observational. And not all observation is perceptual. While empiricism might have been a useful philosophy for science when most scientific observation was perceptual, and when there was little solid

[5] Even then versions of the view remained into the nineteenth century.

underlying theory, it is nonetheless a fundamentally flawed philosophy. This is true not only of the more overt empiricism of anti-realists such as van Fraassen but also of the covert empiricism of most realists.

1.2 The structure of modern science

By looking at ancient and early modern science I have introduced two central themes of this book: knowledge (pro) and empiricism (con). In this section I look at three examples of modern science that illustrate some of the ideas and arguments I present in the chapters to come.

1.2.1 *Palaeontology—the dating of Zinj*

On the morning of Friday 17 July 1959, Mary Leakey, working at Bed I in Tanzania's Olduvai Gorge, uncovered a part of cranium and then a jaw with teeth that she regarded as having a 'hominid look'. Over the following three weeks, further cranial fragments were discovered that allowed for the reconstruction of most of a skull. This individual, known as 'Zinj' or 'Nutcracker Man', is now classified as a belonging to the hominin species *Paranthropus boisei*. This was a hugely exciting discovery. Stone tools were found at the site. And so, as Mary's husband Louis held, it looked as though this would be the earliest example of a tool-using human ancestor.

How early? The dating of Zinj was important. It was intrinsically interesting for a cultural anthropologist to know for how long humans and their ancestors have been using tools. It was also crucial to know how old the fossil is in order to know its evolutionary relationship to other human-like species, whose remains had been discovered at numerous locations in East and Southern Africa. In his article 'Finding the world's earliest man', Leakey (1960) estimated Zinj's age at 600,000 years.

In fact, Louis Leakey was wrong on both counts. Zinj was not an early human nor was he 600,000 years old. The geophysicists Garniss Curtis and Jack Evernden were able to date the volcanic tephra deposits from Bed I using a method developed by their Berkeley colleague, the physicist John Reynolds. This gave an age of 1.75 million years, almost three times older than Leakey's estimate. Reynolds had developed techniques in mass spectrometry to determine the ratios of various isotopes in a sample. The element potassium is found in tephra (as in many other rocks). The radioactive isotope of potassium, ^{40}K, decays to yield the stable argon isotope, ^{40}Ar. Since argon is absent from the molten rock from which tephra is formed, the argon found in a deposit will come exclusively from the decay of ^{40}K since the solid rock was formed. The exponential decay of ^{40}K is well understood by physicists. And so by measuring the ^{40}K and ^{40}Ar in a sample from the Olduvai tephra, the age of the rock, and so of the fossils found in it, can be calculated.

Such is the basic idea behind radiometric dating. Radiometric dating of rocks depends on a number of assumptions, some general and some specific to the sample being measured. For example, one general assumption is that the radioactive decay rate of ^{40}K is unaffected by the physical state of the sample. That is, changes in temperature and pressure (which rocks may well experience) do not affect the rate of decay. This assumption is well-founded, theoretically and experimentally. An assumption specific to the sample is that all the ^{40}Ar in it has come from the decay of ^{40}K in the sample. Sometimes this is not true, for example, when a xenolith is incorporated into the magma when the sample was created, or when there is contamination of the sample by atmospheric argon. Geological examination of the rock stratum from which the sample is taken can eliminate the first source of error. The second is handled by measuring the amount of ^{36}Ar (which can only come from the atmosphere). That will quantify the ^{40}Ar in the sample that is atmospheric in origin (since in air ^{40}Ar is 295.5 times more plentiful than ^{36}Ar), which we can subtract from the ^{40}Ar measured in the sample to give the ^{40}Ar that is radiogenic in origin (i.e. comes from decaying ^{40}K).

The use of radiometric dating in order to provide evidence in palaeoanthropology and in palaeontology more generally is an example of the epistemic dependence of one field in science on a quite different field—in this case, the dependence of palaeontology on geology and the dependence of geology on physics. A good palaeontologist will understand the principles of radiometric dating. She will nonetheless have to trust her physicist and geologist colleagues regarding the theory (of radioactive decay) and the details of its application (e.g. the half-life of ^{40}K) as well as the various assumptions made (those discussed above among others). As Hardwig (1991) emphasizes, trust is unavoidable. While this is not blind trust, it is nonetheless also that case that the palaeontologist is not in a position to avoid this trust by verifying the relevant theory, details, and assumptions for herself.

In 1960 Louis Leakey retracted his claim that Zinj was from an ancestral species of *Homo sapiens*. Two years after discovering Zinj, Mary and Louis discovered another hominin, whose species became known as *Homo habilis*. *H. habilis* had a markedly larger brain than *P. boisei* and was a tool user. Indeed it was likely that the tools found near Zinj were in fact made by *H. habilis*. Clearly the dating of these and other fossils is important evidence for the correct understanding their evolutionary relationship. If *H. habilis* and *P. boisei* are contemporary but distinct species and *H. habilis* is an ancestor of modern humans, then *P. boisei* is not an ancestor. If they are not contemporary then one species could be the ancestor both of the other species and of modern humans.

Note that I said the dating of *P. boisei* is evidence for its not being an ancestor of *H. habilis*. What makes the dating *evidence*? As I shall argue it is the fact that radiometric dating is sufficiently reliable that we *know* when *P. boisei* existed. For now, we may reflect on what the perfectly natural claim, that the dating of *P. boisei* is evidence, implies. First, it is *not* a natural way to talk to say that the dating of *P. boisei* is an observation—one does not observe the age of objects, nor the date on which past events occurred. So this is an instance of evidence in science that is not an observation. Is this evidence derived from observations? The dating of *P. boisei* at 2.4 –1.4 mya is inferred from a large number of radiometric measurements of samples from multiple sites. Even one of these measurements is not ordinarily called an observation. Each measurement is calculated from the output of a mass spectrometer. One might call the output of the mass spectrometer an 'observation', but even that is not normal usage. That output is a data set of measurements of electric current (proportional to the number of ions) versus the mass/charge ratio of the ions. Furthermore, even if we do wish to call this an observation, it is not essentially perceptual. The data are typically presented to the researcher in a chart (often simplified as a stick diagram) in which relative intensity (or abundance) is plotted against the mass/charge ratio. This will often be the first and perhaps only perceptual contact the researcher has with the data (the researcher *could* also print out a table of the raw data). Of course, this perceptual contact is not with the sample. Nor is contact with a simple perceptible causal consequence of sample—it is not like looking at a photograph of a solar eclipse or of a cloud chamber. Rather, the recorder chart or stick diagram is a convention-governed representation of numerical data. Furthermore, the data may undergo statistical processing (resampling, normalization, baseline correction) before being represented in a chart. So even if we insist on calling this process 'making an observation', it remains the case that this observation is not perceptual in any epistemologically interesting way. I shall expand on this conclusion, that not all evidence is observational, much less perceptual, in this chapter and in Chapters 5 and 6.

1.2.2 *Meteorology—automating data collection and analysis*

A remarkable feature of modern science concerns the *automated* manner in which data are collected and analysed, a prime example of which is found in meteorology. In brief: the process is this: automatic weather stations generate basic meteorological data; the data are collected centrally via telecommunications networks, where it is then analysed by high-performance computers; the analysed data set is then made available to users.

We may start by focusing on just one part of the Japan Meteorological Agency's (JMA) system for gathering and using meteorological data, the Automated Meteorological Data Acquisition System (AMeDAS). AMeDAS has 1,300 observation stations across Japan, of which 1,100 are unmanned. These stations automatically measure precipitation, and of these 690 also measure air temperature, wind, and sunshine duration. The stations are integrated with 156 more sophisticated meteorological observatories that monitor air pressure and temperature, wind speed and direction, precipitation, visibility, weather, and other factors. Of the latter, 68 are manned and 88 are unmanned and fully automatic.[6]

The automatically obtained data are sent via telephone landlines to AMeDAS's data collection server, and thence (along with any manually entered data) to the principal data storage and sharing server. That data set undergoes processing, including basic quality control checks (such as real-time internal consistency of the data). The processed AMeDAS data are then disseminated through two routes. One route takes the data to a data archive system, which also involves quality control checks (these checks are non-real time checks of the data for spatial and temporal consistency). This archival data and the statistics gained from it are made available to users, including governmental, commercial, and academic users. The second dissemination

[6] This information was obtained from the website of the Japan Meteorological Agency (www.jma.go.jp/jma/indexe.html) in 2012—the scale of automation has increased since then.

route is via the JMA's 'Computer System for Meteorological Services' (COSMETS). COSMETS has two components. The Automated Data Editing and Switching System (ADESS) gathers data from AMeDAS and other JMA data collection systems, such as from satellites, earthquake observation stations, weather radar, and so forth. ADESS provides automatic analysis and editing of the data, which are then made available to the various users, in various formats. Much of the data is available directly over the internet. ADESS data are also fed to the second component of the COSMETS computer system, the Numerical Analysis and Prediction System (NAPS), which is used for weather forecasting. Although nowcasting and forecasting are important uses of the data collected and processed by AMeDAS and the JMA's other systems, another use is the testing of meteorological and climate models.

The processes used by the JMA may be compared with those employed by the United Kingdom's Met Office in its infancy. The first national meteorological service was set up in 1854 under the leadership of Robert FitzRoy, who when captain of the *Beagle* was accompanied by the young Charles Darwin. FitzRoy, 'Meteorological Statist to the Board of Trade', arranged for standardized instruments to be issued to ships' captains, who sent their observations to FitzRoy. In 1861 he instituted fifteen coastal stations, which by the 1870s sent their data using the newly invented telegraph. Those observations were made by human observers, who would use instruments for making some measurements, such as air pressure and wind speed, while others were made by estimating by eye, such as cloud cover or whether conditions at sea were stormy. Likewise, statistical analysis of the data was carried out by human computers.

All the processes carried out by FitzRoy's observers and computers are now performed without human intervention. AMeDAS collects and analyses its data automatically. Nonetheless, some of the principal uses are broadly similar: forecasting and testing weather models. For example, FitzRoy was able use his data to show that Stephen Saxby's lunar weather system was erroneous, while the first published weather report and forecast was printed in the *Times* on 1 August 1861. The automated collection and use of data for testing scientific models, once carried out by humans using their perceptual and intellectual capacities, is now widespread in science. One might wonder, then, which parts of the scientific process require specifically human input. Perhaps, one might think, the process of hypothesis generation still demands human imagination—though some systems are attempting to automate even this in specific, constrained domains (King *et al.* 2009). In any case, the close linkage between scientific evidence and human perception that is essential to the empiricist picture of science is entirely absent from the AMeDAS data generation process and those like it elsewhere in science, as our next example shows.

1.2.3 *Particle physics—the ATLAS experiments*

The Large Hadron Collider (LHC) is an integrated array of experimental devices designed to perform experiments to test various theories in fundamental physics, such as elements of the Standard Model (including, most famously, the hypothesis of the Higgs mechanism). The central elements of the LHC are the particle accelerator (two parallel beam pipes), the supercooled magnets used to control the particle beams, and the seven different particle detectors, designed for distinct experiments. The ATLAS detector, which surrounds one of the intersection points of the two beams, is not a single detector but an array of multiple detectors of several different types: the Inner Detector, itself an array of thee distinct particle detectors/trackers; the Calorimeters (an electromagnetic calorimeter and a hadron calorimeter); and the Muon Spectrometer. (These are supplemented by a set of additional 'forward' detectors, at a distance from the principal ATLAS detector(s).) The different types of detector within ATLAS function in different ways. There are, however, two basic types of detection system at the back end of the detectors: scintillation counters and proportional chambers. Scintillation counters use the fact that particles hitting a crystal of (e.g.) germanium will emit photons. The photon signal can be enhanced by a photomultiplier and the resulting light detected by sensors similar to those in a digital camera, which use the photoelectric effect to generate an electron signal. The latter is used by the detector's electronics to generate data that can be stored and analysed, again just as in a digital camera. Proportional chambers use the ability of particles to ionize a gas, separating an electron from an atom. An electric field accelerates the electron towards the positive plate (anode) ionising further atoms, resulting in an avalanche of electrons, thereby generating a substantial electrical signal from a single initial ionizing event (a Geiger–Müller tube is a basic proportional chamber). The electrical signal is again used by the detector's electronics to generate data. The ATLAS calorimeters are designed to measure the energy of incoming particles by absorbing them in a layer of steel, with which they interact, then using both scintillation tiles and a liquid argon proportional chamber to detect the results of the interaction.

The quantity of data produced by these processes is vast, so vast that it cannot possibly all be stored. Various 'triggers' automatically decide which data are to be retained and which is to be discarded. The first level of triggering is built into the hardware of the calorimeters and detectors. Another level operates further downstream and is implemented by software. The purpose of the triggering processes is to keep and store only the data that might be of scientific significance—these triggering decisions are highly theory-laden. The (by now highly-processed) data that are finally stored, a small proportion of that originally generated, is then available for analysis. CERN's teams of scientists will be able to interrogate that data set and use it to test theories, such as the Higgs hypothesis. Two things may be noted about that process of testing. First, much of that process will itself be automated, for example in the use of statistical software to provide statistical analyses of the data. Secondly, the scientists' interaction with that data is entirely mediated by computers and their software. So an individual scientist's perceptual interaction with the experiment of which she is a part is limited to what she sees when looking at her computer screen (and perhaps what she hears when conversing with her colleagues). And the most important thing that she might see on that screen is not a representation of the original data set itself but a line of text or a spreadsheet displaying the outcome of a statistical analysis of that data.

The entirely electronic processes of the collection and analysis of data by ATLAS, and its use to test hypotheses, stand in stark contrast to the corresponding processes in the experiments that are the ancestors of the ATLAS experiments. The core pieces of equipment in these earlier experiments were cloud and bubble chambers. The cloud chamber was invented in Cambridge in 1911 by Charles Wilson. A particle passing through a cloud chamber ionizes the gas in its path, thereby leaving a trail of vapour condensing around the ions. The bubble chamber, invented forty years later by Donald Glaser at the University of Michigan, contains a liquid at near boiling point—passing through this, a particle will leave a trail of gas bubbles. In both cases the trails must be photographed and the photographs analysed by hand. First, scientists would examine thousands of such photographs by eye trying to find tracks that might be scientifically significant. Then the scientist will measure the curvature of the trail in the photograph. (The curvature is the result of the particle passing through a magnetic field. Measuring that curvature allows the scientist to determine the charge-to-mass ratio of the particle.)

The visual interaction with (the effects) of the particle that was essential to these earlier experiments is entirely superseded in the ATLAS experiments. In August 1932 Carl Anderson first saw in a photograph the track of a positron traversing a Wilson cloud chamber. That moment was both epistemically and, one could argue, perceptually significant. Even if Anderson did not strictly see the positron, his specific perception of the distinctive left-handed curved trail in the photograph was a fairly direct effect of that positron. The content of that perception is directly correlated with its epistemic value in confirming Paul Dirac's prediction of antiparticles such as the positron. What one sees is an image of the cloud chamber divided horizontally by a lead sheet, and a track that is gently curved to the left below the lead divider and more sharply curved above it. One can see therefore that one is looking at a track left by a particle that is travelling upwards and which then loses speed because of the lead plate, spiralling more steeply as a result. One can see that the particle must be positively charged because the curve is to the left, not to the right. It is this key fact (in tandem with others) that confirms Dirac's hypothesis. In contrast, when a scientist at CERN sees the text of her spreadsheet (or perhaps a graphical representation of its statistical analysis) indicating a statistically significant outcome that confirms the hypothesis under test, the content of that report (or graphical representation) is epistemically important—it its evidence that confirms the hypothesis. There is, however, no corresponding distinctive perception, as there was in Anderson's seeing the curve in his cloud chamber photograph. Perceptually speaking, the CERN scientist's experience is uninteresting.[7]

The ATLAS experiments, like AMeDAS, demonstrate the role of automated processes in modern science, in particular in the collection and analysis of data. The

[7] One might reasonably think that the two cases are not quite as far apart as described, concluding that even Anderson's perception wasn't that significant after all—Anderson's exact tasks of identifying and interpreting significant photographs could have been automated if he had equipment and software that is available nowadays. I am sympathetic to that line of reasoning, which I consider in more detail in Chapter 6.

ATLAS experiments also show, in common with identifying the place of Zinj in the hominin family tree, the division of labour in modern science. The latter required the cooperation of scientists from widely differing fields: anthropology and physics. Likewise, the ATLAS Collaboration includes not only theoretical physicists and particle physicists but also solid-state physicists, computer scientists, and statisticians; furthermore, within these sub-fields further specializations are required. The intellectual skills and knowledge (not to mention the technical expertise) demanded by the ATLAS experiments, it is obvious, could not be mastered by any individual. A highly structured division of labour is needed, held together by relations of trust and mutual dependence.

1.3 Knowing science

The two central claims of this book, introduced earlier in this chapter, are the following:

- The key concept we need in order to understand science is *knowledge*. For example: science aims at knowledge; scientific progress is the accumulation of knowledge; evidence is that which can lead to knowledge, and therefore is itself knowledge.
- Even moderate empiricism is too much empiricism. Sense-perception is not fundamental to scientific knowledge. Insofar as observation is crucial for science, observation is not essentially perceptual. Epistemologically more important than observation is *evidence*, which is not limited to the observational let alone to the perceptual.

Further important claims, to be articulated, illustrated, and defended in the remainder of the book, are:

- Science is irreducibly social in a strong way: scientific knowledge, in a collective or social sense, does not even supervene on the mental states of all individuals.
- In order to come to know the truth of an explanatory theory, all plausible alternatives must be ruled out by the evidence: for knowledge, *Inference to the Best Explanation* must be *Inference to the Only Explanation*.
- If an instance of Inference to the Best Explanation falls short of eliminating all plausible alternatives, then we can have a rational preference short of full belief, based on the relative explanatory virtues of the theories. A scientist's sense of these virtues is acquired by exposure to successful exemplars.
- These rational preferences can be expressed as probabilities: the Bayesian calculus formalizes the assessments of plausibility-in-the-light-of-the-evidence.
- Metascientific knowledge is knowledge about the status of science in the light of evidence regarding its successes and failures. We have very limited metascientific knowledge. Arguments such as the No Miracles Argument and the Pessimistic Meta-Induction do not trump the first order scientific arguments. There is no global metascientific knowledge to be had from these arguments. Any metascientific knowledge is limited and local in nature.

1.4 Synopsis

Chapter 2 argues that the aim of science is the production of scientific knowledge. The focus of the chapter is on showing that the epistemological status of that aim is knowledge rather than truth. It is important first to say what exactly it is for science to have an aim. I draw an analogy between science and belief, an analogy supported by the work of Durkheim and Parsons, and so I look at the literature on the aim of belief. I defend a teleological account of the aim of belief. But that account finds teleology not in the intention of a believer, but in the function of belief. Strictly, the function of belief is to produce true belief, but our cognitive capacities are functioning correctly only when they are producing true belief in the right way. And true belief produced in the right way is knowledge. So the norm for belief that flows from the functionalist account is that a properly functioning cognitive system is one that produces knowledge. Applying the analogy leads to the conclusion that the aim of science is knowledge.

Chapter 3 concerns scientific progress. If the aim of science is knowledge, and an activity makes progress insofar as it achieves its aim, then scientific progress is a matter of accumulating more scientific knowledge. I provide an independent argument for this conclusion by contrasting it with two other possible aims for science: truth and problem-solving (as the latter is understood by Kuhn and Laudan). I show that in episodes where there is accumulation of truth but not of knowledge or where there is accumulation of problem-solutions but not of knowledge, then we do not regard such episodes as progressive. Our judgments of progress track gains in knowledge.

The picture of science developed in Chapter 2 says that science is an institution, and that scientific knowledge is some kind of social or collective knowing. How should we understand this idea of science as social knowing? Chapter 4 provides an answer. It argues that social knowing is not reducible to individual knowing. It is not even supervenient on individual knowing or indeed on any set of individual states. In particular, social knowing, in the case of science at least, is not a matter of a kind of group commitment to a proposition. Rather, a better model for science's social knowing is distributed cognition. I develop the connection between distributed cognition and the Durkheimian conception of science articulated in Chapter 2.

Having articulated a framework whereby we understand science as an institution that aims at producing knowledge, I then turn to the epistemological details of how that end can be achieved. Put very bluntly, science uses evidence to make inferences concerning the truth of propositions of scientific interest. So we need to take an interest in evidence and in the inferential processes applied to that evidence. What then is evidence? This is the question addressed by Chapter 5. The way to understand evidence is to see it as performing a function in inference. We have seen in Chapter 2 that the aims of belief and of science are knowledge. Correspondingly, an inferential process is successful when it produces knowledge. That depends both on the inference pattern being satisfactory and also on its input, the evidence, being satisfactory. So we can characterize evidence functionally—that which when given as input to a satisfactory inference pattern delivers knowledge. I show that only knowledge satisfies this functional characterization of evidence, confirming Williamson's E=K (Williamson 1997).

The importance of E=K is that it says that any proposition can be evidence, if it is known. The picture we gain from that is this. Science produces knowledge, and that knowledge can be used as evidence in some further process of scientific inference, which produces knowledge, and so on. At one stage of inquiry a proposition *p* is the object of inquiry, and not an evidence proposition. As science progresses we can come to know that *p*, and so *p* may be used as evidence in some further inquiry. For Kepler, the elliptical orbit of the planets was not evidence but was inferred from evidence. But for Newton, the elliptical orbits were evidence (for his theory of gravitation). I show that the way that scientists use 'evidence' conforms to this pattern.

In particular evidence is not limited to any particular type of content or origin. It is not the case that all evidence is observational. More importantly, it is not the case that all evidence is perceptual. The idea that observation itself is perceptual is an enduring legacy of empiricism that even scientific realists typically buy into. I reject these ideas in Chapter 6. Instead I propose a functional characterization of observation: observation of *x* supplies the basic evidence regarding *x*. So there is still a connection between observation and evidence, but none between observation and perception. And this again accords with the scientists' claims that they observe such imperceptible things as magnetic field strengths and collisions between black holes.

In Chapters 5 and 6 I will have looked at one component of a successful scientific inference, evidence. In Chapters 7 and 8 I therefore look at the other component, a satisfactory inference pattern. It is widely held that Inference to the Best Explanation (IBE) is an important form of inference in science, especially for the confirmation of explanatory theories that introduce new entities and properties. IBE ranks explanatory hypotheses according to their explanatory goodness—what Lipton (2004) calls 'loveliness'. In certain conditions the loveliest of the hypotheses is inferred to be the actual and therefore true explanation. In Chapter 5 I argue that this is insufficient for knowledge. For knowledge we need our evidence to eliminate all but one of the hypotheses. I explain why this 'Inference to the Only Explanation' (IOE) is possible (refuting the Duhem–Quine thesis along the way). Nonetheless, we do use IBE to give ourselves a rational preference between hypotheses. That this should be a truth-conducive practice depends on loveliness having some correlation with the truth. I use Kuhn's concept of an exemplar to explain how this can be the case. The rough idea is that learning with exemplars gives scientists a quasi-intuitive grasp of what counts as a lovely explanation in a field. Should we have learned from true exemplars, then the standards of loveliness thus acquired will tend to be truth-conducive. This will be enough for a rational preference, although not enough for knowledge.

In those circumstances where there is insufficient evidence for an IOE and we have to fall back on IBE, the top ranked (i.e. loveliest) hypothesis cannot be regarded as known but must be accorded some lower epistemic status. It is appropriate then to give that hypothesis a probability of less than 1 to represent this. In Chapter 8 I look at various problems in reconciling IBE with Bayesianism. My version of what Lipton calls the 'heuristic' approach to this reconciliation parts with standard Bayesianism in certain respects. It rejects subjective Bayesianism in favour of a naturalistic super-objectivism. In the latter, probabilities represent plausibility in the light of the evidence. Scientists' priors regarding hypotheses are given by a quasi-intuitive sense of loveliness that they acquire from exemplars, as described in the previous chapter.

I have argued that science aims at knowledge. But does it in fact achieve that aim? Realists give a blanket positive answer to that question and anti-realists give a blanket negative answer. While particular realists and anti-realists will add various caveats and restrictions to their otherwise blanket answers, their answers are intended to be general philosophical-cum-historical theses about science as a whole. In my final chapter, I reject all such global, metascientific theses. For example, I reject the global argument that the success of science is best explained by the truth of its theories. On inspection this argument adds nothing to the first-order scientific arguments adduced in favour of each individual theory. The Pessimistic Meta-Induction and a more sophisticated version of the No Miracles Argument both require there to be a single scientific methodology shared across all of science for their global arguments to gain traction. But the idea of a single scientific method is a myth.

2

THE AIM OF SCIENCE

This chapter argues that the aim of science is the production of scientific knowledge. I set up a parallel between the institution of science and the cognitive faculty of belief. The aim of belief, I argue, is knowledge, and likewise the aim of science is knowledge. The 'aim of belief' and 'the aim of science' are to be understood not in terms of intentions but instead in terms of 'norms of correctness'. These norms of correctness stem from the fact that belief and science each have a constitutive function—true belief and true accepted hypotheses. A belief or a hypothesis meets the norms of correctness when it is the product of a properly functioning cognitive or scientific system. A system that has the function of producing truth may malfunction and produce falsehood or accidental truth. When such as system is functioning properly the product is knowledge.

2.1 Introduction

I began the preceding chapter with a quotation from Aristotle's *Metaphysics*: 'All people by nature desire to know', and proposed that this desire is embodied in the institution of science. In this chapter I put more flesh on this connection.

I argue that the constitutive aim of science is the production of (scientific) *knowledge*.[1] More precisely, I argue that the aim of science is knowledge rather than truth. The core argument for this knowledge-aim of science draws a parallel with the aim of belief. Later in the chapter, I discuss what it is for belief to have an aim. In my view, belief has an aim because it is governed by norms of 'correctness', and these norms arize from the fact that our belief-producing (i.e. cognitive) systems have a *function*. A correct belief is one produced by a properly functioning cognitive system. And a belief produced by a properly functioning system is knowledge. So beliefs aim at knowledge. In order to draw the same conclusion for science, I first propose, following Émile Durkheim and Talcott Parsons, that we should understand institutions such as science as having functions. Science, in particular, has a cognitive function.

In passing I note that linguistic evidence and our intuitions about particular cases also support the conclusions of the functionalist arguments I develop. For example, our intuitions about which developments constitute successful, progressive science

[1] Clearly not just any old knowledge falls within the purview of science. I will not pursue here the question of what distinguishes scientific knowledge from non-scientific knowledge. Paul Hoyningen-Huene (2013) has a plausible and interesting account of the difference in terms of systematicity. My concern in this chapter is with the epistemological status of the aim of science, principally whether science aims at (scientific) knowledge or (scientific) truth.

Knowing Science. Alexander Bird, Oxford University Press. © Alexander Bird 2022.
DOI: 10.1093/oso/9780199606658.003.0002

(thereby indicating what amounts to meeting the aim of science) suggest that knowledge but not anything less than knowledge meets the aim of science. Likewise, our intuitions about correctness of belief also do not take beliefs that fail to amount to knowledge to be correct.

Before commencing I note that one may reasonably complain that the options of knowledge and truth I have mentioned as the possible aims of science are not exhaustive. For one might think that science aims at understanding. Or that it aims at solving problems (where that is understood in a manner that does not imply truth). Or one might hold that science aims not at truth but at something related to truth, such as empirical adequacy (truth regarding the empirical consequences of a theory) or verisimilitude (nearness to the truth). In Chapter 3 we will look at these views of the aim of science in connection with the progress of science.

2.2 The aim of science

I claim that the aim of science is the production of knowledge:[2]

(Aim-K) Science aims at knowledge.

This claim has two elements. First that science has a single aim at all. And secondly that the aim of science must be characterized epistemically, specifically as *knowledge*—science does not aim at anything less than knowledge. In particular *truth* is not enough, so I reject:

(Aim-T) Science aims at truth.

(I consider other possible aims for science—problem-solving and understanding—in Chapter 3.)

The argument of this chapter is for the most part orthogonal to a discussion of the *scope* of science's aim. Van Fraassen (1980) argues that science aims at the empirical adequacy of its theories. This in effect restricts the scope of science to the observational. Nonetheless, my question still arises. Does science aim at theories with observational consequences that are true? Or does it aim at theories whose observational consequences are known to be true? My discussion applies to deciding between those alternatives, should one be sympathetic to constructive empiricism. (In Chapter 6 I explain why one should *not* be sympathetic to constructive empiricism, or any other kind of empiricism for that matter. But for now that is an distinct discussion.)

I will first argue that it is plausible to think of science as having a single aim and that this is an epistemic aim. I then note the suggestive linguistic evidence that science is concerned with producing knowledge rather than true belief. I next ask, what counts as success in science? Science is successful when it produces knowledge, not merely true belief. Hence, science has the aim of producing knowledge.

This will motivate and support the conclusion of Section 2.3, in which I elaborate on the idea of science as an institution and argue that it has a cognitive function.

[2] Others who endorse the knowledge aim of science include Barnes (1991) and Cohen (1980). I discuss Barnes's view in Chapter 2.

2.2.1 *The multiple aims of science and of scientists*

The aim or aims of science are not to be read off directly from the aims of individual scientists. The aims of the latter are varied. Some scientists are motivated by the nature of the work itself. Others may seek the fame of a Nobel Prize. Yet others might be looking for the practical applications of their research and for lucrative patents. Some but not all may have lofty ideals of expanding our knowledge of the world. None of this conflicts with the claim that there is a single constitutive aim of science. Institutions have aims that may well differ from the aims of those working in them. The armed forces have the aim of protecting the nation from its external enemies; the health service has the aim of maintaining and restoring the health of citizens. Yet soldiers and sailors, nurses and doctors may have a multitude of intentions in taking on these roles, whether it is seeking camaraderie, an exciting profession, personal fulfilment, or a reliable income. A commercial enterprise might have the aim of making profits for its shareholders, without any of its employees having that aim—so long as it is structured in such as way as to incentivize employee behaviours that in combination are disposed to be profit-making.

The complexities of the institution of science and its interactions with other institutions are nevertheless consistent with its having a single constitutive aim. The army puts on parades for visiting heads of state, the health service builds hospitals, and both employ large numbers of people. But none of these is a constitutive function of the institution in question. An army that didn't parade would still be an army. But if it did nothing towards defending the nation, because it lacked troops for example, it would not be an army any longer. The health service builds hospitals because these are necessary for treating the sick. If a health service built hospitals as an end in itself and didn't treat any patients in them, then it would be a health service in name only. So organizations and institutions can have constitutive aims or functions, even though there are many different kinds of thing that they actually do. The things that they typically do are means to the ends that are their constitutive aims (building hospitals), though they need not be (parading for dignitaries). Likewise science and scientific organizations can engage in activities that are not constitutive of science and its aim. Much science produces technology for various social or commercial purposes. So not only have scientists studied the mechanisms of photosynthesis, but they have also sought to replicate photosynthesis artificially in order to generate energy from sunlight in an environmentally sustainable way. Although science in fact does much of the latter kind of thing, it could fail to do anything of the sort and still be science. Most ancient science and a large proportion of current science has no practical or technological application.[3] Although it was part of Bacon's depiction of Salomon's House that through its acquisition of knowledge it contributed to the economy of Bensalem (the mythical nation he describes in *The New Atlantis*), it was a

[3] Douglas (2014), however, argues that dissolving the pure/applied distinction for science allows for a better understanding of scientific progress.

complaint levelled at the Royal Society that it failed to live up to this expectation.[4] But that failure was not thereby a failure of the Royal Society to do science. On the other hand, had the Royal Society made no effort to extend knowledge, then it would not have been a scientific society. The Royal Society could have instructed Robert Hooke to buy equipment and to devise experiments purely for the ephemeral entertainment of its fellows, with no intention of understanding why such things happened or of generalizing from them to propositions about nature. Had it done so then whatever it would have been doing would not have been science.

2.2.2 *Science as knowledge—the linguistic evidence*

We should first note the linguistic evidence in favour of the view that the nature and aim of science concern knowledge. The *Oxford English Dictionary* gives a definition of science as 'The kind of organized knowledge or intellectual activity of which the various branches of learning are examples'. This definition is a disjunction of two ways of conceiving science: as a set of propositions and as an activity. Regarded as a special set of propositions, it is notable that the *OED* describes science as organized knowledge, not as organized belief. With this propositional conception in mind, it is much more common to use 'knowledge' than 'belief' to describe science. 'Science is organized knowledge', also asserted by Herbert Spencer (1911: 61–2), has a naturalness not shared by 'science is organized belief' or even 'science is organized true belief'.[5] If science, under the propositional conception, is a body of knowledge, then how should we think of science when conceived of as an activity? The intuitive answer is that science-as-an-activity aims at achieving or augmenting science-as-a-set-of-propositions. That is, the activity of science aims at producing or adding to the body of scientific knowledge.

Etymology also provides linguistic evidence for the knowledge view of the aim of science. In English and the Romance languages, the relevant terms are derived from the Latin *scire* meaning *to know*: *science* (English), *la science* (French), *la scienza* (Italian), *la ciencia* (Spanish). In Germanic (including Nordic) languages, the terms for science all relate to those languages' words for *know*: *wetenschap* / *weten* (Dutch), *Wissenschaft* / *wissen* (German), *vetenskap* / *veta* (Swedish), *vísindi* / *vita* (Icelandic);

[4] Bacon (1620) writes, 'We have three [fellows of Salomon's House] that bend themselves, looking into the experiments of their fellows, and cast about how to draw out of them things of use and practise for man's life.'

[5] Such instances may be multiplied, as exemplified by these definitions: from the UK's Science Council 'Science is the pursuit and application of knowledge and understanding of the natural and social world following a systematic methodology based on evidence' (https://sciencecouncil.org/about-science/our-definition-of-science/); from the American Physical Society 'Science is the systematic enterprise of gathering knowledge about the universe and organizing and condensing that knowledge into testable laws and theories' (https://www.aps.org/policy/statements/99_6.cfm); *Encyclopaedia Britannica* 'Science, any system of knowledge that is concerned with the physical world and its phenomena and that entails unbiased observations and systematic experimentation. In general, a science involves a pursuit of knowledge covering general truths or the operations of fundamental laws' (https://www.britannica.com/science/science) (the above all accessed on 10 August 2018). While we may quibble about the details of any of these definitions, it is notable that how natural it is to use the term 'knowledge' to characterize science, and, in the last definition, the goal of science.

likewise in the Finno-Ugric languages: *tiede / tietää* (Finnish), *tudomány / tudás* (Hungarian); and in Irish science is *eolaíocht* and I know is *Tá eolas agam*. We find the same pattern beyond Europe: *bilim / bilmek* (Turkish), *uloom / ilm* (Arabic). The Farsi (Persian) word for science, *elm*, derives from the Arabic for knowledge, *ilm*. The words *madda* and *episteme* denote science in modern Hebrew and modern Greek respectively, but refer to knowledge in the ancient versions of those languages. Not all languages make the link between science and knowledge directly—in most Slavic languages the word for science is linked to the word for teaching or for learning, though Czech (*věda / vědět*) and Croatian (*znanost / znati*) are like the Germanic languages in this respect. Like Polish, Russian, and Serbian, Chinese links science to learning, rather than to knowledge *per se*. But in no language of which I am aware is there a link between the words for science and for truth.

Linguistic evidence provides no knock-down argument. But it is highly suggestive. Supporters of the truth view of the aim of science need not deny that science delivers knowledge (although some do). They can maintain that those who aim at truth will want to use reliable methods of achieving that truth and when they do, their beliefs will be knowledge. That makes knowledge a kind of by-product of science. If that is the case it is odd that our words for science are so closely related to our words for knowledge, but never to our words for truth. Why should our words for science derive from its by-product? It is much more plausible that those words link to knowledge because knowledge is the essence of science.

2.2.3 *Progress and knowledge*

When is science successful? What counts as good science? Because the aim of an activity sets conditions for the success of that activity, answering those questions will provide an insight into the aim of science.

Imagine a scientific community that tests its theories by an unreliable method (such as astrological divination). By using that method it comes to believe theory T; coincidentally T is true. Does that count as successful science? In 1903 and René Blondlot and a good number of other French scientists believed that he had discovered a new kind of ray, hitherto unknown to physics—N-rays.[6] In due course it became clear that these scientists were subject to a form of self-deception, in large part fostered by a patriotic desire to further the reputation of French science. There are of course no N-rays. Consider this variation on the actual story:

> **Lucky Blondlot** It turns out that there are in fact rays corresponding to Blondlot's theoretical description of N-rays. Call this theory 'T'. But this is a pure coincidence—Blondlot's methods and reasoning gave him no good reason to think that there are such rays. So Blondlot correctly believed significant novel truths of the form: there are rays with the property ϕ.

In this story Blondlot has no justification for those beliefs and so they do not count as knowledge. Would Blondlot's science have been successful under those circumstances? Would it have met the minimum standards for good science?

[6] See p. 47 for more on the Blondlot case.

It seems clear that Blondlot's science would not have been good or successful science in such a scenario. So the truth of T is not enough. Clearly what is lacking is adequate justification. Good science requires not just that our theories be true but that we have adequate grounds for believing them, that our methods are reliable, and that they do not rest on self-deception or other irrational causes. The best explanation for those additional conditions is that they are required for science to be knowledge.

In Chapter 3 I will argue that science progresses when it accumulates knowledge.[7] If scientific progress were a matter of increasing truth or truthlikeness, then, under the imaginary scenario where T is (coincidentally) true, the French scientific community at first made progress by believing in T. Furthermore, when it gave up belief in T, it then regressed. But that is not correct. The community made no progress by coming to believe T on unreliable grounds. And giving up an unreliably tested theory is not regressive—it is, if anything, progressive. Again, the best explanation of this verdict is that progress requires not true belief but knowledge. Because T was not known to be true, adding a belief in T was not progressive and giving up a belief in T was not regressive. So scientific progress is the accumulation of knowledge rather than truth. Progress in some activity is related to the aim of that activity thus:

> If an activity A aims at goal X, then A makes progress insofar as it gets closer to achieving X or does more of X or does X better.

So if science aimed at truth, then science would be making progress by adding to truth or truthlikeness. The latter is false, as we have just seen. By modus tollens therefore, it cannot be that the science aims at truth. On the other hand, if science aims at knowledge, then scientific progress would be the accumulation of knowledge. So we should conclude that science aims at knowledge rather than truth.

2.3 Science as a cognitive institution

Although the institution of science is complex and multi-faceted, I have argued that it is plausible that it has a constitutive aim, the production of knowledge. In this section I examine in more detail what it is for science to have an aim. I first sketch reasons for thinking that science should be regarded as an institution that is itself the possessor of knowledge and furthermore that in virtue of science knowing things, societies, whose members are mostly not scientists, can know the same things. These conclusions support (but are not essential to) the argument that follows, that science has a cognitive *function*. Drawing on sociological theory, I propose that science stands to society much as a cognitive faculty stands to an organism; the accepted theories of science thus correspond to belief. This sets us up for a discussion of the aim of belief in the following sections. The aim of belief, I will argue, is knowledge, thus supporting the conclusion that the aim of science is knowledge,

[7] See also Bird (2007*d*, 2008, 2016); and see Rowbottom (2008, 2010), Cevolani and Tambolo (2013), Niiniluoto (2014), Dellsén (2016) for critical responses.

2.3.1 *Science as a knowing institution*

Groups can be the possessors of knowledge and belief in a way that does not reduce to a summary of the corresponding propositional attitudes of the members of those groups (Gilbert 1987; Tuomela 1992). Indeed the knowledge of a group does not even supervene on the mental states (including knowledge) of the group's members (Bird 2010*c*). Such groups include scientific groups, such as a research team. They can also be larger groups, such as a scientific community defined by field. Groups or collectives that are epistemic agents can be larger still. The institution of science as a whole can be regarded as such a collective entity. As such it is often the subject of knowledge attributions, particularly of the form 'Science knows that *p*'.[8] Furthermore, in virtue of the discoveries of science, we make statements of the form 'We know that *p*'.[9] One might think of the relationship between 'science knows' and 'we know' as a case of testimony. Statements such as 'Science tells us that the universe is over 13 billion years old' would suggest this. But, I think, given certain conditions regarding the close relationship between science and the rest of society, this relationship is constitutive. That is, in virtue of science itself knowing that *p* and science having a certain role in society, society at large knows that *p*.

Here is a related case. We say 'North Korea knows [*or* the North Koreans know] how to build an atomic bomb'. Yet, only a restricted group of scientists and engineers has any direct part in this knowledge.[10] Normal North Koreans and even leading members of the North Korean government are not members of this restricted group. Yet they are included in the collective (the North Korean nation or people) to whom the attribution of knowledge is made. So we have a case where a broader collective can be said to know something in virtue of the fact that some sub-collective knows something, without there being any transfer of knowledge by testimony.[11]

The usual relationship between science and society is not like that between North Korean nuclear weapons engineers and the North Korean people. In the usual case, scientific knowledge is typically made available so that individuals and organizations in wider society can access and make use of it. Of course, most individual members of a society will not be able to understand let alone themselves make use of the

[8] Examples of science as the subject of knowing (or not knowing): James Franklin's (2009) book *What Science Knows and How It Knows It*; 'Why science will never know everything about our universe', article in *Forbes* <www.forbes.com/sites/startswithabang/2017/02/10/why-science-will-never-know-everything-about-our-universe/#76932ff55513>; 'What science knows about why people are gay', article in *Newsweek* <www.newsweek.com/what-science-knows-about-why-people-gay-751927>; '9 things science knows about baseball' article in *Discover* <discovermagazine.com/galleries/zen-photo/b/baseball-science> (the above sites all accessed 12 February 2018).

[9] For example: 'Science, evolution, and what we know: 2018 edition', article in *The Huffington Post* <www.huffingtonpost.com/entry/science-evolution-and-what-we-know-2018-edition_us_5a5bccb6e4b0a233482e0cb7>; 'What we know', American Association for the Advancement of Science website on climate change <whatweknow.aaas.org>; 'How do we know what stars like our Sun are made of?' BBC website <www.bbc.co.uk/earth/story/20160128-how-do-we-know-what-stars-like-our-sun-are-made-of> (the above sites all accessed 13 February 2018).

[10] And even then they only have this knowledge collectively rather than individually.

[11] If there are concerns that this case is about knowing *how* then we can replace 'knows how to build an atomic bomb' with 'knows that *p*' for some proposition *p* that is central to atomic bomb making.

esoteric discoveries of science. Nonetheless, they are connected to those discoveries in numerous ways. They may be consumers of products designed using science-based technology or as citizens subject to government policies devised on the basis of social science research. The North Korean example can be seen a special case where the North Korean people (in the view of its government) benefit from the social need for military power that is met by possessing knowledge of nuclear technology. The connections are not just through technology and applied science. We draw on science as consumers of popular science in books, newspaper articles, and television programmes. And in principle those with sufficient resources and aptitude can study science so that they can directly access the knowledge that science produces. Given the multiple ways in which the knowledge that science produces affects or is available to wider society, for many purposes we can regard the 'we' in 'we know that *p*' as encompassing all or most of humanity.

2.3.2 *Science as an institution with a cognitive function*

According to a conception of institutions derived from Émile Durkheim (1893, 1894) and Talcott Parsons (1961), we should see society as analogous to an organism. Institutions are like the organs and physiological systems of an organism. And like the organs of an animal or plant, institutions have functions. And such functions can be constitutive of the organs in question. A wing has the function of enabling flight. And it has that function essentially. That function determines which part of an animal is its wing. We can identify the wings of an albatross and the wings of a bee as wings, because they are analogues even though they are not homologues. That is, they are both wings because they perform the function of enabling flight.[12] Likewise, what determines the identity of an institution, such as the law or the army or the higher education sector, is the function it performs in society.

Herbert Spencer (1874) suggested analogies between particular functional systems in animals and in societies, for example between the central nervous system and the government, since both played an overall regulatory function, coordinating the interactions of other organs/institutions. While no one would suggest that every animal organ has a social analogue or vice versa, it is nonetheless possible to see some parallels as sufficiently strong that we may hypothesize that they belong to the same functional type. Below I shall discuss the cognitive faculties of a person, how they are integrated, and how their function is to produce knowledge (or true belief). In parallel, we can say that various social institutions and practices perform cognitive functions. That is, they serve to collect, generate, and distribute information of a certain kind. Governments, for example, need information in order to carry out their functions. Therefore governments have created various organizations whose

[12] It is *not* that they occupy the same locus in the structure of the two organisms: they do not. A bat's wings and a bird's wings and a human arm are homologues, because they are evolved from the same structure in a common ancestor. Whereas a bee's wing has no evolutionary ancestry in common with those wings; it is a wing nonetheless. Penguins, therefore, do not have wings. Zoologists describe their forelimbs as flippers, as in a dolphin, since the function they perform is that of propelling and guiding the animal through water. They are vestigial wings. But vestigial wings are no more wings than a retired postman is a postman or someone's ex-spouse is still (thereby) a spouse.

function is to collect that information, from spy agencies to census bureaus. Non-governmental organizations, such as businesses and charities, likewise have divisions whose function is to create or collect relevant kinds of information.

So there are formally created institutions or sub-components of institutions with cognitive functions. Science as a whole, however, is not a formally created institution. Science has grown up more or less informally, across national boundaries, inspired in large part by the curiosity of individuals. That organic development of science does not preclude it from having a social function. Above I characterized science as the institutional embodiment of the human desire to know. The desire to gain knowledge from others, as well as the prestige associated with displaying one's own knowledge, led to the exchange of knowledge and thence, as Kuhn (1962) describes, also to the sharing of methods and values. This leads to the development of an institution, that is to say a group whose members show a degree of social cohesion; they are bound together by mechanisms of what Durkheim called 'solidarity'. On the one hand there is organic solidarity arising from a division of intellectual labour—scientists depend on the results produced by other scientists.[13] And on the other there is mechanical solidarity—scientists share values and methods, and a conception of themselves as scientists (or, earlier, 'natural philosophers' or 'men of science'). Although it is itself an informal institution, science from the very beginning has been pursued in formal institutional settings, from Aristotle's Lyceum and the Mouseion at Alexandria through the madrassas and universities of medieval Europe. And states took an interest, supporting and creating scientific institutions of their own, both for reasons of prestige as well as in the hope that the knowledge generated might be of practical benefit—as increasingly it came to be.

As we shall see, the outputs of a properly functioning cognitive faculty of an organism are varied in the use to which they are put: knowledge as an input into practical reasoning, leading to action; as an input to theoretical reasoning, leading to further knowledge; knowledge to be stored in memory (short term or long term). Likewise the outputs of science are varied in use: knowledge for practical purposes, as evidence for theorizing, and to be added to the store of shared knowledge. Nonetheless, a common nature underlies the variety of the uses of these outputs: the outputs all ought to be knowledge, and will be if the systems in question are functioning correctly.

In conclusion, then, institutions in a society have functions just as faculties and organs within an organism have functions. While there need not be any specific parallels at a greater level of detail, it turns out that there are faculties and institutions which have a cognitive function. The commonality between faculties with cognitive functions and institutions with cognitive functions is not an accident: the fundamentality of information collection, production, and processing for any complex biological or social entity makes this inevitable (Burgin 2009). In the next section I therefore look more closely at cognitive faculties in an individual organism.

[13] One does not want to overstate parallels between science and commerce. Nonetheless, it is intriguing that commerce is another largely informal, supra-national institution characterized by division of labour, linking individuals many thousands of miles apart, that developed early on through a process of exchange.

2.4 Belief has an aim

In this section I articulate an approach to answering the question 'what is the aim of belief?'[14] In the next section I continue the argument to show that the aim of belief is knowledge.

The key claims of this section are these:

- The normativity implicit in 'the aim of belief' is teleological: the teleology of belief sets 'correctness' conditions for belief.
- That teleology derives from the fact that cognitive systems have a function.
- The function of a cognitive system is to produce true belief.

Note that I do not claim that the normativity deriving from the functional teleology of belief has any implications for a deontology of belief. That is, they do not tell us (not directly anyway) what we ought to believe.[15]

2.4.1 *Teleology and the aim of belief*

Since Bernard Williams (1973) first used the phrase 'the aim of belief' philosophers have sought to articulate precisely what it might mean. On the one hand, one may take the term 'aim' seriously, as implying a goal or intention. Such is the teleological (or 'teleologist') interpretation. On the other hand one may prefer to understand 'aim' more metaphorically, as referring to a *norm* of belief (Wedgwood 2002*a*; Gibbard 2005). This is the normativist interpretation.

While these are the standard opposing views about the aim of belief, that should not be taken to imply that the teleologist must deny that there are norms of belief. The ordinary term 'aim' implies both normativity and teleology. If a darts player aims at the triple twenty, then he intends the dart to hit that part of the board; the purpose of his throw is to get that score. That is the teleology of 'aim'. If he hits the triple twenty, his throw was successful; if he hits the triple five, then his throw failed. This is the normativity of 'aim'. The two are related, since what counts as success or failure (normativity) of the throw depends on the intention of the player (teleology). Had the player intended to hit the triple five, then hitting the triple five would have been a successful shot.

While the teleology of aim generates norms of success and failure, not all norms come from teleology. A person's behaviour is governed by ethical norms, even if they have no intention of acting ethically. The same goes for constitutive norms—norms that govern the nature and existence of some process. The rules of chess, for example, tell us what chess is and determine whether some activity constitutes playing a game of chess (Wedgwood 2002*a*). Such norms, like ethical norms, allow us to evaluate an outcome independently of a subject's intentions. If a player is bribed to throw a match, they still count as winning if they do what the rules say constitutes winning, despite lacking an intention so to do.

[14] I have previously sketched this functionalist approach in Bird (2007*b*, 2010*b*). Papineau (2013) holds that the 'aim of belief' can be understood in terms of biological function, while Sullivan-Bisset (2017) takes a biological functionalist approach to epistemic normativity. For a related view see Millikan (1993).

[15] McHugh (2014) and Kiesewetter (2017) have argued that a functionalist account of the kind I offer cannot account for the normative demands of rationality.

The norms governing belief *qua* belief set the conditions for the 'correctness' of a belief.[16] So when it is claimed that a belief is 'correct' precisely when it is true, that claim is not as trivial as it sounds (indeed, I maintain that it is false). One might think that belief *per se* does not generate any norms.[17] Or one might think, as I do, that it generates norms beyond truth. If there are such norms these may not only be constitutive of belief—they may also individuate belief (Fassio 2018). That is, they may not only be necessary conditions of belief; they may also be sufficient to distinguish belief from other attitudes. I take constitutive and individuative norms to describe essences. It is the essence of belief that it is governed by these norms. I shall argue that a belief is not correct (does not satisfy all the norms of belief *qua* belief) unless it is also knowledge. I note at this point that the truth norm does not seem to be individuative, since guessing is also subject to the truth norm (Owens 2003). This criticism does not affect the knowledge norm for belief, since guessing is not subject to the knowledge norm. A guess is successful if the proposition guessed to be true is in fact true; it does not need to be known—indeed a guessed proposition cannot be knowledge.

So the debate between the teleologists and the normativists may be framed as one regarding the source of norms—whether they derive from something teleological, such as an intention, or from some other non-teleological source (or are basic).

Intention, however, does not seem to be the source of the norms of belief. I have mentioned that the most widely favoured norm for belief is truth whereas I think the correct norm is knowledge. Often neither truth nor knowledge is the intended outcome of believing. For one thing, our beliefs are not typically subject to our will, as Williams (1973) pointed out; they usually just occur, most obviously in ordinary perception. Since one cannot ascribe an intention to a subject's believing in such a case, it cannot be that a norm is generated by such an intention. Furthermore, insofar as we can influence the contents of our beliefs, we might not always intend them to be true. If Jane fears that she might be captured and subjected to an interrogation with a reliable lie-detector or truth drug, she may intend that certain beliefs she has about sensitive matters should be false, and she might try to bring that about. Wedgwood (2002*a*) holds that even in such cases, where falsity is intended and may even be morally good, it remains the case that the truth norm still applies and the false belief fails to satisfy it. Thus those involved in this debate do not think that the norm of belief can be derived from the intentions of the believer.

For this and other reasons, most discussions of the 'aim of belief' have assumed that the normativity of belief's aim does not arise from teleology, and so have preferred normativism. Nonetheless, teleology may still be the source of the norms of

[16] One might draw a distinction between *standards* of correctness and *norms*, as does Sullivan-Bisset (2017), on the ground that norms generate oughts whereas not all standards do (see also Bykvist and Hattiangadi 2013). If that is a valid distinction, then I am concerned with standards of correctness when I talk of norms.

[17] Papineau (2013) denies that there are any distinctive norms of belief. But he takes a norm of belief to be something that generates an obligation on a potential believer, whereas I take a norm of belief to be a standard of correctness, generated by the (biological) function of belief-producing systems. The latter view is neutral on whether such norms also lead to obligations on believers.

belief, for not all teleology is a matter of intention. Velleman (2000) asserts the thesis that 'belief is an acceptance regulated in ways designed to ensure that its content is true'.[18] We can understand this as expressing a teleology that is independent of intention. Consider someone who has accepted Pascal's conclusion in his famous wager. Like Jane, he intends to believe something that he in fact strongly doubts (that God exists). Pascal's advice was to ensure that he is exposed only to the evidence (including the testimony of theists) that favours belief in God. So although his intention aims at falsity, he exploits the fact that his belief-forming mechanisms are, as Velleman says, oriented towards ensuring that their contents are true (e.g. they are determined by the available evidence). And, clearly, the mechanisms of perception may be regarded as regulating the resulting belief in a way designed to ensure a true perceptual belief. Here 'designed' is not a matter of intention, but a matter of a function brought about by natural selection.

2.4.2 *A functionalist account of belief*

Teleology can be a source of normativity without deriving norms from intentions. The teleology of functions also generates norms. If the pineal gland does not produce melatonin (leading to sleep disorders) then the pineal gland is not functioning correctly; it is not doing what it ought to do. While Aristotle also saw teleology in nature, a naturalistic understanding of biological function is available in the light of our knowledge of evolution, as I discuss below. This fact is, perhaps, under-appreciated by those who reject a teleological approach on the ground that we can fail to intend our beliefs to be true. With the teleology of functions in mind, we should instead interpret 'X is the aim of belief' as meaning 'X is the product of a properly functioning belief-producing (i.e. cognitive) faculty'. On this view the use of the term 'aim' is not entirely metaphorical. For the teleological aspect of 'aim' is genuinely present in the function of a cognitive faculty. (I should note briefly, as I explain in more detail below, that it is not entirely correct to say that the aim of belief is simply identical to the function of a cognitive faculty.)

Organisms have systems that cause them to respond to environmental stimuli. Examples include the various tropisms of plants and simple animals. Chemotropism in plants causes their roots to grow towards beneficial soil minerals. While such organisms can be said to carry information about their environments, just as a simple thermostat carries information about temperature, they cannot be said to process that information, since the information is not stored in a manner that makes it available for any purpose other than the stimulus-response system of which it is a part. More complex animals, by contrast, are able to integrate information from multiple stimuli. In the most sophisticated instances, most notably in humans, but also in higher primates, some other mammals, and some birds, this integration constitutes practical reasoning—using information (consciously or unconsciously) to make decisions about how to achieve goals. Such integration via reasoning requires that the

[18] Velleman's earlier view has developed in the light of criticism by Shah (2003), leading to their joint view in Shah and Velleman (2005). The latter leans more towards normativism. For an alternative teleological view which resists Shah's arguments see Steglich-Petersen (2006).

information in question is represented in a common fashion. Practical reasoning typically involves putting together novel combinations of inputs in the form of perceptual information from several senses and information stored in memory with desires of different sorts to produce an action or intention to act. This would not be possible if the contents of all these states were represented in different ways—if the different systems produced outputs of entirely distinct types. We can therefore regard these information-generating systems as having a common function—the function of producing a certain kind of output that is the raw material for the processes of reasoning. Having the same function, these are systems of a common type: *cognitive* systems or faculties.

Reasoning in this sense may be conscious or unconscious, reflective or unreflective. Even near instantaneous responses to new stimuli can be the results of reasoning, if they are suitably responsive to background states of the subject (it is this that distinguishes them from reflex actions). In other cases the reasoning may be complex and need not lead directly to action. In these cases a process of reasoning may lead to an output that is of the same kind as its input. That is to say, it is an output that can itself be used as an informational input for practical reasoning. Such processes are instances of theoretical reasoning. The capacity for theoretical reasoning is thus another cognitive faculty. The outputs of theoretical reasoning need not be used only as inputs for practical reasoning, since they can be used as inputs for a further process of theoretical reasoning. And the outputs of theoretical reasoning can be stored in memory, either because the process of reasoning is extended in time (short-term memory) or because those outputs might be useful as inputs at some later time (long-term memory). Memory is thus another cognitive faculty—although it does not generate information, its function is also to output information for the processes of reasoning.

The concept of 'function' in use here is the concept of biological function. Just as it is the function of the pineal gland to produce melatonin, it is the function of our cognitive systems to produce a certain kind of beneficial output. Accounts of biological function (Wright 1973; Millikan 1989; Neander 1991; Walsh 1996) link a system's function to biological fitness. The function of a system (organ, faculty, ...) is ϕ if by ϕ-ing that system confers an advantage in a regime of natural selection.[19] By producing *true* beliefs in the manner described above, human and animal cognitive systems give their possessors selective advantages over organisms that lack such systems or which have similar systems but which do not produce truth.[20]

[19] Debates among philosophers of biology focus on whether the selective regime in question is located in the distant past, the recent past, or the present. Which view is correct need not trouble us here.

[20] Sullivan-Bisset (2017) argues that cognitive faculties also have functions other than truth production. Some belief-forming mechanisms, such as those produced by self-enhancement bias and self-deception, can be adaptive because they assist the effective functioning of the person in a different way, aiding 'self-organization'. This does not detract from the claim that the primary function of a cognitive system is truth. Biological systems with secondary functions are common. For example, a secondary endocrine function of the heart is the production of atrial natriuretic peptide (ANP), which instructs the kidneys to reduce sodium reabsorption when there is a sudden increase in blood pressure or volume. Pumping blood is still the constitutive function of the heart—producing ANP is tacked on to this function. Similarly, even if

2.4.3 *An objection: heuristics and biases*

An objection to this argument for true belief as the function of cognitive systems argues that fitness might not be best served by truth. No cognitive system is perfect, and so the fitness conferred by a cognitive system will be determined not just by the value of accurate beliefs but also but the costs of inaccurate beliefs. Furthermore, there are costs associated with having or implementing highly accurate systems, and these too will have an effect on what an optimally fit system does. For example, an important use of cognition for our ancestors (in many ancestral species) will have been to tell us whether a predator is nearby. The costs of a type-II error (a false negative) will be considerable (death, serious injury), whereas the costs of a type-I error (false positive) will be rather less (effort spent in moving away, some opportunity costs in feeding, etc.). Consider circumstances in which a creature's current sensory stimuli are of a kind that is marginally better correlated with the absence of a predator than with its presence. Given the imbalance of costs, fitness will be served by the creature's cognitive system recording the presence of a predator. On the other hand, the system would maximize truth by recording the absence of a predator (since that verdict will be right more often when the stimuli are marginally better correlated with absence). Additionally, fitness will not be served by employing additional time and effort to get the most accurate verdict, since the predator might seize its chance before this process is complete—better to respond quickly to a few salient indicators and beat a hasty retreat if they are present. The heuristics and biases research programme suggests that the demands of fitness lead to belief-forming dispositions that do not maximize the probability of our beliefs being true.[21]

Before responding in detail, note that it is consistent with the claim that our cognitive, belief-producing systems have evolved under pressure from natural selection to deliver true beliefs, that sometimes false beliefs are more advantageous than true ones. The evolutionary claim rules out only its being very frequent that false beliefs are advantageous. If the true belief that one is a slow runner would cause one to freeze in the face of a predator, then the false belief that one is a fast runner will be advantageous. But a system that made most of one's beliefs (and those of one's conspecifics) false would not be advantageous. If a system were to deliver a large proportion of false beliefs, there would be, overall, selection pressures to improve it, or, if it remains unchanged, to disconnect it from behaviour, in which case it would no longer be a system producing *beliefs*.

It is important to distinguish a heuristic from a biased mechanism.[22] A heuristic assists the believer in getting to the truth—I describe the recognition heuristic (Goldstein and Gigerenzer 2002) on p.197. A heuristic is imperfectly reliable because

some belief-forming mechanisms have a non-epistemic function, it remains the case that the function of cognitive faculties is truth.

[21] The nature of these heuristics and biases is contested. See, for example, Tversky and Kahneman (1974), Gigerenzer (1991), Baron (2014).

[22] Although the concepts are distinct, a mechanism can both be a heuristic and a bias. Indeed one might expect heuristics to be sometimes biased. A heuristic will accept a moderate rate of error. If errors of one sort are more costly than errors of another sort, there may be benefits from those errors being biased.

the costs of greater accuracy may outweigh the benefits. That fact does not contradict the proposition that the heuristic aims at truth. Other things being equal (e.g. costs of implementation), a more accurate heuristic has greater fitness than a less accurate one. Nonetheless, an objector might propose that the function of the heuristic is not 'to produce true belief' but is something like 'to produce true beliefs subject to cost constraints'. That is not correct, since functions should be defined in terms of their output (or input-output relationship), and efficiency constraints are not elements of the output. Note in any case that any function is subject to such constraints. Any organ could be better at its function. In many cases it will be that the better version has not evolved because although its better functioning would *per se* be an addition to fitness, its resource needs would reduce fitness elsewhere. For example, it might be that a human heart twice as big would be much better at pumping blood, which might make one better at running. But the large heart might therefore require more calories to operate than (in our ancestors at least) could be gained by the improved ability to hunt that better running provides. Even so, one should not deny, on that ground, that it is strictly correct to say that the heart has the function of pumping blood *simpliciter*.

On the face of it, the objection from biased cognitive mechanisms cannot be dismissed in this way. For a biased mechanism may be more fit than an unbiased one, even when the costs of implementation are the same. The biased mechanism has evolved the way it has, not despite its inaccuracy, but in order to be inaccurate in a certain way. The cognitive mechanism I described at the beginning of this section, which has evolved to produce Type-I errors more frequently than Type-II errors, is just such a biased mechanism.

The answer to this objection starts by reminding us that it is always the *truth* of the animal's belief that contributes to its success. It never benefits from falsely believing that a predator is present when it isn't. And so on biological conceptions of function, which link function to what contributes to fitness, the function of a cognitive system must be to produce true belief. For example, Denis Walsh (1996: 564) proposes that, 'The/a function of a token of type *X* with respect to selective regime *R* is to *m* iff A's doing *m* positively (and significantly) contributes to the average fitness of individuals possessing *X* with respect to *R*.'[23] Since believing what is true will contribute to fitness and believing falsely does not, a cognitive system has the function of producing true belief.

If the function of a cognitive system is to produce true belief, then how has it evolved to have a bias? As we saw, the value of bias arises because a cognitive system is imperfect, delivering false belief on some occasions. So bias is valuable only because cognitive systems may fail in their functions. Some physiological systems are evolved as fail-safe systems, which come into play once a failure of the primary system is detected. For example, the heart has a natural pacemaker in the sinoatrial node; if the signal from the sinoatrial node fails, then cells in the atrioventricular

[23] Walsh's view is a dispositional theory, in that it defines function in terms of current contribution to fitness. Other accounts will emphasize past contribution to fitness. That difference, important though it is, does not bear on the current point.

node will act as a backup pacemaker. The bias of a cognitive system is not a backup, but is a pre-emptive 'modification'—though strictly it is not a *post hoc* modification but a pre-existing feature that has been retained.

Cognitive systems have evolved from stimulus-response systems. The latter do not produce beliefs, just behaviour. In a sense they are biased—such a system might produce flight behaviour out of proportion to the number of occasions on which predators are present. But this is not biased belief, just 'biased' behaviour. The cognitive systems that have evolved from particular stimulus-response systems have in some cases retained the 'biases' of the latter, in particular if they are biases that might be beneficial to fitness for the reasons under discussion. Insofar as these cognitive mechanisms are belief-forming mechanisms they aim at truth; insofar as they are behaviour-influencing mechanisms, they may be biased. The point of the evolution of cognition, the production of belief, over a suite of stimulus-response systems is to decouple the information contained in a stimulus from behaviour. It is decoupled so that this information may be integrated with information from multiple sources, such as the various senses and memory and processes of theoretical reasoning—which is to say that it allows for practical reasoning. If the only influence on the production of beliefs were the benefits generated by this process, then a mechanism would not produce biased beliefs because its influence on behaviour depends on being integrated with other beliefs. The benefits of such a system arise precisely because one cognitive system does not 'know' what outputs another is producing. Nor does it know all the circumstances under which the belief might be deployed. While the belief that one is a fast runner might be advantageous on an occasion when one might otherwise be paralysed by fear when unexpectedly facing a predator, the same belief might lead to dangerous over-confidence when knowingly approaching a predator. Specifically *belief* forming mechanisms evolve to enable us to optimize behaviour in epistemically novel situations which are not well suited to stimulus-response mechanisms. These are precisely the conditions under which there is no benefit to be gained from bias.

2.4.4 *Cognition and the aim of belief*

Belief is subject to norms of correctness. These conditions are constitutive and individuative of belief. So a belief *qua* belief is what it ought to be if and only if it satisfies these norms. The normativist takes this to say all there is to say about the aim of belief. For the normativist the claim 'A belief is correct if and only if it achieves the aim of belief' is trivial. For the teleologist this is a more substantial claim because 'the aim of belief' is construed realistically—the norms arise from the teleology of belief. The teleology of belief, according to my argument, is found in the function of a cognitive system (rather than the intention of an individual believer). Functions are teleological and the source of the norms of belief.

A cognitive system has a function just as other systems of an organism have functions. The concept of function in use here is biological; it is nonetheless also normative. The heart's function is to pump blood. If it fails to do so, it is malfunctioning, it is not doing what it ought to do. Metabolic pathways are biological systems for the synthesis of biologically important molecules. The porphyrin synthesis pathway has the function of producing haem, a key component of blood. If that pathway fails (for example because of an enzyme deficiency), then it will not produce haem, and a serious disease may result. The product of the porphyrin synthesis pathway *ought* to be haem, but sometimes it is not.[24]

So to identify the correctness conditions of belief, we need first to identify the function of cognitive systems. As we have seen, that function is the production of true beliefs as the input into reasoning and action, since it is the truth of beliefs that makes them adaptive. Simply producing a true belief, however, does not exhaust the norms of correctness arising from the function of producing true beliefs, since a true belief might not be correctly produced. A fully correct belief, one that satisfies all the norms of correctness, is one that is produced by a properly functioning cognitive system, as I explain in Section 2.5.2 below.[25]

2.5 Belief aims at knowledge rather than truth

In the preceding section I argued that 'the aim of belief' is to be understood in terms of correctness conditions for belief, and that these are set by belief-producing (i.e. cognitive) systems having a function. This function is to produce true beliefs. It is important to note, however, that this does not mean that the aim of belief and the norm of correctness for belief is true belief. That is because a malfunctioning system can bring about the appropriate sort of output by accident—producing the functionally appropriate output, true belief in this case, is not enough for a system to be functioning correctly. So, I shall argue in this section, the correctness conditions for a belief (the norms of belief *qua* belief) are that it fulfils the function of cognitive systems in the right way. Those correctness conditions are fulfilled by knowledge but not by states less than knowledge.

I precede the main argument concerning the proper functioning of a cognitive system with considerations from the epistemology of Moorean and lottery beliefs.

2.5.1 *The aim of belief—epistemological considerations*

Above, in Section 2.2.3, I argued that our conception of 'correct' science is that it produces knowledge, and that we do not regard accidentally true science as successful science. Likewise, I first argue that our conception of correct believing is not satisfied

[24] For example, a failure in the pathway may lead to sideroblastic anaemia. The pathway produces erythroblasts that are abnormal—megaloblasts—rather than normoblasts. Erythroblasts *ought* to be normoblasts but sometimes, in cases of sideroblastic anaemia, they are not.

[25] This functionalist approach will be attractive especially to naturalized epistemologists. Neta (2007) promotes an explicitly functional account of knowledge, while a similar view is articulated by Kornblith (2002). Bishop and Trout (2004) also develop a naturalistic approach to epistemology, but seek to bypass questions concerning the nature of knowledge. Neta appeals, as I do, to the function of cognitive systems, saying that knowledge is a matter of achieving the 'goal-state' of a cognitive system. He says that the goal-state of a cognitive system can differ between inquisitive and non-inquisitive creatures. It is not clear to me that this distinction is justified by a satisfactory account of function. In any case, I don't think that the goal-state of a cognitive system includes reliability (non-inquisitive creatures) or being reasonably based on adequate grounds (inquisitive creatures). The goal-state is truth; knowledge is a matter of the goal-state being achieved by the proper functioning of a cognitive system.

merely by true belief, whereas it is satisfied by knowledge. This is revealed by certain epistemological phenomena surrounding Moorean beliefs and lottery beliefs.

Consider Elliott who does not know that dogs bark. He believes (Whiting 2013: 188):

(M) Dogs bark, but I don't know that dogs bark.

Elliott's belief (M) is true. But there is something wrong with it. He should not believe (M). If the norm for belief is truth, then since (M) is true, Elliott does satisfy the norm, and so there should be nothing to criticize about the belief (M). On the other hand, if knowledge is the norm for belief, then (M) cannot meet the norm, for the norm would require him to know the first conjunct, that dogs bark. But he does not know that dogs bark. Furthermore, meeting the norm for the first conjunct (knowing that dogs bark) is inconsistent with meeting the norm for the second conjunct (knowing that he doesn't know that dogs bark). So the truth norm does not explain why it is wrong to believe (M) whereas the knowledge norm does explain this.

Now consider Stanley who has one ticket in a million ticket fair lottery. The draw has taken place but Stanley has not been informed of the outcome. Stanley believes (Whiting 2013: 191):

(L) My ticket didn't win.

which is in fact true. Nonetheless Stanley shouldn't outright believe (L). Again the truth norm does not explain this, since believing (L) meets that norm. Whereas the knowledge norm does explain why Stanley should not outright believe (L) since that belief does not amount to knowledge.

If the incorrectness of believing (L) is not apparent, then consider a time shortly before the lottery draw. Stanley correctly thinks:

(P) My ticket might win the lottery.

He also believes:

(L*) My ticket will not win the lottery.

Both are true propositions. However, it would be wrong for Stanley to believe the true conjunction:

(P&L*) My ticket might win the lottery but it will not win the lottery.

If believing (P) and (L*) are both correct in this context, then believing their conjunction should also be correct. But it is not. Since it is clearly alright to believe (P), the problem with (P&L*) must be (L*). If it is wrong to believe (L*) ahead of the draw it is wrong to believe (L) after the draw since Stanley has received no new information in the meantime. Moreover, (P&L*) is itself another case of a true proposition that it is wrong to believe, which the truth norm cannot explain but the knowledge norm can.

If Stanley believes (L) or (L*) he might act on that belief, for example by throwing his ticket away or not making the slightest effort to find out the result of the draw, even if the prize is significant. But those would be irrational actions. Imagine that the lottery is organized by a philanthropist who offers a €1,000,000 prize but sells the tickets for €0.50 each. So Stanley could reasonably buy a ticket on the ground that its expected value is greater than the cost of buying it. If it is reasonable to believe (L)/(L*) and so to act on it by throwing the ticket away, then he is guaranteed to lose money in a sequence of reasonable beliefs and actions.

In conclusion, we have two cases that are explained by knowledge being the norm for correct belief that cannot be explained by truth being the norm for correctness.[26]

2.5.2 *Knowledge and properly functioning cognitive systems*

I have been arguing that belief aims at knowledge. One might argue that it is prima facie just as plausible that belief aims at truth, because, as I have acknowledged, the function of our cognitive systems is to produce true belief. True belief, it would seem, would have the same selective advantage as knowledge.

Timothy Williamson (2000) would deny the last assertion, on the ground that knowing has different behavioural consequences from true belief. For true belief is consistent with belief being formed unreliably. In such cases true belief is less robust than knowledge. While I am sympathetic to Williamson's position on this point, I pursue here a different line of argument.

If the function of a biological (e.g. cognitive) system S is X, there is more to the correct functioning of S than bringing about X. The normativity of function includes both what the function produces and how it produces it. Consequently, the corresponding norm is not merely that the system should produce X. For example, what is the function of the heart? To pump blood. That answer, although correct, does not tell us when a heart is functioning properly. Consider Elsa the electrician. While on a job, her heart stops beating in the normal way. She falls over and her chest touches some live cables she is working on and the electric shock she receives stimulates her heart muscles so that it pumps her blood. In this case her heart has brought about the supposedly defining activity of the heart's function, pumping blood. Yet, this is not a case of a properly functioning heart but is a case of a malfunctioning heart. A properly functioning heart is one whose pumping is brought about in a normal way. (It is up to physiologists to tell us what this is—the electrical stimulation comes from the sinus node, not from accidentally touching live cables.) For this reason, the norms of correctness for belief are not merely that the belief should be true. Those norms include that the belief should be produced as a result of the subject's cognitive system functioning properly

So while it may not be wrong to say 'belief aims at truth', like 'the function of the heart is to pump blood', such statements fail to capture the full normativity of 'aim' and 'function'. When epistemologists discuss the 'aim of belief', it is this normativity they are trying to capture and account for. They are asking for the conditions under which a belief can be said to be 'correct', which is to say that it satisfies all the norms it is subject to *qua* belief. The functionalist construal of 'aim' shows that a correctly functioning heart and a correctly functioning cognitive system are also subject to norms concerning the way in which pumping and true belief are brought about. As a result a correct pumping of the blood is one brought about by the heart's being

[26] But see Whiting (2013) for a defence of the truth norm against these arguments.

stimulated by the sinus node, and a correct belief is a true belief brought about by a similarly proper, normal process.

What then is a true belief brought about by a similarly proper, normal process? An unjustified true belief and a justified true Gettier belief are beliefs not brought about in the normal way. In such cases our cognitive systems are malfunctioning despite producing true beliefs, or are functioning properly but the true belief is not produced by that proper functioning. According to Plantinga (1993) and Boyce and Plantinga (2012), *knowledge* is true belief brought about by a properly functioning cognitive system in an appropriate environment. For our purposes that seems right.[27] If unjustified beliefs and justified true but Gettier beliefs are examples of the products of malfunctioning cognitive systems or a well-functioning cognitive systems in an inappropriate environment, then it is highly plausible that to be the product of a properly functioning system in an appropriate environment, the belief must amount to knowledge.[28]

My view can therefore be summarized thus: it is the essence of a properly functioning cognitive system that, in an appropriate environment, it produces knowledge as its output. I note that this coheres with other views concerning knowledge. For example, John Hyman (1999) argues that S knows that *p* precisely when the fact that *p* can be used by S as a reason for something. For example, knowing that it is raining is a matter of being able to use the fact that it is raining as a reason for my taking an umbrella when I go outside, or as a reason for thinking that this is the third wettest June since 1900. According to my view, the function of the cognitive faculties is just that, to provide a link between the subject and the relevant facts so that they may be used as the inputs (reasons) in practical and theoretical reasoning. It also implies (and explains) the truth of Williamson's (2000: 47) claim that (mere) belief is botched knowing. Knowing is what a properly functioning cognitive system produces. False belief and true belief that isn't knowledge are what is produced by a cognitive system that is not functioning properly, when it botches its job. Finally, since the output of a properly functioning cognitive system is knowledge and such outputs are themselves the inputs into reasoning, it follows that the appropriate input into reasoning is also knowledge. In Chapter 5 I expound and defend the view that we can give a functional account of evidence—evidence is that which is the appropriate input into a knowledge-producing inference (an instance of theoretical reasoning). So the

[27] One does not have to take proper functioning to provide an analysis of knowledge to hold that to meet the norms belief is subject to as the output of a cognitive system, a belief must amount to knowledge. The proper functioning view has been criticized on the ground that proper functioning is not necessary for knowledge, as shown by the Swampman case (Sosa 1993). The right answer to this, if you think that Swampman can in due course have knowledge, is to adopt a current fitness account of biological function (Walsh 1996). Although Swampman's 'biological systems' are not themselves evolved they can participate in evolution, and they do have current fitness and this determines their function. For more on proper functioning accounts of knowledge, warrant, and justification see Boyce (2018).

[28] The proper functioning account of knowledge has also been criticized for being insufficient for knowledge. Supposedly it is subject to its own Gettier-style cases (Boyce 2018). This depends on the details of one's version of proper functionalism. See Bergmann (2006) for another approach.

account I am giving here of the role of knowledge in a cognitive system supports the view that evidence is knowledge (Williamson 1997; Bird 2018).

2.5.3 *Demandingness*

Properly functioning cognitive systems in normal circumstances produce knowledge. The caveat 'in normal circumstances' acknowledges that to produce knowledge even a properly functioning cognitive system requires that the world should cooperate rather than present a well-executed *trompe l'oeil* or elaborate fraud, for example. Now consider conditions that are propitious in this respect, but in which knowledge regarding some proposition is not possible because relevant evidence is limited or contrary. A properly functioning cognitive system will not be disposed to produce belief in these conditions. A person with normal vision presented with a red tomato in good lighting conditions just cannot believe that it is green. A competent scientist who starts with an open mind on whether smoking causes cancer cannot just believe it does not solely on the basis of knowing about a healthy nonagenarian who has smoked heavily for seventy years.

The fact that a properly functioning cognitive system produces knowledge therefore explains the phenomenon that McHugh (2011) calls *demandingness*: the fact that 'you cannot, deliberatively and in full awareness, form a belief in a proposition if you regard your evidence for that proposition as less than *sufficient*, where sufficiency involves more than having better or stronger evidence for the proposition than for its negation' (see also Owens 2003). McHugh is not claiming that beliefs with insufficient evidence are never possible. The exceptions, however, are pathological and clearly involve incorrect means of belief formation, such as self-deception. In cases where the preponderance of evidence is against a proposition, cannot the belief aim account for demandingness? After all, a belief in such circumstances is more likely to violate such a norm than satisfy it. McHugh (2011: 373, fn 9) defends demandingness even for such cases. In fact, I think that a stronger statement of demandingness can be maintained, according to which one cannot form beliefs (deliberatively and in full awareness etc.) when the evidence favours a proposition but only very slightly. I subscribe to this stronger form.

Let us say you have the aim of achieving X. You can do action A with the intention of achieving X even if A has a relatively small chance of achieving X. But you cannot do A with the intention of achieving X if you know that A will not achieve X. For example, you want to go out of your house to the shops and back and you aim to do so without getting wet. You see that the clouds are dark and that rain is quite possible. On the other hand there are brighter clouds on the horizon and these seem to be coming in your direction. If the intention of avoiding getting wet is important you can wait until the skies are clear of the darkest rainclouds, reducing the chances of being rained on. If getting wet is not quite so important you can leave sooner. On the other hand, if it is actually raining now, so that it is clear that leaving your house will inevitably lead to you getting wet, then you cannot leave with the intention of not getting wet. Likewise, if the aim of belief were truth, then one could believe that *p* with that aim even if the evidence is only marginally in favour of its being true that *p*. On the other hand, if the aim of belief were knowledge, then one could not believe that *p* with that aim if the evidence is only marginal. For in such a situation it is clear that belief could not achieve that goal. Belief on the basis of limited evidence, even if the evidence is favourable and the belief is true, does not amount to knowledge. The truth aim of belief cannot explain demandingness whereas the knowledge aim does explain it.

2.6 Conclusion

I have proposed—and this is an idea on which I shall expand in Chapter 4—that we can identify a broad class of 'cognitive systems'. Sophisticated cognitive systems have propositional outputs. An advantage of such systems is that such outputs can be put to multiple uses, including being stored for future use. The cognitive faculty in individual humans that produces beliefs is one such cognitive system, one that is largely (though perhaps not entirely) biological. A functionalist view of social institutions says that institutions are like biological systems in having functions. Science is a social cognitive system.

These systems have functions. And functions generate norms of correctness. A system S may have producing F as a function. That it brings about F is a norm of correctness for the functioning of S. But satisfying just that norm is not the same as S functioning correctly, since S might be malfunctioning and the output's being F might be an accident. The full set of norms for the correct functioning of F require that S produces F as a direct result of S's functioning properly.

What, then, are the conditions under which a cognitive system is functioning properly? When it produces *true* belief *in the right way*. True belief produced in the right way is knowledge. So the norm of correctness for a belief or a scientific hypothesis is that it constitutes knowledge.

I follow that standard view that to ask for 'the aim of belief' is to ask for the norms of correctness of belief. Likewise to ask for the 'aim of science' is to ask for the norms of correctness for scientific hypotheses. While we might intend many things for our beliefs and for science, those intentions do not generate norms of correctness that are constitutive of belief or of science. It is the functional analysis given above that does that.

The close connection between science or belief and knowledge is revealed in a number of ways, as we have explored above. One of these concerns the nature of scientific progress. We make progress in some enterprise to the extent that we fulfil or get closer to fulfilling the aim of that enterprise. As briefly mentioned, the goal of science being knowledge accords with progress in science being a matter of acquiring scientific knowledge

3

SCIENTIFIC PROGRESS

The preceding chapter argued that the aim of science is the production of scientific knowledge. This chapter argues for the corresponding claim about scientific progress: scientific progress is the accumulation of scientific knowledge. This exemplifies the epistemic approach to progress, according to which an episode must meet certain epistemic standards for it to be progressive. I contrast this account with the views that hold that progress is increasing verisimilitude or increasing problem-solving power. These lack an epistemic element (at least when understood externalistically), which make progress too easy. I also contrast the cumulative knowledge account with the noetic account, which links progress to understanding, where the latter is taken to concern explanatory and predictive power. This view is over-demanding, since there are contributions to knowledge that do not involve addition of understanding (or explanatory or predictive power).

3.1 Introduction

Francis Bacon's *New Organon* (1620) has a frontispiece in which ships are seen sailing between two columns into the endless ocean beyond. The columns are the Pillars of Hercules, representing the limits of the Mediterranean sea at the Straits of Gibraltar, whence it opens out on to the Atlantic. Just as Columbus, who embarked at Palos de la Frontera just a short distance from Gibraltar, had discovered new lands unknown to the ancient and medieval geographers, scientists too could discover startling new facts, by taking the route laid down in Bacon's book. The motto on the frontispiece (taken from Daniel 12:4) states:

Multi pertransibunt & augebitur scientia
Many shall go about and knowledge shall increase.

Bacon's motto provides a simple and obvious answer to the question: What is scientific progress? Science, or some particular scientific field or theory, makes progress precisely when it shows the accumulation of scientific knowledge; an episode in science is progressive when at the end of the episode there is more knowledge than at the beginning.

Sir William Bragg (1936: 41), who with his son Lawrence won the 1915 Nobel Prize in Physics for work on X-ray crystallography, endorsed this cumulative knowledge view of scientific progress thus:

Knowing Science. Alexander Bird, Oxford University Press. © Alexander Bird 2022.
DOI: 10.1093/oso/9780199606658.003.0003

> If we give to the term Progress in Science the meaning which is most simple and direct, we shall suppose it to refer to the growth of our knowledge of the world in which we live.

Bragg was far from alone in his opinion—scientists in general take knowledge to be at the core of scientific progress (Mizrahi 2013).[1]

Despite its appeal to scientists untainted by philosophy, philosophers of science have almost entirely ignored this conception of progress in science. Anti-realists have been impressed by Kuhn's demonstration that science does not develop in a smoothly cumulative way. Deprecating truth, they have preferred an account of progress in terms of increasing problem-solving power. Realists too have regarded the cumulative knowledge view as just too naive. Realists have instead sought an account of progress in terms of increasing verisimilitude (truthlikeness, nearness to the truth). More recently, it has been proposed that scientific progress is a matter of advances in scientific understanding.

There are broadly speaking four approaches to understanding scientific progress, of which the cumulative knowledge view and the three alternatives just mentioned are examples:

> (i) The *functional-internalist* approach. Progress is to be understood relative to a certain function of science (such as solving problems), where fulfilment of that function is something that can be straightforwardly assessed from within science, by the scientists involved.
> (ii) The *semantic* approach. Progress is to be understood in terms of truth or some related notion such as verisimilitude.
> (iii) The *understanding* approach. Progress concerns gains in scientific understanding.
> (iv) The *epistemic* approach. Progress is an epistemic phenomenon; it should be understood in terms of knowledge or justification or related concepts.

To help appreciate the difference between these views consider the simple cumulative truth account of progress:

> (CT) An episode in science makes progress precisely when it shows a growth in the stock of scientific truths accepted by the scientific community.

This is the simplest incarnation of the semantic approach. (CT) does not exemplify the epistemic approach because it makes no demand that the scientific community be in any particular epistemic or cognitive relation to the truth of those theories. It would still be progress according to (CT) if the community came to believe a new and true theory even without any justification and so without any knowledge. (CT) would also not satisfy proponents of the functional-internalist approach, because, they will say, it is not straightforward for the scientific community to tell, when it accepts a

[1] And beyond practicing scientists. For example, according to the author Michael Shermer (2002: 31) 'Scientific progress is the cumulative growth of a system of knowledge over time' while Auguste Comte (1970: 13) wrote 'Its [science's] character will be henceforth unchangeable, and it will then have only to develop indefinitely, by incorporating the constantly increasing knowledge that inevitably results from new observations or more profound meditations'.

hypothesis, that it has thereby accepted a true hypothesis. (CT) would not satisfy the understanding approach, since (CT) will accept as progress additions of new facts to those that scientists accept even if those facts do not add to understanding.

In this chapter I promote the naive and simple conception of scientific progress as the accumulation of scientific knowledge, (CK), endorsed by Bacon and Bragg. In addition to this appeal to authority, I present two arguments in favour of (CK). The first draws on the premise that the aim of science is the production of scientific knowledge. The second considers some actual and hypothetical developments in the history of science and shows that (CK) gives a more accurate assessment of whether those episodes are progressive or not than the principal competing theories of progress. The accounts in competition are:

> (PS) (problem-solving): An episode in science makes progress precisely when the relevant scientific community has accumulated more solutions to its scientific problems. Here the criterion for solving a problem is one whose satisfaction (or not) can be straightforwardly assessed by the scientific community. (PS) exemplifies the functional-internalist approach.

> (VS) (verisimilitudinarian): An episode in science makes progress precisely when the scientific theories accepted by the relevant scientific community show an increase in verisimilitude (they are increasingly near to the truth). (VS) exemplifies the semantic approach. (CT), as mentioned, also exemplifies the semantic approach. I concentrate on (VS) since that is the version favoured in the literature.

> (NA) (noetic): An episode in science is progressive precisely when scientists grasp how to correctly explain or predict more aspects of the world at the end of the episode than at the beginning (Dellsén 2016: 72). (NA) exemplifies the understanding approach, *if* understanding is a matter of grasping how to explain and predict.

> (CK) (cumulative knowledge): An episode in science makes progress precisely when it shows an accumulation of scientific knowledge in the relevant scientific community. (CK) exemplifies the epistemic approach.

3.2 Progress and the aim of science

How does progress relate to the aims or goals of an activity? As Niiniluoto (2011) puts it, 'Debates on the normative concept of progress are at the same time concerned with axiological questions about the aims and goals of science.' As we shall see, Laudan and Kuhn think that science aims at solving problems; thus success in so doing is their standard of progress. Oddie (2014) accepts that truth is the aim of inquiry, motivating accounts of verisimilitude and, thereby, the 'verisimilitudinarian' view of progress (see Cevolani and Tambolo 2013). My own view, as I have argued in Chapter 2, is that the goal of science is knowledge; accordingly I hold scientific progress to be the addition of scientific knowledge.

Because a goal can have subsidiary goals as means, those who hold that science aims at truth or verisimilitude can agree that science aims at problem-solving, since solving a problem will be a route (albeit a fallible one) to the truth about the matter in question. Likewise the view that knowledge is the goal of science will imply that problem-solving and truth are legitimate goals for science, since pursuing these will further the pursuit of knowledge. So the difference between the views must turn on differences between these goals as ultimate or constitutive goals.

A simple view of the relationship between having a goal and progress would be:

> (A) If an activity A aims at goal X, then A makes progress insofar as it achieves X.

In Chapter 2 I considered two possible constitutive goals for science: truth and knowledge. To these I add a third, one that is available to the anti-realist: problem-solving. In the light of (A), we can relate those aims to conceptions of progress as follows:

> (Aim-P) Science aims at solving scientific problems. Therefore science makes progress when it solves such problems (PS).
> (Aim-T) Science aims at truth. Therefore science makes progress when when it accumulates more truth (CT).
> (Aim-U) Science aims at understanding. Therefore science makes progress when it gives greater or more understanding (NA).
> (Aim-K) Science aims at knowledge. Therefore science makes progress when it adds to the stock of knowledge (CK).

(A) is, however, just too simplistic a view of the relationship between a goal and progress. A journey may have the goal of reaching a destination; one makes progress by reaching intermediate places. So a more nuanced account says:

> (A′) If an activity A aims at goal X, then A makes progress insofar as it achieves X, achieves more of X, achieves X better, gets closer to achieving X, or promotes the achievement of X (depending on what X is and how it is specified).

Not just *anything* that promotes the aim of X constitutes progress. The large grant awarded to a research team may well promote knowledge, but is not itself scientific progress. On the other hand developing a new technique, such as X-ray crystallography, or refining an old one, so that we have new experimental tools for testing theories and investigating phenomena, does count as a contribution to progress. The difference is that the latter are directly connected to the cognitive goal of science whereas the former is only indirectly connected.

So while the conclusions about progress above, (AIM-P) etc., do give sufficient conditions for progress (according to the proposed aims) they may not give necessary conditions. In particular, in the light of (A′) we may also say:

(Aim-T) Science aims at truth. Therefore science makes progress when it gets closer to the truth (VS).

I argued in Chapter 2 that (Aim-K) gives the correct account of the aim of science. One reason among others was that it best explains scientists' preferences. For example, (Aim-T) does not explain aversion to false belief. When supplemented by an aim of avoiding falsity, the problem arises of balancing the twin aims of gaining truth and of avoiding falsity. (Aim-K) neatly explains why scientists do seek truth (since it is necessary for knowledge), but not in ways that might easily give a false belief (since true beliefs that might easily have been false are not knowledge). (Aim-K) is also motivated by the view that *belief* aims at knowledge. (Aim-K) together with (A) tells us that (CK) is the correct account of scientific progress. (A′) in place of (A) yields:

(CK′) An episode in science makes progress precisely when it shows an accumulation of scientific knowledge in the relevant scientific community, or improves the quality of that knowledge, or gets that community closer to adding to knowledge, or promotes its acquisition of knowledge.

That said, the differences between (CK) and (CK′) are marginal, because the developments that lead to additions to knowledge themselves typically involve adding to knowledge. For example, gathering evidence is important for coming to know the truth of a theory and so by (CK′) is a contribution to progress. In Chapter 5 I argue for this characterization of evidence: evidence is that from which knowledge can be gained by inference. I show that only known propositions can satisfy this characterization. So the acquisition of scientific evidence will count as a contribution to progress by (CK) also.

It is worth noting that an activity can be said to show progress along dimension D, even if that activity does not have any goal, let alone a consciously entertained or publicly agreed one, to which D is related. Progress in the arts, as mentioned in Chapter 1, is like this. We might judge that political institutions have made progress if they are more responsive to the desires and needs of citizens, even if we do not suppose that this is always the aim of those institutions. In such cases we judge progress by some appropriate standard, typically concerning a good that relates to the nature of the activity in question. For example, an artist might aim at producing a certain kind of aesthetic experience in an audience. An appropriate standard for judging progress in the arts is therefore the expansion of the possibilities for producing aesthetic experiences. In the preceding chapter I noted that a collective enterprise may make progress and may have a goal even if it is not the intention of any individual to contribute to that progress or to promote that goal. The artist may be focused solely on the experience of his or her audience and have no specific intention of widening the expressive possibilities of the art form. Nonetheless, if the work in fact does broaden the range of expressive possibilities, then he or she may thereby be contributing to artistic progress. A business may have the aim of making a profit, even though most of its employees (perhaps even all of them) do not have this as an individual goal.

Thus we can talk of progress in science even if we do not attribute an aim to science and do not attribute the same aim to scientists. In the remainder of this chapter I pursue a different line of argument in favour of (CK). I show that when considering certain actual and hypothetical developments in science, (CK) gives the intuitively correct evaluation of their contribution to progress whereas (PS) and (VS) do not. Since (VS) is the standard view of progress among scientific realists, I go further into the problems and advantages of (VS). Following Barnes (1991) I show that a key problem for the concept of verisimilitude (language-relativity) is avoided by the epistemic approach. And I argue that an alleged advantage for (VS), that it can account for progress in theories that are not strictly true, is in fact no advantage. I finally look at the understanding approach and the noetic account (NA). Not all contributions to scientific progress add to understanding (acquisition of important data, falsifying of previously believed theories, discovery of puzzling phenomena). On the other hand, a gain in understanding is sufficient for scientific progress since, in my view, understanding is a species of knowledge.

3.2.1 *The Pessimistic Induction*

Here I briefly note and address a motivation for (PS) and (VS)—the Pessimistic Induction. Stated baldly, the Pessimistic (Meta-)Induction (PMI) maintains that the eventual failure of past theories that were considered successful and which were widely endorsed in their day is a reason to doubt that our current successful and widely endorsed theories are true. Famously, Poincaré (1943) articulated the view thus:

> The ephemeral nature of scientific theories takes by surprise the man of the world. Their brief period of prosperity ended, he sees them abandoned one after another; he sees ruins piled upon ruins; he predicts that the theories in fashion to-day will in a short time succumb in their turn, and he concludes that they are absolutely in vain. This is what he calls the *bankruptcy of science.*

The PMI is one argument put forward by anti-realists in favour of their conception of the aim and progress of science (e.g. Kuhn 1970; Laudan 1981*a*). If science does not aim at truth, then falsity is not then a failure of science. Science can progress, despite refutation and change of favoured theories, if it grows in its problem-solving power.

Others have attempted to recover some element of realism from the PMI. A sequence of false theories can nevertheless increase in some truth-related property. Poincaré himself responded to the argument, and commented regarding the fact that Maxwell's theory of light superseded Fresnel's:

> Then if Maxwell's theory is to-day preferred, does that mean that Fresnel's work was in vain? No; for Fresnel's object was not to know whether there really is an ether, if it is or is not formed of atoms, if these atoms really move in this way or that; his object was to predict optical phenom-

> ena. This Fresnel's theory enables us to do today as well as it did before Maxwell's time. The differential equations are always true ....

So Poincaré's response was to retreat from full-on realism to a view that is partly realist, partly anti-realist. His claim that Fresnel aimed only to predict optical phenomena points in the direction of van Fraassen's (1980) constructive empiricism. His point about the preservation of the differential equations eventually develops into structural realism (Worrall 1989; Ladyman 2014). Even so, the focus on the mathematical, 'structural' parts of theories is not enough, for these are not always preserved by later theories. The relativistic revolution not only rejected the non-structural elements of Newtonian mechanics, but also dispensed with its mathematical core. Nor were the mathematical details of Maxwell's theory preserved in the subsequent development of quantum mechanics. But as Einstein himself observed, as have many others since, the earlier equations approximate to the later ones, so that one could see the sequence of theories as approximating ever more closely to the truth (Agassi 2011). The idea of verisimilitude is first found in the work of ancient sceptics, who used it to temper their radical scepticism (Niiniluoto 1987: xi). Popper (1963), who fully endorsed the conclusion of the PMI, if not its inductive argument, was nonetheless the first among modern philosophers to develop the idea of verisimilitude as another way of articulating the thought that scientific theories, though born refuted, are nonetheless increasingly getting some things right. Even those less sceptical than Popper accept enough of the force of the PMI to feel the need to appeal to the concept of verisimilitude. Niiniluoto (1980: 430–1) ascribes this motivation to Nicholas of Cusa and to Peirce, while the point is also endorsed by Psillos (1999: 276): 'In our interactions with the world, the exact truth cannot generally be had, especially concerning the unobservable, and spatiotemporally remote aspects of the world. A perfect match between theories and the world is impossible.'

So (PS) and (VS) are at least partly motivated by the PMI and related sceptical concerns about the ability of science to get to the (full) truth. I address the PMI in greater detail in Chapter 9, focusing in particular on the question whether there can be any such argument that encompasses *all* of science in its conclusion. Without anticipating this argument in detail, we ought to note the sheer implausibility of a global pessimistic conclusion (i.e. one correct for all fields of science). While one might well be willing to accept a high probability of major changes in the future for our best cosmological theories or theories of fundamental physics, or in relatively young fields such as social psychology that are experiencing some degree of methodological turmoil, there are other areas of science where there is not the remotest scientifically plausible possibility of rejecting the fundamental theoretical ideas. Examples include: blood circulates pumped by the heart; chemical substances are constituted by atoms; water is a compound of hydrogen and oxygen; light is electromagnetic radiation; electrons are negatively charged; smoking causes cancer; the tides are caused by the gravitational influence of the moon; the continents have moved over time; mankind has evolved from ape-like ancestors; DNA has a double-helical structure; the sun's energy comes from nuclear fusion. Nor are these claims approximations. Subsequent science has added detail to them but not challenged their truth. Large parts of modern science do not seem to undergo refutation of their most successful theories; their development is cumulative.

The pessimist may retort that the reference to *modern* science just begs the question, for it is these current theories that the PMI predicts will likely fail in due course. If we look at older science we can find many theories that were successful and were accepted for extended periods before eventually succumbing to refutation. Laudan (1981*a*: 33) lists the following as successful but not only false, but as lacking reference for their central theoretical terms: the crystalline spheres of ancient and medieval astronomy; the humoral theory of medicine; the effluvial theory of static electricity; 'catastrophist' geology, with its commitment to a universal (Noachian) deluge; the phlogiston theory of chemistry; the caloric theory of heat; the vibratory theory of heat; the vital force theories of physiology; the electromagnetic aether; the optical aether; the theory of circular inertia; theories of spontaneous generation.

This, however, is misleading. For the success of a theory is not a matter of the length of time during which it is accepted; nor is the latter a good proxy for success. Rather success depends on the quantity and degree of testing or other evidence gathered that bears on its truth or falsity. And the exponential growth of science means that success in this sense is not proportional to duration (Fahrbach 2009; Fahrbach 2011*b*; Fahrbach 2011*a*). About half of all science ever carried out has occurred within this century. Theories devised before 1980 fall within the initial quartile of science's history if the latter is measured not in years but in quantity of experiment, testing, and observation. Whereas all the theories listed by Laudan were refuted before the first centile of science was complete. That means that the currently accepted theories of modern science developed in the twentieth century come from fields subject to vastly more evidence than any of Laudan's theories. As Fahrbach points out, if the PMI were correct one would expect ever increasing refutation of theories. But that is not what one observes. I find it striking that of all the Nobel Prizes awarded in the sciences (physics, chemistry, and physiology or medicine) only one has been awarded for work that we now regard as fundamentally mistaken (to Johannes Fibiger in 1927, in physiology or medicine—more on Fibiger's work later). The PMI seems to have run out of steam.

Consequently, we should not regard the PMI, nor arguments with similar conclusions, as a good reason for abandoning the cumulative knowledge view of progress, (CK), for the problem-solving account, (PS), or for verisimilitudinarian account, (VS). The evidence of history (properly quantified) shows that much science is cumulative. Successful theories of modern science in many fields are not refuted and replaced, neither by radically different alternatives, nor even by better approximations to the truth; i.e. much science is not beset by revolutions, nor need it be seen as converging on (but not reaching) the truth.

That is not to say that scientific changes of these kinds never occur. If a field shows frequent radical changes in its leading theories, then we should question whether it really is progressing (instead of adopting a different theory of progress). On the other hand, if a field does show replacement of theories by more precise theories to which each predecessor is an approximation, then, as I argue below, this can be accommodated by (CK), as I will show in Section 3.4.3.

3.3 The functional-internalist approach to progress

In this section I look at what I call the *functional-internalist* approach to progress. The principal representatives of this approach are the puzzle- and problem-solving views of Kuhn and Laudan, which I bundle together as (PS). (Much of what I have to say will apply to Lakatos's methodology of scientific research programmes also.) (PS) is 'functional' because it takes progress to be a matter of the success a scientific field has in fulfilling a function—that of solving problems. Why it is *internalist* I shall come to shortly.

When compared with the verisimilitudinarian view, (VS), there is a superficial resemblance between (PS) and the cumulative knowledge view, (CK). While (CK) sees scientific progress as the accumulation of scientific knowledge, (PS) sees progress as the accumulation of solutions to scientific problems. The two views would coincide if (i) solving a problem is a matter of acquiring a certain sort of knowledge; and (ii) all scientific knowledge is knowing some problem-solution. However, the views do not in fact coincide since neither (i) nor (ii) is true, according to the conception of 'problem-solving' I am dealing with here. Furthermore, where the views differ, (CK) account has the advantage over (PS).

Considering (ii), it is clear that not all scientific knowledge is a matter of knowing the solution to some problem. Astronomers and naturalists of the eighteenth and nineteenth century spent lives collecting data on stars and comets, or on new species and habitats. These were contributions, albeit not dramatic, to scientific progress. At the same time there are serendipitous discoveries (such as X-rays) that progress science, the importance of which was clear without their being solutions to any problems.[2,3]

More significant is (i). Kuhn and Laudan do not think of solving a problem as involving knowledge, when knowledge is understood as requiring truth. For Kuhn a puzzle or problem is solved when a proposed solution is sufficiently similar to a relevant paradigmatic puzzle-solution. The latter is an instance of puzzle-solving that the scientific community holds to be an exemplar of what a scientific puzzle is and how it should be solved. Laudan's account is superficially similar to Hempel's D-N model of explanation—a problem phenomenon P is solved by T when one can deduce P from T. But importantly Laudan does not require that T is true. He does not even require that P really exist: 'A problem need not accurately describe a real state of

[2] Yafeng Shan (2019) has recently offered a version of the functional-internalist approach that seems to me to be an improvement on the Kuhn–Laudan version. Nonetheless, I think it still fails on the point raized in this paragraph, that not all progress involves solving a problem.

[3] On p. 64 below I describe the interest that X-rays generated among scientists who had for some time only a very limited understanding of the phenomenon. The discovery of X-rays was progress—a gain of knowledge of their existence and some basic facts—that generated a new puzzle without solving any old ones.

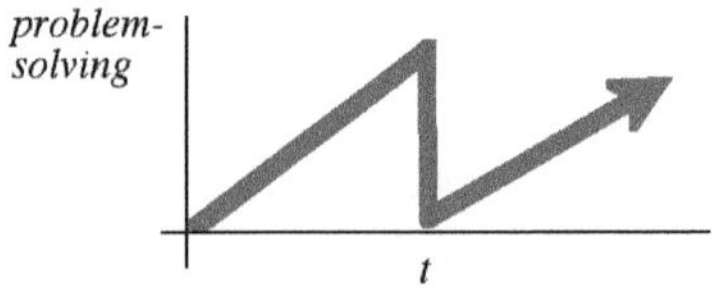

Fig. 3.1

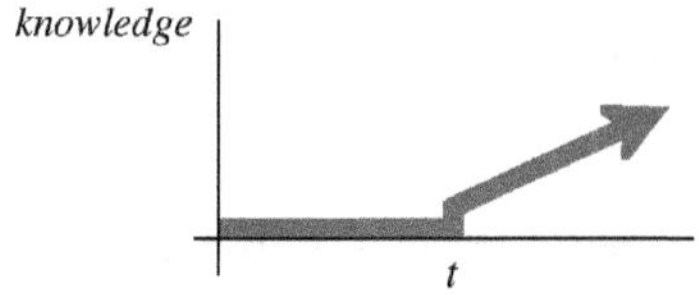

Fig. 3.2

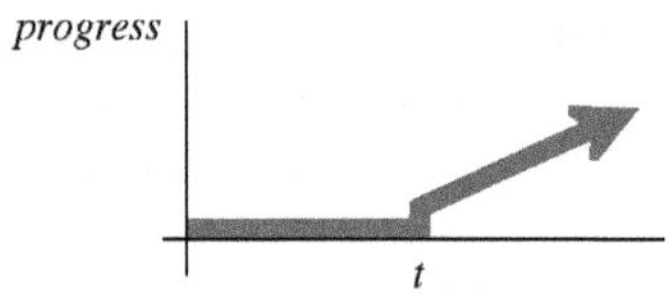

Fig. 3.3

affairs to be a problem: all that is required is that it be *thought to be* an actual states of affairs' (Laudan 1977: 16; Laudan's emphasis).

Now consider this following (actual) case:

> **Goat's blood** The ancient authors Pliny and Solinus believed that goat's blood would soften and split diamonds.[4] Christian writers such as Albertus Magnus, Cyprian, and Augustine explained this by noting that the blood of the Saviour on the cross softened the hard hearts of his enemies (such as Longinus, the Roman soldier who oversaw the crucifixion). An analogy between the goat and Jesus is licensed by scripture, since Jesus was a scape-goat, sent into the wilderness and ultimately slain to expiate our sins. Therefore, just as Jesus' blood softened the hardest of hearts, goat's blood would soften the hardest of substances, diamond. Later Sir Thomas Browne (1646) demonstrated that goat's blood does not have the power to split diamonds, noting the empirical evidence provided by lapidaries (gem cutters).

According to Laudan's account, the Christian writers provide a solution to the problem of why it is that goat's blood can soften diamonds and allow them to be split, since the proposition in question can be deduced from their theory. It therefore constitutes a contribution to scientific progress. Furthermore, Laudan and Kuhn not only do not require the problem to contain any truth, they do not require the solution to contain any truth either: progress can heap falsehood upon falsehood. Furthermore, consider the consequences of Browne's work in 1646. Whereas we had a problem and a solution before, we now have no problem and no solution. Thus problem-solving power has been reduced; by Laudan's and Kuhn's standards that would mark a regress. Hence in Fig. 3.1, showing problem-solving capacity over time,

[4] See Browne (1646) for this history. Laudan (1977: 16) ascribes the belief to Nicole d'Oresme, though, as Browne says, it is much older. Thorndike (1914: 287) notes that Roger Bacon rejected the idea.

the line dips downward at t (=1646). Now consider changes in knowledge. There was little or no knowledge in this area while those concerned were in the grip of a false problem and false solution. Browne provided some knowledge—that hot goat's blood does not split diamonds—which was one may assume a better basis for subsequent generation of knowledge. Hence in Fig. 3.2 the curve rises at t. Our question now is, was there regress or progress in 1646? When earlier, the medieval thinkers developed their explanation in terms of the scape-goat analogy, they thought they were making a contribution to progress. But they were surely mistaken. So undoing their solution was no regress. On the contrary, it is clear that Browne did contribute to progress, albeit in a small way, and he did so by giving us knowledge that something previously thought to be true is in fact false. So in Fig. 3.3, tracking progress coincides with the graph of changes in knowledge in Fig. 3.2, not the changes in problem-solving in Fig. 3.1.

We can see the same pattern in an episode from more recent history of science:

> **The Blondlot Case** In 1903 René Blondlot believed that he had discovered a new form of radiation, which he named 'N-rays' ('N' for Nancy, where he was born and worked). N-rays attracted much interest until the American spectroscopist R. W. Wood visited Blondlot's laboratory at the behest of the journal *Nature*. Wood surreptitiously removed a crucial prism from Blondlot's experiment. Blondlot nonetheless continued to see the rays. After Wood's intervention, reported in *Nature*, it became accepted that N-rays simply did not exist; scientists gave up research in N-rays and the community repudiated the earlier 'discoveries'. During the intervening period, 1903–5, more than 300 articles on the topic were published (Lagemann 1977), discussing the nature of the rays and their properties (e.g., which materials could generate or transmit the rays). While the scientists engaged in this research were mostly French, some Irish and English scientists also reported seeing N-Rays.

Until Wood's intervention, N-ray research was carried out as normal science; research problems were posed and solved, by the criteria of Kuhn and Laudan. Yet it would be absurd to suggest that progress was being made. Since Wood's disclosure led to the loss of those problems and solutions, it must count as regressive according to (PS). Again that it absurd. Surely, if anything, by helping science get back on the path of genuine as opposed to apparent knowledge, Wood aided progress.

More generally, scientists can be 'successful' in solving problems by the standards of their day. In some cases, their solutions will not stand the test of time. It is implausible that individual scientists aim at solving problems only in the internalist sense. It is true that scientists want to solve scientific puzzles and problems. But the notion of 'solving' here is an objective, externalist one. A scientist wants to uncover the *correct* answer to the problem, not merely an answer that meets the standards of her peers. Likewise it is implausible that the institution of science aims at (internalist) problem-solving, (Aim-P), and that it progresses when it succeeds at (internalist) problem-solving, (PS). Above I mentioned the sole case of a Nobel Prize known to be awarded for a mistaken piece of research. Johannes Fibiger won the 1926 Nobel Prize in Physiology or Medicine 'for his discovery of the Spiroptera carcinoma'. In research that was published in 1913 Fibiger was the first to induce cancer in rats. He observed that the wild rats with which he was working and which were infected with *Spiroptera* nematode worms also had stomach cancers. He claimed to have found that experimental rats went on to develop stomach cancers if they were fed with larval cockroaches that were infected with the nematodes. Fibiger hypothesized that the nematode worms secreted a poisonous substance that caused the cancer. Later it came to be understood that this was not the case and that the cause is not specific to the worms; it is probable that Fibiger's rats were suffering from vitamin A deficiency. No contemporary questioned the quality of Fibiger's research; indeed Fibiger was highly regarded as a meticulous experimentalist. Later his work was criticized for lacking a control group of rats. But at the time scientists were not expected to use controls (Stolt *et al.* 2004).[5] So by the standards of the day Fibiger solved various problems sufficiently well to deserve a Nobel Prize: he 'solved' the problem of what caused the cancer in the rats, the problem of whether cancer could be caused by a parasite, the problem of inducing cancer artificially, and thereby the problem of developing an animal 'model' for cancer. Yet no scientist would accept that this constitutes achieving the aim of their research, or of the aim of science generally. Likewise, Fibiger's research did not constitute scientific progress.

3.3.1 *Anti-realism and progress*

Kuhn and Laudan countenance contributions to scientific progress that do not involve any knowledge. In Laudan's case this is because he thinks that we never have scientific knowledge—he accepts the Pessimistic Induction as sound. In Kuhn's case this was initially because he wanted to give an account of scientific change that is neutral over questions of truth and knowledge. In the postscript to the second edition of *The Structure of Scientific Revolutions* Kuhn adds to this an attack on the concepts of truth and verisimilitude (Bird 2012: 875).

Because they believe that as matters are normally understood knowledge and truth are problematic, Kuhn and Laudan adopt the classic strategy of the anti-realist. The anti-realist is impressed by sceptical arguments that on the basis of evidence E one cannot get to know facts of kind K, where 'K' is conceived of in a standard, natural way. But rather than be an out-and-out sceptic the anti-realist reconceives 'K' in such a way that 'K' can be inferred from E. And the easiest way of doing this is to reconceive 'K' in such a way that it is more or less identical to E (just as Berkeley reconceives 'corporeal substances' as identical to sensations). Similarly, it is natural to think that success in problem-solving is *evidence for* the progress of science, when the latter is understood as the accumulation of knowledge. If one is sceptical about such an inference, one might conclude that we do not have scientific progress, or at least that we do not know whether we do. But one may prefer instead to reconceive of scientific progress so that it is identical to success in problem-solving. In which case the sceptical worry just evaporates and one is able to share the talk of the vulgar who

[5] Indeed, Fibiger himself was one of the first to use controls in a clinical trial (Hróbjartsson *et al.* 1998).

do think that there is scientific progress.[6] This move is clear in Laudan (1977: 126–7), who asserts, '...what I am suggesting is that we apparently do not have any way of knowing for sure (or even with some confidence) that science is true, or probable, or that it is getting closer to the truth. Such aims are *utopian*, in the literal sense that we can never know whether they are being achieved.' Laudan (1977: 127) uses this sceptical admission to justify his account of progress:

> The workability of the problem-solving model is its greatest virtue. In principle, we can determine whether a given theory does or does not solve a particular problem.... If we have had to weaken our notions of rationality and progress in order to achieve this end, we are at least now in a position to decide whether science is rational and progressive—a crucial necessity denied to us if we retain the classical connections between progress, rationality, and truth.

It is this that makes Laudan's account (and similarly Kuhn's account) epistemologically internalist. Whether or not a scientific community has made progress can be judged and known by that community, and does not depend on features that may be unknown to them (such as whether their problem is genuine and whether their solution is true or known).

An upshot of Laudan's and Kuhn's view of problem-solving and progress is that revisionary scientific changes (such as scientific revolutions) are not progressive in a straightforward way. Kuhn (1970: 160) says that there is progress through revolutions, because the new paradigm solves more problems than its predecessor. But while the later paradigm may solve more problems it need not solve (or dissolve) all the problems that were previously solved. For example, Descartes's vortex account of planetary gravitation accounted for the fact that the orbits of the known planets are co-planar and have the same sense of rotation. Yet this account of gravitation was superseded by Newton's theory which gave no explanation of the phenomenon. Similarly the geocentric view of the planets made it quite natural that the Moon should present the same face towards the Earth while the heliocentric view does not. According to Laudan and Kuhn the move to the Newtonian and geocentric theories was accompanied by a loss of problem solutions in this respect. This phenomenon—often called 'Kuhn-loss'—certainly raises interesting questions of why the transitions were made. That is not our current concern. Our current concern is whether we can describe the changes as progressive. For Laudan and Kuhn to be able to describe the changes as progressive we have to take progress to occur despite loss of problem-solving power in some areas. Thus we have to take an increase in problem-solving power in other areas to outweigh the losses described. That in turn requires that we are able to quantify problem-solving power which means that we need an idea of how to individuate and count problem solutions and how to weigh their relative importance. This is a difficult matter, and Collingwood (1956: 329, 332) for one thought

[6] By referring to the talk of the 'vulgar' I am referencing Berkeley's (1710) view that one can maintain an anti-realist metaphysics while being willing to make what appear to be realist assertions, 'in such things we ought to think with the learned, and speak with the vulgar.'

the problem is insoluble. Laudan (1981*b*: 150; 1977: 31–40, 64–6) nonetheless thinks that we can weigh the importance and number of problems so as to measure whether gains in problem-solving outweigh losses.

The problem may not be impossible; even the epistemic, cumulative knowledge account of progress (CK) may invite supplementation with an account of which additions to knowledge are more significant and progressive than others. But for present purposes what deserves notice is the fact that (CK) does not need such supplementation in order to describe the Kuhn-loss transitions as progressive. For the problem 'solutions' that are lost are not knowledge. Since gravity is not explained by vortices, Descartes did not know what explains the co-planar orbits of the planets. He believed that the co-planar orbits are explained by vortices, but this belief is not knowledge since it is false. Since Kuhn-loss is not loss of knowledge it presents no difficulty, not even a prima facie one, for (CK). The theoretical changes that involved Kuhn-loss brought with them gains in knowledge but no losses of knowledge (only losses of false belief). So (CK) can deliver the verdict that these changes are progressive without any need to weigh gains against losses.[7]

Although Kuhn and Laudan avoid scepticism about progress, their ability to speak with the vulgar is very far from perfect, since it is part of the vulgar concept that there is a gap between problem-solving (in their sense) and progress, the gap between evidence and what it is evidence for. Either their accounts of progress are attempts to capture the everyday concept of scientific progress, in which case their attempts are unsuccessful. Or they are seeking to reform the concept (the passage quoted from Laudan suggests that he might take such a view), in which case that reformation needs strong motivation. Anti-realist scepticism may supply some motivation. But realists will be unmoved.

3.4 The semantic approach to progress

(CT) is the version of the semantic view of progress that says that scientific progress is the accumulation of true scientific beliefs. (CT) bears a structural similarity to my version of the epistemic approach, (CK), with truth instead of knowledge. Philosophers of science are not always careful about the difference. Richard Boyd (1983: 45), in characterizing scientific realism, moves without comment from one to the other: 'The historical progress of mature sciences is largely a matter of successively more accurate approximations to the truth about both observable and unobservable phenomena. Later theories typically build upon the (observational and theoretical) knowledge embodied in previous theories.' In good science instances of gains in truth and verisimilitude should indeed also be instance of additions to knowledge

[7] Laudan (1977: 147) notes that several philosophers have promoted cumulative accounts of progress, including Whewell, Peirce, Duhem, Collingwood, Popper, Reichenbach, Lakatos, and Stegmüller. That an account of progress is cumulative is consistent with its conforming to any of the epistemic, semantic, functional-internalist, and understanding conceptions. What will vary among the cumulative versions of such views will be what is accumulated (knowledge, truth, problem-solutions, understanding). But since knowledge is harder to come by and more difficult to shift once acquired, the problem of a corresponding loss will be rarer for the epistemic conception than for the other conceptions.

(I say more about this later in this chapter). Nonetheless, the difference between a semantic concept and an epistemic one is a deep difference indeed.

(CT) has had few explicit supporters; the more popular version of the semantic approach takes scientific progress to be a matter of increasing verisimilitude.[8] Verisimilitude or nearness to the truth has been the realists' weapon of choice in their battles against anti-realists. Popper, somewhat forlornly, hoped that his adherence to verisimilitude would differentiate him from his anti-realist contemporaries. More recently, Ilkka Niiniluoto (1987, 1999) has sought to build a sophisticated and systematic philosophy of critical scientific realism on the foundations of his own account of verisimilitude. And a central component of Niiniluoto's critical scientific realism is the idea that progress is a matter of increasing verisimilitude (Niiniluoto 1980: 428; 1984: 76; 1999: 201).[9]

In this section I shall show that the semantic conception yields a verdict about progress in certain kinds of case that is at odds with our intuitions. Given that science is an epistemic activity it seems almost tautologous to suggest that its success and so progress should be measured by epistemic standards. I shall argue that our intuitions concerning progress in possible episodes of change do imply that epistemic characteristics are essential to progress.

The semantic and epistemic accounts diverge when it comes to considering beliefs with insufficient epistemic support to count as knowledge. Imagine a sequence of beliefs that show the accumulation of truth. (And, *a fortiori*, increasing verisimilitude.) Let the truth (or the increasing verisimilitude) of these beliefs be entirely accidental. This is a sequence of lucky guesses or lucky irrational beliefs. Such a sequence will be progressive on the semantic approach but not on the epistemic approach. The latter gives the correct verdict here. Progress and rationality cannot diverge that easily.

> **Flukish astrology** A scientific community C has formed its beliefs using some very weak or even irrational method M, such as astrology. But by fluke this sequence of beliefs is a sequence of true beliefs. These true beliefs are believed solely because they are generated by M and they do not have independent confirmation. At time *t* an Archimedes-like scientist in C realizes and comes to know that M is weak. This scientist persuades (using different, reliable methods) her colleagues that M is unreliable. This may be C's first piece of scientific knowledge. Community C now rejects its earlier beliefs as unsound, realizing that they were formed solely on the basis of a poor method.

[8] Niiniluoto (1984: 76–7) briefly discusses the cumulative truth view, (CT). He says the view is 'nowadays widely regarded as giving a naïve and oversimplified picture of the development of science.' He rejects the view because, among other reasons, it cannot, he says, accommodate strictly false theories that are nonetheless idealizations that are close to the truth. I reject this reason below.

[9] See also Kuipers (2000). Modern accounts of verisimilitude were developed simultaneously by Niiniluoto (1987) and Oddie (1986). For a comprehensive overview, see Oddie (2014).

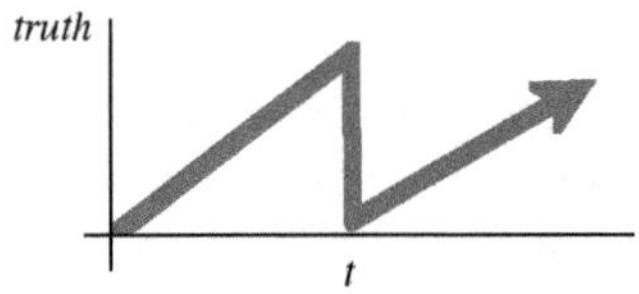

Fig. 3.4

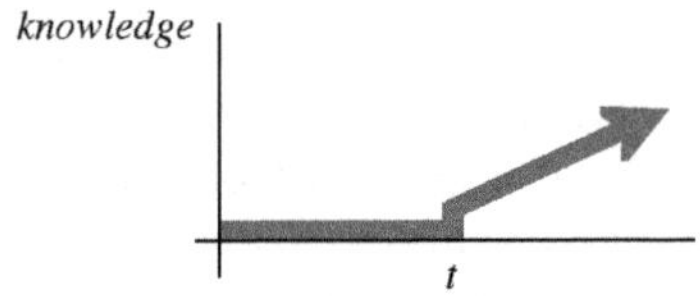

Fig. 3.5

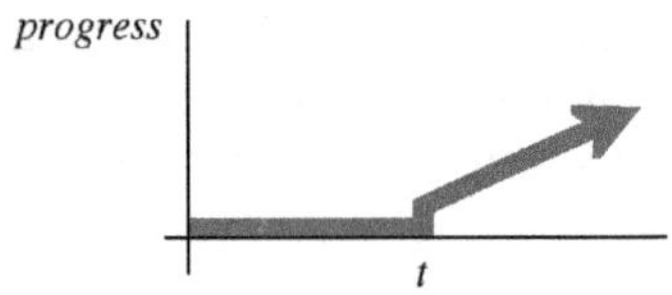

Fig. 3.6

According to the semantic approach this community was making progress until time *t* (it was accumulating true beliefs) and then regressed (it gave up those beliefs). This, it seems, contradicts the verdict of our intuitions about this episode. The acquisition of beliefs by an unreliable method cannot be genuine scientific progress, even if the beliefs so acquired are, by accident, true. Far from being a regressive move, giving up those unreliably produced beliefs, because of a now well-founded belief that they were unreliably produced, is a positive, progressive step. So the semantic view yields a description in terms of progress and regress that conflicts with what we are intuitively inclined to say.

The epistemic approach, in contrast, says that the community made no progress at all until *t* at which time it did start to make progress. This, by contrast with the semantic view, does accord with the verdict of intuition. Fig. 3.4 and Fig. 3.5 show the change in true scientific belief and scientific knowledge respectively. Intuition tells us that Fig. 3.6 charts scientific progress. And so I conclude that progress does not match change in true belief but does match change in knowledge. Clearly the same argument may be run where the society forms a sequence of beliefs with increasing but accidental verisimilitude. Hence the semantic approach does not capture our intuition about scientific progress, whereas, in this case at least, the epistemic approach does.

In Chapter 2 I considered an imaginary variation on the Blondlot case mentioned above, in which it turns out that by coincidence there are rays corresponding to Blondlot's description of N-rays. Their existence does not explain the evidence in the way that Blondlot supposed. And it remains the case that Blondlot's thought processes were dominated by self-deception and national pride. So, in this case, the core of Blondlot's theory was composed of true propositions that he believed, but these did not amount to knowledge. This imaginary version of the Blondlot case is like Flukish Astrology—there is first an addition to true belief and then a reduction in true belief after Wood's intervention became known. Thus, if the semantic approach is right, this episode ought to show progress until that intervention and regress thereafter. But that is not correct. Blondlot's views, even if correct, were not a contribution to the progress of science. That is explained by the epistemic approach, since they did not amount to knowledge.

Genuine cases in the history of science that show the acquisition of truth by unreliable reasoning are rare. Even so, they are not entirely absent. One well-known example is Johann Friedrich Meckel's prediction that human embryos would be found to have gill slits, which is a consequence of his theory that embryonic development mirrors the great chain of being, from the least perfect animals to the most perfect (humans) (O'Connell 2018). The prediction was accurate, but the theory that generated it is false. In the Renaissance period it was believed that infants and young children had more body water as a proportion of total body mass than older children and adults. The source of this belief was a physiological theory based on the doctrine of the humours. The body was supposed to be governed by four humours (bile, black bile, blood, and phlegm); ill-health was held to be a matter of an imbalance or corruption of the humours. The natural balance was supposed to change with time, and since the humours were characterized by their proportions of the four basic qualities (moist, dry, warm, and cold), the changing balance of the humours implied change in the natural proportion of these qualities with age. The prevailing view was that the predominant humour in childhood is blood, which is moist and warm (Newton 2012). Children are moist because they are formed from the mother's blood and the father's seed, both of which are moist and warm. Childhood is also influenced by the moon, which was held to be warm and moist. In youth, however, the dominant humour is choler, which is warm and dry. And so, in growing to maturity, a child loses moisture, and indeed this drying continues with age, explaining the wrinkled skin of the elderly in contrast to the soft skin of babies and infants. When the humoral theory was dismissed in the late eighteenth and nineteenth centuries, these opinions fell into abeyance. It turns out that modern physiology has shown that body water is proportionally highest in neonates and then declines (Friis-Hansen *et al.* 1951; Schoeller 1989). So the Renaissance doctors, basing their opinion in the false humoral theory, fortuitously had correct opinions about the variation of moisture in the body with age.[10] According to (VS), when these true implications of humoral theory were drawn, science thereby made progress, but it regressed when these beliefs were dropped along with the humoral theory. On the contrary, even the fortuitously true consequences of humoral theory were no contribution to scientific progress insofar as they were based principally on that theory and not on appropriate evidence.[11]

[10] Another example of a true belief in science acquired by unreliable reasoning is Herschel's view that the surface of the Sun is cooler than its atmosphere. This has been known to be true since the middle of the twentieth century, from analysis of spectral emission lines (Grotrian 1939; Edlén 1943; Parnell and Moortel 2012). The surface of the Sun is around 6,000°C, whereas its atmospheric temperature reaches 1,000,000°C. Herschel (1795) recognized that the Sun's atmosphere is hot but gave reasons for thinking that its surface was cool and habitable, indeed habited.

[11] For defences of the semantic approach and the verisimilitudinarian view in particular against my arguments, see Rowbottom (2008, 2010), Tambolo (2011), Cevolani and Tambolo (2013), Niiniluoto (2014).

3.4.1 *Truth versus falsity*

Darrell Rowbottom (2008) objects to regarding my case of imaginary Blondlot as a counterexample, on the ground that the semantic approach is not obliged to regard it as showing progress. That is because, although the example attributes new true beliefs to Blondlot, he would also have had many false beliefs (e.g. that N-rays were causally responsible for the perceptual experiences he was having when he thought he was observing them). If the false beliefs outweigh the true beliefs, then perhaps the semantic approach will allow that, all things considered, there is no progress.

A detailed version of the semantic approach that allowed for this answer would have to count beliefs. If we consider an episode that has both true content and false content, then if we adopt the logical fiction that the belief set includes all of its own logical consequences, then such an episode will have infinitely many true beliefs and infinitely many false beliefs. Trying to accommodate this will lead to the insuperable objections that faced Popper's (1963) account of verisimilitude (Miller 1974; Tichý 1974). Even if we drop the logical idealization, the counting of beliefs remains problematic, especially when, presumably, we need to give additional weight to the more significant ones. Almost any scientific episode will have some false content, and so we cannot restrict ourselves to those instances where there is an addition of true belief with no addition of false belief.

Another problem concerns the relative value of truth and falsity. (CT) says that progress is a matter of accumulating truth, and the associated aim of science, (Aim-T), says that science aims at truth. This aim is best achieved by science accepting as many propositions as possible, even when there is no evidence. Accepting at random is better than withholding acceptance. For (CT) and (Aim-T) say nothing about the disvalue of falsity. So long as truth is achieved and accumulated the aim of science is met and progress occurs. But acceptance at random cannot be an acceptable way of achieving the aim of science. So the semantic approach not only needs to promote truth acquisition, it should also encourage falsity avoidance. Such an account will be a hybrid view, amalgamating these two aims. In which case the question arises, how should they be balanced? Is the value of a true belief that *p* equal to the disvalue of a false belief that *q* (where *p* and *q* are equally significant propositions)? In 1805 John Playfair famously summarized the different characters of his friends the chemist Joseph Black and the geologist James Hutton thus:

> Ardour and even enthusiasm, in the pursuit of science, great rapidity of thought, and much animation, distinguished Dr Hutton on all occasions. Great caution in his reasonings, and a coolness of head that even approached to indifference, were characteristic of Dr Black .... Dr Black dreaded nothing so much as error and Dr Hutton dreaded nothing so much as ignorance: the one was always afraid of going beyond the truth and the other of not reaching it. (Playfair 1805: 95, quoted in Jones 1996: 120)

Is the choice in weighting truth and falsity a mere matter of temperament? Or is there some rationally optimal balance? (Note that an equal balance would still mean that it would be a good strategy to accept every proposition that has even the flimsiest of evidence in its favour.)

The epistemic approach does not suffer from such problems. (Aim-K) says that science aims to add to knowledge. So a strategy of adding beliefs on no evidence or on weak evidence, even if that makes them more likely to be true than false, will not achieve (Aim-K). (Aim-K) endorses a strategy of believing a proposition when and only when there is sufficient evidence for it to be known when believed. The issue of weighing truth against falsity just does not arise.

Does that mean that one or other of Hutton and Black is irrational (or both are)? I think that the verdict is that one or both might be imperfectly rational, but not downright irrational. So it is possible that Hutton was ready to believe a proposition before enough evidence was adduced to make it known whereas Black needed more evidence than was necessary for knowledge in order to accept a proposition. Hence one or other or both were not pursuing an optimal strategy. On the other hand, it is not easy to know exactly what the optimal strategy is, i.e. what for any proposition counts as sufficient evidence for knowledge. The KK principle states that if a subject knows, then she knows (or is in a position to know) that she knows. If the principle were true, then when a subject had gathered enough evidence to know that *p* then she would be able to know that and then could stop gathering further evidence. She would always be in a position to know that she had enough evidence to fulfil (Aim-K) but not more than is required. However, I regard the KK principle as false. The subject will often not be in a position to know that she knows, and so may be in a state of reasonable doubt as to whether her evidence is sufficient for knowledge. Hutton-like individuals may be more sanguine when in such a position, Black-like individuals may be less optimistic.

I now return briefly to Rowbottom's objection. Rowbottom considers two worlds, with changes brought about by a *malin génie*. In one the community wakes up with an array of new, true scientific beliefs. In the other, the community wakes up with the same new, true scientific beliefs plus a lot of fake evidence (e.g. lab records) that justifies those beliefs. Rowbottom says that the epistemic approach should regard the latter world as having experienced more progress than the first, since its scientists benefit from an epistemic good, justification, that the scientists in the first lack. Yet, he says, things are the other way around. For in the first world, the scientists will realize that they have no evidence for these beliefs they have and so will test them and thereby science will progress. Whereas in the second world, because of the spurious justification of their beliefs, the scientists will make no such efforts.

The most important thing to say in reply is that although a version of the epistemic approach might value justification in itself, my preferred account, (CK), does not. (CK) says that progress requires knowledge. What the scientists in Rowbottom's second world have is, at best, justified true belief. And as Gettier showed, justified true belief is not always knowledge. And clearly this is a Gettier case. So (CK) denies that more progress is made in the second than in the first world.[12]

[12] See Bird (2008) for further discussion.

3.4.2 *Verisimilitude and Gettier*

In this section I present a problem for the semantic approach in its (VS) form. (Some supporters of (VS) may claim that a good account of verisimilitude should be able to address the problem of the trade-off between truth and falsity discussed in the preceding section.) I start by outlining the approach to verisimilitude developed by Tichý (1974) and others (Hilpinen 1976; Niiniluoto 1977) following problems facing Popper's original attempt to give a formal account. This approach aims to understand the verisimilitude of a theory in terms of how much a world described by that theory is like the actual world.

Tichý imagines a language with three atomic propositions to describe the weather: 'it is hot', 'it is rainy', and 'it is windy'. As it happens the actual weather is such that all three propositions are true. Smith believes: 'it is not hot and it is rainy and it is windy' whereas Jones believes 'it is not hot and it is not rainy and it is not windy'. We can say that Jones is further from the truth than Smith since there are three respects in which a world may be like or unlike the actual world and the possible world described by Jones's theory differs from the actual world in all three respects (so the distance from the actual world is 3) while the possible world described by Smith's theory differs from the actual world in one respect (so distance = 1). The world of Jones's theory is thus more distant from the actual world than the world of Smith's theory, and so the latter theory is closer to the truth than the former.

The weather according to Smith and Jones (original)

truth	hot	rainy	windy
Jones	not hot	not rainy	not windy
Smith	not hot	rainy	windy

The weather according to Smith and Jones (redescribed)

truth	hot	Minnesotan	Arizonan
Jones	not hot	Minnesotan	Arizonan
Smith	not hot	not Minnesotan	not Arizonan

Miller (1974) however considers an alternative language with the atomic propositions 'it is hot', 'it is Minnesotan', and 'it is Arizonan'. 'It is Minnesotan' is true precisely when 'it is hot $\leftrightarrow$ it is rainy' is true, and 'it is Arizonan' is true precisely when 'it is hot $\leftrightarrow$ it is windy' is true. In this language the truth is expressed by 'it is hot and it is Minnesotan and it is Arizonan', while Smith's beliefs are expressed by 'it is not hot and it is not Minnesotan and it is not Arizonan' and Jones's beliefs are expressed by 'it is not hot and it is Minnesotan and it is Arizonan'. Given this language for the description of possible worlds, the world described by Smith's theory differs from the actual world in three respects (distance = 3) and the world of Jones's theory differs in just one respect (distance =1). So Jones's theory is measured to be closer to the truth than Smith's, when using this language. It would thus appear that this approach to capturing verisimilitude makes verisimilitude relative to a language. And that is a problematic conclusion for the scientific realist, who wants progress and verisimilitude to be objective. There are various ways in which one might tackle this problem. For example, one might say that some languages are objectively better at capturing the natural structure of the world. (While I think that is correct—and one should reject the nominalism that drives Goodman's New Riddle of Induction—it seems insufficient to solve the problem entirely.) Or one might try quite different approaches to verisimilitude. (See Oddie (2014) for a clear and detailed survey of the options.) Eric Barnes (1991) proposes a more radical response that notes an epistemic asymmetry between the beliefs expressed in the two sets of vocabularies. I agree, and take this as a reason to prefer (CK) to (VS).

In Miller's version of the story, they inhabit a windowless, air-conditioned room. So they are making guesses: Jones guesses *it is not hot* and *it is not rainy* and *it is not windy* while Smith guesses *it is not hot* and *it is rainy* and *it is windy*. The truth is *it is hot* and *it is rainy* and *it is windy*; so none of Jones's guesses are correct whereas two of Smith's are correct. If we further assume that Smith's guesses are made after Jones's, then that would look like progress according to (VS). But, as argued in the preceding section, lucky guesses are no contribution to progress. Now let us imagine, as does Barnes (1991: 315), a variation whereby Smith and Jones form their beliefs not by guessing but by investigating indicators of temperature, precipitation, and airspeed in their vicinity, using methods that are usually very reliable by normal scientific standards. However, things go awry in the case of all Jones's investigations, leading to his false beliefs. They go awry for Smith also in the case of temperature, but his methods for measuring precipitation and airspeed are reliable and working fine. Accordingly, Smith knows *it is rainy* and *it is windy*. Now consider the beliefs of Smith and Jones when expressed using the language of 'hot', 'Minnesotan', and 'Arizonan'. The oddity, recall, is that now Jones has two correct beliefs: *the weather is Minnesotan* and *it is Arizonan*, even though, when expressed in ordinary English, he has no true beliefs. Let us have a closer look at how it is that Jones has these two true beliefs. The belief *it is Minnesotan* is logically equivalent to the belief *it is hot* $\leftrightarrow$ *it is raining*, which in turn is equivalent to the belief **DIS**: *either it is both hot and rainy or it is neither hot nor rainy*. Jones believes this proposition because he believes *it is not hot and not rainy*.

Why Jones believes that the weather is Minnesotan

	Jones believes	
	it is Minnesotan	
	which is equivalent to	
	DIS:	
it is hot and rainy	or	it is not hot and it is not rainy
true		*false*
not believed by Jones		*believed by Jones*

So he believes the second disjunct of **DIS**. Note that this disjunct of **DIS** is false. Since **DIS** is a simple consequence of that second disjunct, Jones believes the whole of **DIS**. However, the whole of **DIS** is true, since the first disjunct is true—a disjunct that Jones believes to be false. So the position is that Jones believes a true proposition, **DIS**, because it is a disjunction and he believes one of the disjuncts—the one that happens to be false. As Barnes (1991: 317) points out, Jones's belief has the structure of a standard Gettier case: a case of a true and justified belief, but where the truth and justification come apart. The justification (in this case Jones's normally reliable investigation that has gone awry) is attached to the false disjunct of **DIS** whereas the truth comes from the other disjunct of **DIS** for which Jones has no justification (in fact he thinks it is false). As a standard Gettier case, we must deny that Jones has knowledge in this case, although he has fortuitously got a true belief.

The problem we faced was that it looked as if verisimilitude depends on choice of language. Combined with (VS), that implies that Smith makes progress compared to Jones when we express their beliefs using normal language and that Jones makes more progress when their beliefs are expressed using the terms 'Minnesotan' and 'Arizonan'. Barnes's important observation shows that the change of language makes no corresponding difference regarding *knowledge*. The shift of language reveals that Jones has some true beliefs, but the new language does not render those beliefs epistemically successful—they are not knowledge. So that in turn means that while (VS) faces problems, (CK) can simply bypass them.[13] Fortuitously true beliefs notwithstanding, Jones knows nothing substantive about the weather whereas Smith does, and so Smith thereby contributes to progress and Jones does not. A suitable response will therefore propose that progress should be seen in terms of increasing knowledge, not nearness to the truth.

3.4.3 *Progress with approximate truth*

The cumulative knowledge account of progress faces the following objection:

> (O) Theories are very often at best only approximately true; they rarely attain full truth. Since knowledge entails full truth, these theories cannot be the objects of knowledge. Consequently we cannot be accumulating knowledge in such cases. Hence (CK) cannot explain the sense in which our theories are getting better. (VS) can explain this and so is a better account than (CK).

Note first that this criticism would also hold of (CT), the view that progress is a matter of accumulating truths. (O) seizes on the truth component of knowledge; since knowledge entails truth, only true propositions can contribute to progress, according to (CK). Trivially the same is true of (CT). (O) does not say anything about the epistemic element of (CK). Accordingly my strategy will in effect be to show how we can defend (CT) against this criticism. That is, I will show that when there is a sequence of false propositions that get closer to the truth, and so is progressive according to (VS), there is also a related sequence of true propositions that is progressive according to (CT). Of course that does not show that we have progress according to (CK). For that requires epistemic conditions to be met also. But (O) has not raised any problems for the epistemic side of things. The relevant true propositions can

[13] Barnes himself thinks that we can use these epistemic features of such cases to identify an appropriately privileged language—appropriate to that epistemic circumstance. The language issue strikes me as a red herring and that Barnes's contribution is more effective without the detour via language.

be known and so the sequence can be progressive according to (CK), should the additional epistemic conditions be met.

The idea is to argue that once we have granted a notion of approximate truth, then certain relevant propositions will have full truth (e.g. those of the form 'approximately *p*').[14] Since such propositions are fully true, they are potentially knowable (should the right epistemic conditions be met) and the objection in (O) does not apply.

We can accept that in many cases we need to use a notion of approximate truth without agreeing that there are no relevant propositions that can be characterized as fully true, as (O) implies. If *p* is strictly false but approximately true, then the proposition *q*, that *p* is approximately true, is itself strictly true, not merely close to the truth. This is legitimate, since if 'planets travel in ellipses' is a scientific proposition, then so is 'approximately, planets travel in ellipses'. Even if *p* is not true and so not knowable, q ($q =$ approximately, p) might well be knowable.[15]

One might have a conception of scientific theories whereby 'approximately, *p*' is not a theory even if '*p*' is. After all it is '*p*' that will be used in explaining and predicting, not 'approximately, *p*'. An objector may also suggest that whereas '*p*' is about the world, '*p* is approximately true' is a higher-order proposition about *p*, and for this reason is not a theory.

First, this is irrelevant to the defence of (CT) and (CK). These tell us that an episode is progressive when the stock of scientific truths, (CT), or of scientific knowledge, (CK), grows. They do not require that the true or known propositions be theories. Not all scientific propositions are theories, and so our conception of theories is irrelevant.

Secondly, propositions such as 'approximately, *p*' can be used in explanations and predictions. For example, the fact that it is approximately true that the only force acting on the Earth is a central force directed towards the Sun explains why the Earth's orbit is approximately an ellipse. This is a better explanation than one involving the 'non-fact' that (precisely) the only force acting on the Earth is a central force. Of course, explanations involving 'approximately, *p*' are often parasitic on what *p* would explain if *p* were true. But not always. That the motion of a pendulum is approximately simple harmonic is explained by the fact that for small θ, $\sin\theta$ is approximately equal to θ. This is not best understood in terms of what would be explained if $\sin\theta$ were exactly equal to θ. For if $\sin\theta = \theta$ (exactly) then $\theta = 0$, and we do not have any pendulum motion at all.

Thirdly, theories of truth acceptable to realists will regard propositions such as '*p* is true' and '*p* is approximately true' as at least partly about the world, while minimalists about truth will happily regard such propositions as being entirely about the world.

Having granted that the concept of approximate truth has a use, we can see that cases that show increasing nearness to the truth will also show the accumulation of (full, strict) truth. Those cases can in principle show the accumulation of knowledge

[14] Barrett (2008) gives an account of approximate truth for cases of the kind discussed here.

[15] 'Approximately, *p*' may be read as '*p* is approximately true'.

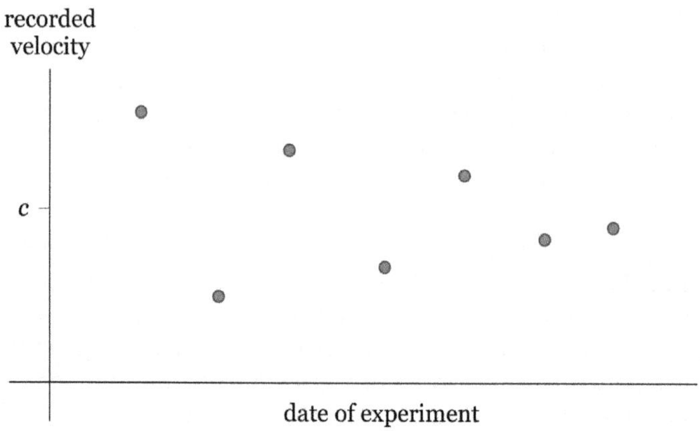

Fig. 3.7: The points represent a sequence of experimental results for the value of the speed of light, converging on the correct value, *c*.

also, should the required epistemic conditions be met. They therefore present no special reason for preferring an account of progress in terms of verisimilitude. Let $< p_1, \ldots, p_k >$ be a series of hypotheses, accepted in that order over time, that monotonically get closer to the truth. This sequence shows progress according to (VS). We can see that there is also a distinct but related sequence of propositions, entailed by the first, that exhibits the accumulation of truth and (potentially) the accumulation of knowledge. Let $A\ldots$ be a propositional operator whose meaning is given thus: Ap iff approximately p. Assuming first, for simplicity, that all the p_i are approximately true, the sequence of propositions $< Ap_1, \ldots, Ap_k >$ will be a sequence of propositions each of which is fully true and adds to the truth provided by its predecessors. So the sequence is progressive according to (CT). If these propositions are believed, then we have an accumulation of fully true propositions; and if they are sufficiently well supported by the evidence, the propositions will be known and we will have progress according to (CK). The improving precision of our approximations can be an object of knowledge. As Bragg (1936: 41) states 'The first purpose of scientific enquiry is to add to the extent and accuracy of our knowledge.'

The latter is one specific kind of case. Sometimes the approach to the truth may not be monotonic. We will now consider a case of non-monotonic convergence. Let c represent the speed of light, and Val(x) represent (de re) the numerical value of x (with appropriate units). (So the sentence '$c = \text{Val}(c+\delta)$' where $c = 2.99793x10^8 ms^{-1}$ and $\delta = 0.00002 \times 10^8 ms^{-1}$ expresses the proposition: $c = 2.99795 \times 10^8 ms^{-1}$.) Researchers might over time attribute best values to c, numerically equivalent to:

$$\text{Val}(c+\epsilon_0), \text{Val}(c-\epsilon_1), \text{Val}(c+\epsilon_2), \text{Val}(c-\epsilon_3), \ldots,$$

where $\epsilon_0 > \epsilon_1 > \epsilon_2 > \ldots > 0$. The improving approximations fall on either side of c, converging from both sides, as in Fig. 3.7. Following the approach of the preceding paragraph, we would form the following sequences of true propositions using the 'approximately' operator:

$$A(c = \mathrm{Val}(c + \epsilon_0)), A(c = \mathrm{Val}(c - \epsilon_1)), A(c = \mathrm{Val}(c + \epsilon_2)), A(c = \mathrm{Val}(c - \epsilon_3)), \ldots.$$

A problem arises however. For if the first two propositions in the sequence are true, then so are all the others. For example, if someone is both approximately 180cm tall and also approximately 185cm tall, then it follows that they are approximately 182cm tall. Adding the later members therefore does not add any truth. And so knowing them does not add any new knowledge. So the argument employed above that where there is verisimilitude there will be accumulation of truth (and so potentially knowledge) will not apply here.

The problem arises because a single 'approximately' operator is insufficiently discriminating. We use multiple such operators in ordinary English, from the less precise 'very roughly' to the more precise 'almost exactly'. As our estimates get more accurate, we can use more precise approximation operators and get a truth. Let p and q both be propositions expressing false estimates of the value of a constant, where p is more accurate than q. Then 'roughly, p' and 'roughly, q' may both be true, whereas 'almost exactly, p' may be true while 'almost exactly, q' is false. So we consider a sequence of increasingly demanding approximation operators: A_0, A_1, A_2, ... For example, A_0 = very roughly, A_1 = roughly, A_2 = to a fair approximation, A_3 = very nearly. Or the A_i could be more formal and precise approximation operators suitable for the appropriate field (e.g., margins for error: ±15%, ±10%, ±5%, ±2%, etc.). For some such sequence of approximation operators, the propositions:

$$A_0(c = \mathrm{Val}(c + \epsilon_0)), A_1(c = \mathrm{Val}(c - \epsilon_1)), A_2(c = \mathrm{Val}(c + \epsilon_2)), A_3(c = \mathrm{Val}(c - \epsilon_3)), \ldots$$

will be a sequence of true propositions whose later members do add to the truths entailed by the earlier members. Since researchers can see that the later values are within the bounds of earlier ones, they can have reason to believe that the later approximations are better than the earlier ones (in addition to whatever reason they may have for thinking that their techniques are more accurate). Hence they are likely to have beliefs where the later ones involve attributions of better approximation (even if the belief is merely of the form '$(c = \mathrm{Val}(c + \epsilon_2))$ is a better approximation than $(c = \mathrm{Val}(c + \epsilon_0))$'). Given the assumption of appropriate confirmation, the relevant sequence of beliefs will also represent the accumulation of knowledge. Although there is still a margin for error, we know the value of the velocity of light with greater accuracy than previously.[16]

The problem addressed by this section is raised by the kind of circumstance that motivates use of the idea of verisimilitude: propositions (theories, etc.) which are not true but which have some kind of nearness to the truth. Theories can improve without being strictly true so long as they get nearer to the truth. This might arise with grand theories: the general theory of relativity is nearer to the truth than Newton's

[16] A modern graphical representation would show shrinking error bars. In many such cases the information will be probabilistic. The relationship between probability and knowledge is discussed below, in Chapter 8.

gravitational theory which is nearer to the truth than Aristotle's account of falling bodies. Or with focused hypotheses, e.g. regarding the value of some constant. This motivates thinking that (CT) cannot be right: progress is not the accumulation of truth, because developments such as these show progress but no increase in truth (all the propositions are false). If (CT) is not right, it needs to be replaced by (VS). Furthermore, it follows that (CK) cannot be right, since false propositions cannot be known.

In effect I have argued that (CT) can be defended against that criticism. If (VS) says that a sequence of propositions S shows increasing verisimilitude, then there is a related sequence S* that shows increasing truth. S* is obtained from S by use of appropriate approximation operators. So (CT) does not have a problem on this score. That is not to say that (CT) is correct—it is not correct because it lacks an epistemic component. Rather than (CT) we need (CK). The importance of the partial defence of (CT) is that if we have a sequence of true propositions, each of which adds to the truth of its predecessor, then we have a sequence of propositions that can be known, if the correct cognitive conditions are in place. In the case of convergence on the truth, it remains possible that we will find the accumulation of knowledge.

3.5 The understanding approach

I began the preceding chapter with a quote from Aristotle: All people by nature desire to know—*πάντες ἄνθρωποι τοῦ εἰδέναι φύσει*. While 'know' is the most suitable translation of '*εἰδέναι*', Lear (1988) and Burnyeat (1984) argue that Aristotle in fact takes the object of inquiry to be *ἐπιστήμη*', and that '*ἐπιστήμη*' denotes *understanding*.[17]

If Lear and Burnyeat are correct, then Aristotle's considered view, the first sentence of the *Metaphysics* notwithstanding, is that humans by nature desire understanding, and that understanding is thus the aim of science.[18]

So I now address a third alternative to the epistemic approach, one which says the scientific progress is a matter of increasing understanding. I will first argue that this cannot be correct, since understanding is not necessary for progress—we can find many example of progress that do not involve an addition to understanding (but which do involve additions of knowledge).

That said, while understanding is not necessary for progress, I agree that science without understanding is an attenuated kind of science. Progress with understanding is generally more valuable than progress without it. (CK) is consistent with such claims because, in my view, understanding is a species of knowledge: knowledge of explanatory relations. Knowledge of causes, mechanisms, laws, and the like is particularly significant scientific knowledge and is frequently (but not always) the goal of scientific inquiry.

[17] Lesher (2001) disagrees, holding '*ἐπιστήμη*' to mean something more like 'expert knowledge'.

[18] All are agreed that '*ἐπιστήμη*' denotes a reasonably sophisticated kind of knowledge/understanding, close to what we would call 'scientific knowledge/understanding'. I note, bearing in mind what follows, that Lear and Burnyeat do not deny that *ἐπιστήμη* is one kind of *ειδήμων*—that is, that understanding is a kind of knowing.

As I say, I will show that understanding is not necessary for progress. But is it sufficient? Given my view of understanding as a species of knowledge, trivially any episode that is regarded as progressive by the understanding approach ought also to be progressive according to the epistemic approach. The relationship between understanding and knowledge is, however, controversial; there is a lively debate between those who hold, as I do, that understanding is a species of knowledge (Achinstein 1983; Lipton 2004; Grimm 2006; Bird 2007*d*; Sliwa 2015) and those who deny this (Elgin 1996; Elgin 2004; Elgin 2007; Kvanvig 2003; Hills 2016; Dellsén 2016).[19] Rather than attempt to resolve that debate here, I directly address Dellsén's *noetic* account of progress which contends that scientific progress is the acquisition of understanding while denying that understanding entails knowledge.

3.5.1 *Progress without understanding*

The understanding approach and the epistemic approach diverge on cases that are additions to scientific knowledge but which make no contribution to scientific understanding. Science is full of such cases and, I shall argue, they clearly contribute to progress. In the late eighteenth century and early nineteenth century, Caroline Herschel discovered multiple new comets and with her brother William produced catalogues that recorded thousands of new stars and nebulae, while also correcting errors in Flamsteed's earlier catalogue. William himself discovered the planet Uranus, the first new planet to be observed since ancient times. These additions to scientific knowledge were important contributions to progress in astronomy—their contemporaries certainly thought so. Their work was published in the Royal Society's *Philosophical Transactions* and both brother and sister were honoured by scientists and scientific organizations in Britain and overseas. Yet it is difficult to see that such discoveries added anything to scientific understanding. No new explanations were proffered nor even any novel predictions made. Similar remarks might be made about the many naturalists of the same period, such as Gilbert White, who recorded the habits of earthworms, discovered new species of beetle, made censuses of birds, recorded details of weather and climate, and so on. While undoubtedly some of these were additions to understanding, many of these systematically recorded facts were not. Even so, they were a central element in the progress of the life sciences of the time.

The accumulation of knowledge of facts just described contributes to progress because they add detail and accuracy to our scientific picture of the world, which they can do without adding understanding. Other additions to knowledge of this kind include the determination of the value of constants, such as the measurement of the gravitational constant G by Henry Cavendish's torsion balance experiment in 1798 through to Paul Heyl's work in 1930 and beyond (and likewise for the speed of light, the fine structure constant, etc.). It is difficult to see that such work adds to our

[19] My own view (Bird 2007*d*: 84), like Lipton's (2004: 30), is that understanding is knowledge of causes and other explanatory relations. The view of understanding as knowledge of causes is developed in detail by Grimm (2014).

understanding—it is the theories that incorporate these constants that do that. Yet this work contributes to progress by adding to our knowledge of the world.

Discoveries are often important contributions to scientific progress because they open up new avenues for scientific research and do so precisely because they are *not* accompanied by understanding. Röntgen's discovery of X-rays in 1895 is a case in point. This was immediately recognized as important, and Röntgen himself studied the various properties of the new rays. Yet neither Röntgen nor his contemporaries had much understanding of their nature (Assmus 1995). Röntgen and Philip Lenard held that the rays were disturbances in the aether (which does not exist). They surmised that they were a form of light. But this correct guess was undermined by their falsely thinking that they were qualitatively different from other more familiar forms of light (visible light, infrared radiation, ultraviolet light) by being longitudinal waves rather than transverse, and by classing them together with cathode rays, which they wrongly thought were also vibrations in the aether. Other leading scientists had other theories and speculations, that X-rays were vortices in the aether (Michelson) or impulses in the aether (Stokes and Thomson) or an acoustic phenomenon (Edison) or a gravitational phenomenon (Lodge). Despite this lack of understanding, the discovery of X-rays was a significant moment in the progress of physics.

3.5.2 *No progress without knowledge*

According to the Dellsén's (2016) noetic account of progress, understanding is a matter of being able to predict and explain aspects of the world, hence the account of progress states:

> (NA) An episode in science is progressive precisely when scientists grasp how to correctly explain or predict more aspects of the world at the end of the episode than at the beginning (Dellsén 2016: 72).

Note that the requirement of *correct* explanation and prediction parallels the truth (or verisimilitude) requirement in the semantic approach. Neither approach has an epistemic component such as justification or knowledge.

The counterexamples discussed above in Section 3.5.1 also refute (NA). None of these examples of progress add explanatory power to science. Some do add predictive power, by collecting projectible generalizations. But that is not what makes them additions to progress—true but accidental correlations do that too.[20] Nonetheless Dellsén does provide an example which, he argues, supports (NA) but refutes (CK).

In his famous 1905 paper 'Investigations on the Theory of the Brownian Movement', Einstein derived predictions from the molecular theory of matter and kinetic

[20] Dellsén might amend his account so that progress requires correct predictions to be produced reliably, not accidentally—the scientists' 'grasp' enables them to generate correct predictions in similar but non-actual circumstances. To require that kind of modal reliability is tantamount to requiring that the predictions amount to knowledge. I note also that correct prediction, even reliably correct prediction, is too weak to count as understanding. For one can know, for example from an observational study in medicine, that two factors are correlated, so that one can predict the second from the first. But since correlation is not causation one might have this ability without any understanding of why these factors are correlated; one might not know whether one factor causes the other or instead both are effects of some unknown common cause.

theory of heat to show that microscopic particles will perform movements that are observable with a microscope. 'It is possible that the movements to be discussed here are identical with the so-called "Brownian molecular motion" ', wrote Einstein, 'however, the information available to me regarding the latter is so lacking in precision, that I can form no judgment in the matter'. It looks as if Einstein's evidence was insufficient for him to have justification for a belief that the motions he predicts are Brownian motion. Without that justification, he could not *know* that the molecular-kinetic theory explains Brownian motion, even though that theory does in fact explain Brownian motion. Furthermore, the kinetic theory itself was not known at that time to be true. Dellsén contends that Einstein's work was nonetheless a contribution to scientific progress; it did constitute understanding of Brownian motion; so the case refutes (CK) while confirming (NE).

While this is a challenge to (CK), the defender of the latter has good grounds for declining to accept it as a refutation. Dellsén's case rests on Einstein's reticence in the light of what he says is imperfect evidence. Einstein's declared reticence may be understandable—especially as his results contradicted the opinions of leading scientists such as Ostwald and Mach, whom Einstein held in high regard. The title of the paper nonetheless gives away what he thought really was the case. As it was, other scientists, particularly those who knew the experimental work on Brownian motion rather better than Einstein, did accept this as an explanation of Brownian motion, as Einstein noted in a paper the following year.[21] The Polish physicist Marian Smoluchowski simultaneously came to similar conclusions as Einstein, with stronger empirical backing. And shortly afterwards Jean Baptiste Perrin gave incontrovertible experimental confirmation of their results. So it is not entirely clear that Einstein did not know that he had explained Brownian motion, even if his reticence is genuine. Lack of confidence is consistent with knowledge.[22] More importantly, it is possible that *the scientific community* or an important section of it had this knowledge as a result of Einstein's paper. What determines whether there is progress in science is not typically a matter of what an individual knows but what the community of scientists knows.

We need to consider a case where it is clear that neither Einstein nor anyone else was in a position to know that the motions that Einstein predicted were Brownian motion, but believed they were. Imagine that all the available experimental data on Brownian motion mistakenly gave it a magnitude very different from that predicted by Einstein. This is a result of an experimental error, but one that could not be detected at the time. So Einstein and everyone else have strong reasons to think that the predicted motions are different from the Brownian motion. Yet Einstein irrationally believes that his motions are Brownian motion (perhaps because he has spent so much intellectual energy in trying to explain Brownian motion). And the

[21] Einstein was characteristically uninterested in the details of experimental work.

[22] The case of the unconfident exam candidate or quiz show contestant who nonetheless answers the questions correctly, because she knows, is well known. This is one reason among others to reject the KK principle, that if a subject knows something, then she knows that she knows it. Einstein might well have known what explains Brownian motion without knowing that he had that knowledge.

scientific community goes along with him.[23] And so, although it is true that Einstein's predicted motions are Brownian motion, no one in fact knew this. Do Einstein and his contemporaries, in this scenario, understand why Brownian motion occurs? It seems to me that they do not have understanding. Whether or not we judge there to be understanding, it is clear that this case does not show progress in science, and so (NA) is mistaken in saying that it does.

Returning to the original, actual Einstein case, we should note that even if it does show progress without knowledge, the defender of an epistemic approach to progress can respond by adopting the more nuanced position, (CK′). The latter allows that something may contribute to scientific progress by directly promoting the growth of knowledge without necessarily itself being an addition to knowledge. There is no doubt that Einstein's paper did this. In the wider epistemic context of his day and in particular by stimulating the work of Perrin, that work led directly to knowledge of the correct mechanism of Brownian motion and knowledge of the kinetic theory and of the molecular constitution of matter, which were accepted almost universally by 1909. In general, true theories that are insufficiently well-evidenced to be known are contributions to progress in this way. If you are on a journey from Paris to Amsterdam and are passing through Brussels, then you have made progress to your goal but have not reached it (you are a little over halfway there). Likewise, proposing a true but as yet unjustified or only partially justified theory is not to reach the goal of knowledge but it is to make important progress towards that goal. In particular the theory may stimulate further investigations that will generate the evidence that will lead to the theory's confirmation and its becoming known. So we can have progress without knowledge in such special cases.

More significantly, Einstein's paper did add important *abstract* knowledge. This knowledge concerns the *relationship* between the kinetic-molecular theory and predictions of observable motion by microscopic particles. The paper (Einstein 1956: 1) tells us this explicitly in its opening sentence: 'In this paper it will be shown that according to the molecular-kinetic theory of heat, bodies of microscopically-visible size suspended in liquid will perform movements of such magnitude that they can be easily observed in a microscope'. So the central claim to knowledge is not a categorical claim that the world is some way but is a conditional or hypothetical claim, that if the world is a certain way (such that the kinetic theory of heat is correct), then there will be observable motion by suspended microscopic particles. And there is no doubt that *this* claim is correct. And, furthermore, it is scientific knowledge, and important scientific knowledge at that.

[23] I find that some philosophers of science do not like discussing such counterfactual examples. To them I say: (i) If our judgments of actual examples are thought to be useful in helping answer questions of the relationship between progress, understanding, and knowledge, why should our judgments regarding counterfactual examples not be useful? (ii) It is possible to find real examples that have some of the relevant features—the story of N-rays is a real example of a scientific community having beliefs that were not well founded in evidence. But the detail of the real examples can detract from our understanding, whereas fictional examples, being cleaner and less complex, can help identify the relevant features. In a way they are like *in vitro* experiments compared to *in vivo* observations.

Dellsén (2016: 77) anticipates this move and suggests that I cannot make it without contradicting myself. Above, in Section 3.3, I discuss Laudan's problem-solving approach to progress, which says that if one can deduce a statement of a problem phenomenon from a theory, then that theory solves the problem, so long as problem statement and the theory are both believed to be true—even if they are in fact false. I pointed out that medieval thinkers such as Oresme believed that hot goat's blood could split diamonds—call this alleged phenomenon 'P'. If Oresme were to develop a theory, T, that predicts P, *that* clearly could not count as progress, I argued, *pace* Laudan, because there is no knowledge (T would be false, of course). Dellsén notes that in rejecting this case I am rejecting the idea that there is progress because Oresme knows that T predicts that P. If I allow relational knowledge to count towards progress in Einstein's case (the relational knowledge = that the kinetic theory would explain microscopic motion), then I ought to do the same in the Oresme case also (the relational knowledge = that T would explain/predict P).

I disagree. Einstein's derivation involved important scientific work. It uses sophisticated and correct mathematics, important and true auxiliary hypotheses (such as the theory of thermodynamics and the value of Avogadro's number), and its reasoning is both novel and impeccable by objective standards. We saw that the standard medieval explanation for the splitting of diamonds by goat's blood appealed to an analogy between the diamond and the hard hearts of the enemies of Christ, and between the goat's blood and Christ's blood. Unlike in Einstein's paper, there is no scientific work in the derivation of the explanandum from the explanans. One might well question the truth of the auxiliary assumptions here and the reliability of the analogy and so on. Let's say one were to couch the explanation in such a way that splitting of diamonds by goat's blood follows from it by an unarguable deduction (e.g. by making all the auxiliary propositions into axioms of the revised explanatory theory). Even so the logical derivation, which can be known to be valid, would be of no scientific value; it would almost certainly be a very trivial piece of logic; so even if an object of knowledge, that derivation would not be an object of scientific knowledge. Not everything one comes to know in the course of a scientific investigation is thereby a matter of *scientific* knowledge.

There are cases where knowing the logical/mathematical relationship between a false theory and a non-existent phenomenon *is* a contribution to scientific progress because knowledge of that relationship is scientific knowledge. In 1905 James Jeans and Lord Rayleigh, and, independently, Einstein, showed that the equipartition theorem of classical statistical mechanics predicts a phenomenon known as the 'ultraviolet' (or 'Rayleigh–Jeans') catastrophe. A black-body oscillator emits energy at all frequencies, and the Rayleigh–Jeans law derived from classical mechanics relates the energy of the radiation and its frequency. While the observed relationship for low frequencies is close to predictions, for high frequencies the predicted energy approaches infinity (this is the catastrophe)—which patently is not observed. In making the connection between classical mechanics and the ultraviolet catastrophe, Jeans, Rayleigh, and Einstein were relating a false theory and a phenomenon that they knew in advance did not exist. So the derivation does not count as a contribution to understanding, according to Dellsén's view of understanding which requires the correctness of the explanation. And so this work cannot be a contribution to progress according to (NA). Jeans, Rayleigh, and Einstein did do important scientific work, it led to knowledge, knowledge that classical mechanics predicts an ultraviolet catastrophe. The case for regarding this work as progressive is strong. It was an important step in the development of the quantum theory. For in showing that something important was wrong with classical mechanics it led directly to Planck developing the quantum concept by proposing an alternative to the equipartition theorem.[24]

3.6 Conclusion

The central part of this chapter has argued that our judgments concerning whether an episode in science is progressive are best explained by our taking progress to be a matter of accumulating knowledge, not increasing problem-solving power nor greater nearness to the truth. This conclusion matches conception of progress of scientists, according to Mizrahi (2013). The same conclusion is reached by starting from our earlier account of the aim of science as knowledge, and taking progress to be a matter of achieving, promoting, or getting closer to that goal. Furthermore, the view that scientific progress is increasing verisimilitude, popular among philosophers of science, suffers from the besetting problem of accounts of verisimilitude, that verisimilitude is language-relative. That problem is neatly bypassed by the epistemic approach, and (CK) in particular. On the other hand, the fact that our preferred theories can be progressive without being true, and so without being known, looks to support (VS) over (CK). But this is not an obstacle to the epistemic approach after all, for we can find a set of related propositions which show the accumulation of full truth and so which can be known. The problem-solving account, (PS), within the functional-internalist approach, suffers from analogous problems, because that account and approach imply the absurd consequence that genuine progress can be made by badly erroneous science.

I also considered the noetic account, (NA). This fails to account for contributions to scientific progress from the collection of important data and from the discovery of new but poorly understood phenomena. The noetic account also fails to accept that

[24] P. D. Magnus (2007) holds that we cannot find a goal for science, and so the characterization of science as distributed cognition (see Chapter 4) is problematic. He notes that the goal of knowledge does not allow for the speculative activity of exploring a hypothesis. He also argues that the aim fails to discriminate between worthwhile and pointless knowledge. Both points can be answered. Some instances of 'exploring a hypothesis' do generate knowledge—for example knowledge of the mathematical or logical relationships between the hypothesis and other hypotheses, models, etc. More importantly, exploring a hypothesis is scientifically worthwhile because it can lead to knowledge—for example by allowing us to see what evidence might refute or confirm the hypothesis. If the aim of science is knowledge, then many activities (such as designing an experiment) will count as scientific because they are conducive to that end even if they do not in themselves achieve that end. Magnus is right that the kind of knowledge that science aims at needs to be circumscribed. He himself points to Kitcher's (1993) notion of significant truth. There are problems, as Magnus says, in saying what exactly 'significant' amounts to. I suggest that significance is relative to a scientific tradition of a Kuhnian kind. But the existence of such difficulties in articulating precisely the goal of science is not a strong reason for denying that there is such a goal—the commonplace idea that there is such a thing as scientific progress implies that science does have a goal.

the correction of scientific errors can be a contribution to progress, whereas (CK) can hold that knowledge of negative truths or knowledge that some theory is false can sometimes be valuable contributions to progress. Instances of such progress will include theoretical advances that show that an accepted theory turns out to have false consequences. (NA) is intended to be an exemplar of the understanding approach to progress. I have briefly outlined my own view that understanding is a species of knowledge—which differs from the view of understanding as being able to give correct explanations and predictions. The examples just given show that any reasonable version of the understanding approach will provide an incorrect necessary condition on progress.

Because knowledge is more difficult to achieve than truth or problem-solving (understood internalistically), progress according to (CK) is more difficult to achieve than it is according to (PS) or (VS). But it is also more robust, in that it is less easily lost in the light of new evidence. So even if our beliefs are not cumulative, our knowledge might be. This suggests one way of articulating realism about science (or, better, about a particular scientific field): science (or that field) is progressive.

4

SCIENCE AS SOCIAL KNOWING

Science is an institution. It is a social entity, as are the communities of specialists in a particular field, learned societies, and research groups. To these we often attribute knowledge. What does it mean to say that a group knows something? How does the epistemology of groups relate to the epistemology of individuals? I contrast two general approaches to social knowing: the supervenience approach—social states supervene on the states and behaviour of individuals; and the analogy approach—social states are analogues of individual states. The exemplar of the supervenience approach is the commitment model, so called because a group believes a proposition when the individuals in the group are jointly committed to letting that proposition stand as the view of the group. The exemplar of the analogy approach is the distributed model. According to this model, there are social structures that have cognitive functions—social cognitive systems—within which individuals typically have distinct functional roles, roles that are often mediated by physical infrastructure. I argue in favour of the analogy approach and of the distributed model in particular. Social knowing neither reduces to nor even supervenes on the epistemic or other mental states of individuals. Its supervenience basis includes the material means of information production, storage, and exchange. I extend the Durkheimian sociology introduced in Chapter 2 to show that science, characterized by distributed cognition, really does consist of social entities capable of possessing cognitive states.

4.1 Introduction

In Chapter 2 I characterized science as an institutional embodiment of what Aristotle says is our natural desire to know. It is therefore both an epistemic activity and a social activity. This was also assumed in Chapter 3, where I applied arguments from epistemology to scientific communities as well as to individuals.

Yet we—philosophers especially—tend to think of epistemology as primarily concerned with the individual. That is in part a matter of convenience and tradition, for which reasons much of the epistemological discussion in this book is couched in the individual idiom. Yet, over-emphasis on the individual leaves the social nature of scientific knowledge under-explored. This chapter aims to correct that. It will thereby also reinforce the arguments surrounding knowledge ascribed to groups and communities in the preceding chapters. It will argue that a scientific community can be the possessor of knowledge; that knowledge does not supervene on the mental states

Knowing Science. Alexander Bird, Oxford University Press. © Alexander Bird 2022.
DOI: 10.1093/oso/9780199606658.003.0004

(or actions) of individuals; and that when a scientific community knows something that is best understood in terms of distributed cognition.

4.1.1 *Social–social epistemology*

The individualism of traditional epistemology has two aspects. First, as mentioned, it has focused on attributions of knowledge to individuals. Secondly, it has ignored an individual's social environment or has treated it as akin to an individual's physical environment. What is known as 'social epistemology' broke initially with traditional epistemology as regards the second of these respects. Epistemologists began to focus on social causes, testimony especially (e.g. Coady 1992; Lipton 1998), of an individual's state of knowledge as being particularly important or even as disanalogous to the physical causes. 'Individual–social epistemology' retains the focus on individual knowers while socializing their epistemic context.[1,2]

A second kind of social epistemology, 'social–social epistemology', diverges from traditional individualistic epistemology by considering groups or social structures themselves as legitimate possessors of epistemic states.[3] It is common to assert the existence of an impersonal species of knowledge. And this is especially common in relation to science. The following are typical examples:

(I) The pure maths group in our department hopes that within five years it will know whether the Riemann hypothesis is true.
(II) The North Koreans know how to build an atomic bomb. (*Or* North Korea knows how to build an atomic bomb.)
(III) We now know that peptic ulcer is often caused by the bacterium *Helicobacter pylori.*
(IV) The world knows that global warming is a major threat.
(V) The growth of scientific knowledge has been exponential since the scientific revolution.

Social–social epistemology emphasizes not only the social nature of the epistemic context but also the social nature of the epistemic subject. According to social–social epistemology, reference to a social epistemic subject is not merely an elliptical way of talking about certain individual epistemic subjects.

We are used to individual epistemic subjects—for the most part there is no particular puzzle about their existence. Although, as our examples attest, we often talk of social epistemic subjects, we have a rather less clear idea about which entities can be social epistemic subjects. While there are borderline or unusual cases of putative individual subjects, those cases are atypical and need not trouble the individualist epistemologist. On the other hand, the existence of social subjects at all may be

[1] Individual-social epistemology is exemplified by Alvin Goldman's (1999) *Knowledge in a Social World.*

[2] Richard Pettigrew, in correspondence, notes the parallel shift in rational choice theory from decision theory, which treats other individuals just as aspects of the agent's environment, to game theory, which gives them a distinctive theoretical role.

[3] Important contributions to social–social epistemology include Gilbert (1987, 1994, 2004), Tuomela (1992, 2004), Corlett (1996), Pettit (2003), Tollefsen (2006), Palermos (2020). This chapter draws on Bird (2010*c*, 2014).

questioned. Which groups, collections, institutions, or social structures, if any, can be social epistemic subjects? This ontological question is pressing for the social–social epistemologist. So an important part of what follows will address the metaphysics of groups, aiming to give a general answer that will allow us to see which scientific social entities there are, to which epistemic states can correctly be attributed. I shall argue that amongst these are not only formal research teams, but also disciplines and specialist sub-fields; furthermore the global scientific community as a whole is a social epistemic subject, as is (thanks to globalization, in particular as regards communication) the large majority of the human species collectively.

In what follows I shall use 'social knowing' to refer to a state of knowing possessed by some social entity, such as a nation, a government, a corporation, a community of scientific research specialists, a research team, and so forth. I shall use 'group' to refer to any such social entity, which is to say to any social entity correctly treated as such by social science. And so in this context 'group knowing', 'group knowledge', and 'social knowing' mean the same.

4.1.2 *Two approaches to social–social epistemology*

In this chapter I contrast two approaches to social–social epistemology. One accounts for the existence of group knowledge in terms of the mental states of the individuals constituting the relevant group. What the group knows depends on what those individuals know or believe or intend or are committed to and so on. Typically this approach will endorse or imply a supervenience thesis: the epistemic states of a social entity supervene on the mental states of the individuals making up that social entity. So I shall call this the *supervenience* approach to group knowledge.

The most sophisticated version of the supervenience approach is what I call the *commitment model* of the social epistemic subject, so called because a group believes a proposition when the individuals in the group are jointly committed to letting that proposition stand as the view of the group. The commitment model is the model developed by the leading philosophers who study group states, such as Margaret Gilbert and Raimo Tuomela, and it can be said to be the standard view of group belief and knowledge. The working of a formal committee will typically satisfy the commitment model.

A different, more radical, approach to social–social epistemology rejects the supervenience thesis. It sees the social epistemic subject not as supervenient on individual subjects but rather as analogous to them. So I call this the *analogy* approach. An important version of the analogy approach is the *distributed model* of social cognition, which I have co-opted from cognitive anthropology. According to this model, there are social structures that have cognitive functions—social cognitive systems—within which individuals typically have distinct functional roles, often mediated by physical infrastructure. Edwin Hutchins (1995*b*) describes how the cockpit of a commercial airliner knows its correct landing speeds and flap configurations through an interaction of the pilots and their equipment. If individual knowing is unanalysable then the same will be true of social knowing. Nonetheless, one can say that, roughly, a collective entity knows that *p* when it has (or is) a properly functioning social cognitive system, of which the proposition that *p* is a product. As we saw in Chapter 2, a cognitive system is one that has the function of producing true outputs that can be the inputs for deliberation and action.

I shall argue that the analogy approach—in the guise of the distributed model in particular—is the correct way to understand the social nature of scientific knowledge as exemplified in (I)–(V). Social groups to which knowledge can be attributed exist at different scales. The paradigm group in science is a research team led by a Principal Investigator, working in the same laboratory on a well-defined project. There are also much larger research teams, such as the ATLAS Collaboration mentioned in Chapter 1, which has several thousand members from almost 200 institutions in 38 countries. Those groups are 'established' (to use Gilbert's term) in that they came into being as a result of an intentional act, such as the Principal Investigator winning a grant and hiring a team. They will typically have a formal structure and specific goals to which the group is dedicated. Beyond such established groups in science, there are communities of researchers working in the same specialized field. Paleoarchaeologists, entomologists, geophysicsts, and immunologists, as well as plankton researchers, sleep scientists, and econometricians, form communities bound together by shared knowledge and research techniques and by collaborations and other forms of social interaction. Although there are established institutions, such as specialist subject associations, journals, and conferences that give some structure to these communities, they themselves are not established. Beyond these communities, although there is specialization and diversity in science, there is also unity through interdisciplinarity and, to some extent, shared culture and values, and a few methods, such as controlled trials or statistical analysis, that are found in many different fields. In virtue of that we can talk of 'science' in its widest sense as a group or institution. This is what I call 'wider science'. It is this 'group' that is the knower in the case of impersonal uses of 'knowledge' as in (V). I speculate also that there is a global community, of both scientists and non-scientists, that is also the possessor of that knowledge, thanks to the globalization of communications and of access to science and its products.

4.1.3 *The failure of summative reduction*

Before looking in detail at the two approaches to social knowing, and the two particular models thereof, we should first establish that there is a need for an interesting account of group knowledge. Perhaps there is a straightforward reduction of group knowledge to individual knowledge. Anthony Quinton (1976: 19) says the following:

> We do, of course, speak freely of the mental properties and acts of a group in the same way we do of individual people. Groups are said to have beliefs, emotions and attitudes and to take decisions and make promises. But these ways of speaking are plainly metaphorical. To ascribe mental predicates to a group is always an indirect way of ascribing such predicates to its members. With such mental states as beliefs and attitudes the ascriptions are of what I have called the summative kind. To say that the industrial working class is determined to resist anti-trade-union laws is to say that all or most industrial workers are so minded.

Quinton is articulating a *summative* view of group states: a group is in state Φ iff the members of the group are all/mostly in state Φ. The group state thus reduces to the corresponding states of individuals and so is the simplest version of the supervenience approach.[4,5]

None of the leading philosophers of group epistemology think that the summative account is correct. It is nonetheless revealing to look at the reasons given for rejecting the summative account. Supporters of the commitment model reject the summative account, arguing that a more sophisticated version of the supervenience approach is required. I argue that their reasons for rejecting the summative approach are not only mistaken but in fact reveal a significant weakness of the commitment model.

We shall see that the summative view cannot account for the assertions of group and impersonal knowledge in statements (I)–(V) above.

Let us apply the summative view to the predicate 'knows' thus:

> (K-SUM) It is socially known that *p* iff all or most people individually know that *p*.[6]

Universal or nearly universal individual knowing, which is also called 'mutual knowledge' (Vanderschraaf and Sillari 2005), is both necessary and sufficient for social knowing, according to (K-SUM). But clearly mutual knowledge is not necessary for social knowing. The organization of science is characterized by the division of intellectual labour on a massive scale. Not everyone reads *Nature* from cover to cover. So many of the discoveries it publishes will not be known by everyone in science, indeed one expects that pretty well all recent discoveries will not be known by everyone. If the left-to-right implication of (K-SUM), (K-SUM→), were correct then none of these discoveries would be contributions to scientific knowledge (in the usual social sense), but that is just what we think (or hope) they are.[7] In any case, although some major advances get published in *Nature*, most contributions to scientific knowledge appear in more specialized journals that will only rarely be consulted beyond the limited specialism. But the discoveries published in such journals can be contributions to scientific knowledge in the wider sense.

In the brief discussion above, nothing peculiar to the nature of knowledge was at work, and so we may take these examples to refute the left-to-right implications of the analogous claim about social *belief*:

> (B-SUM) It is socially believed that *p* iff all/most people individually believe that *p*.

[4] Strictly, perhaps, Quinton's view is eliminativist, since he thinks that the group ascriptions are metaphorical not literal.

[5] The summative view is sometimes referred to as 'majoritarianism' or 'majoritarian supervenience' in this context (List and Pettit 2006: 92).

[6] The quantifier may be understood as restricted to the relevant group.

[7] The '→' signifies the left-to-right implication of the biconditional in the corresponding principle, and '←' signifies the right-to-left implication.

So it seems that being known by all or many people is not necessary for social knowing, and likewise being believed by all or many is not necessary for social belief. Nevertheless, is widespread or even universal knowledge *sufficient* for social knowing, i.e. (K-SUM←)? (And, *mutatis mutandis*, regarding belief, *viz.* (B-SUM←)?) I shall argue shortly that it is not.

Others have also argued against (B-SUM←), and so, implicitly, against (K-SUM←), but I do not find their reasons compelling. Corlett (1996: 88), for example, argues that if every member of the Zuni tribe believes 'the world exists' for reasons entirely unconnected with their tribal membership it would not be a belief held by the Zuni tribe (see also Schmitt 1994: 261). He thinks this shows that universal individual knowledge does not suffice for social knowing. I disagree. While he is correct that 'this hardly counts as a distinctively Zuni belief', that is no reason for withholding the ascription. The same belief is hardly distinctive of the individuals in question either, but we do not withhold ascription of the belief to them on that account. Schmitt and others also argue that universal individual knowledge does not suffice for social knowing because the latter requires a commitment that is lacking in such cases. I reject their reasoning towards the end of Section 4.2.1, on the ground that it cannot account for the use of common knowledge in social reasoning. In the next subsection I give my own reasons for rejecting (B-SUM←).

4.1.4 *Mutual knowing without social knowing*

Above I mentioned 'mutual' knowledge:

> *Mutual knowledge* Members of a group have mutual knowledge that *p* iff each individually knows that *p* (Vanderschraaf and Sillari 2005).

which we may distinguish from 'common' knowledge, understood in a technical sense:

> *Common knowledge* Members of a group have common knowledge that *p* iff each individually knows that *p* and each individually knows that each individually knows that *p* and each individually knows that each individually knows that each individually knows that *p* and ... etc. (Lewis 1969).

In the last subsection we saw that mutual knowledge within a group is not a necessary condition on that group knowing something (social knowing). *A fortiori*, common knowledge is not necessary for group or social knowing. We raised the question of whether mutual knowing is sufficient for social knowing. And we can raise the further question whether common knowledge is sufficient for social knowing.

Imagine that everyone individually knows that *p* but each person thinks that they are the only one to know it and desires to keep this knowledge secret. So if anyone were to ask whether *p*, all would claim ignorance. It is not correct to say 'in this society it is known that *p*.' Recall Hans Christian Andersen's story:

> **The Emperor's New Clothes** Confidence tricksters have convinced the emperor, the court, and the citizens of Gymnasia that the emperor's very expensive new suit of clothes is made from remarkable cloth that is perceptible only by those who are intelligent and wise—it is invisible to the stupid or foolish. The emperor parades through the streets of Gymnasia

> wearing the 'new clothes'. Not wanting to be thought stupid, everyone claims to see the (non-existent) new clothes. In fact, everyone sees the emperor's bare skin and so comes to know that he has a birthmark on his shoulder. But, of course, no one admits to this knowledge.

The existence of the birthmark is mutual knowledge. But it is not social knowing. All the citizens of Gymnasia know that the emperor has a birthmark. But Gymnasia does not know he has a birthmark.

So mutual knowledge is not sufficient for social knowing. It is true that cases of common knowledge are *typically* cases of social knowing. Common knowledge is usually able to play a useful social function because everyone can rely on other people having the same knowledge, which in turn means that it can guide group action without further discussion. But there can be exceptions: common knowledge without social knowing.

> **Gosplan** Every member of the State Planning Committee of the Soviet Union, Gosplan, knows that the official production figures are grossly inaccurate. And since they know the other members of the subcommittee are intelligent and well-informed they all know that they all know this. Nonetheless, they all act as if the figures were true—such is their commitment to the propaganda centred on the false figures (or fear of being seen to question them) they do not let their common knowledge enter into or influence their decision making.

In such a case there is common knowledge without social knowing.[8] A similar case would occur when a group refuses to use its common knowledge as a matter of policy: a jury follows instructions to ignore evidence that was not legitimately obtained; an appointment committee refuses to take personal information about a candidate into consideration that a member inappropriately revealed to it.[9]

Likewise, in The Emperor's New Clothes the widespread mutual knowledge plays no social role at all—from the social perspective everyone might as well be ignorant. Consider a scientific analogue in which everyone individually believed some hypothesis but pretended not to. That hypothesis could not contribute to further research, it would not be used as an auxiliary hypothesis in scientific reasoning. It would be redundant in science. So social knowing is something other than just mutual knowledge.[10]

[8] My thanks to Alan Weir for the Gosplan example.

[9] '*We* do not know that', the chair might say—my thanks to Michael Potter for this example.

[10] Gilbert (1989: 257–8) makes a related point with regard to belief—it would be incorrect to say that the Zuni tribe believes that *p* if every member of the tribe believes that *p* but is too afraid to share this belief, thinking that he is the only one with this strange view. Gilbert also rejects the sufficiency of widespread belief for collective belief on the ground that it conflicts with the *members' knowledge principle*, that the group *G* believes that *p* only if the members of *G* know that the group believes that *p*. This principle reflects the fact that Gilbert's model of a social group is a highly organized group, one in which the members all contribute to the belief-forming process. The principle is clearly not true for social scientific knowledge, in which contributions to scientific knowledge can be made such that only a few people know that they have been made. Furthermore, it is not obviously true even for groups that satisfy Gilbert's commitment

What lessons may be drawn from the failure of mutual knowledge (or belief) to suffice for social knowing (or belief)? In the next section I consider two models of social cognition that supply different explanations of this failure. In one model, what makes for a social epistemic subject is a certain kind of commitment to the group from its individual members. Since that is lacking in cases of mere mutual belief or knowledge, the latter cannot suffice for group belief or knowledge. The second model takes the processes of social cognition often to be divided and distributed among individuals, with a social structure for transmitting information between individuals as required by the relevant social task. According to this model, mutual knowledge fails to be social knowledge because in cases such as The Emperor's New Clothes there is no social access to the knowledge that individuals have—a newcomer could not come to know what everyone knows individually. As a consequence that knowledge cannot be used by the group or social structure and so has no social effects—everyone acts as if they do not have the knowledge in question. So mutual knowledge, in this case, is sociologically speaking, an idle wheel.[11]

4.2 Two models of social cognition

In the light of the failure of summative account of social knowing, which would reduce it to universal individual knowing, we now turn to more substantial accounts of what such social epistemic states are. In this section I look in detail at two models, the commitment model and the distributed model. The commitment model is an instance of a more general approach to social states which I have called the supervenience approach. The supervenience approach maintains that the social states of kind *K* supervene on the totality of individual states, including individual states not of kind *K*. (The summative account just rejected can be seen as the most simple if also simplistic model within the supervenience approach.) The distributed model can be regarded as exemplifying the analogy approach to social states, which sees them as analogous to individual states.

4.2.1 *The commitment model*

Collective epistemic activity often takes the following form. A committee is formed in order to make recommendations on a particular question. The committee gathers evidence. It discusses the evidence. By some agreed procedure, it reaches a conclusion and submits its recommendations. Committees are governed by certain norms

model (see next section). A committee may form decisions and beliefs on the basis of a quorum that is less than 100 percent. Those absent will be ignorant of the decisions and beliefs of the committee of which they are a part. Gilbert (1989: 311–12) considers explicitly whether such a committee meets the demands of her account of group belief and argues that it does. But she does not notice the inconsistency with the members' knowledge principle. Indeed her account conflicts with the members' knowledge principle even in cases where all members contribute to belief-formation (see (G) in Section 4.2.1 below). It might be that all members are openly committed to a proposition without all members knowing that each has openly expressed this commitment.

[11] So long as it remains merely mutual knowledge. In the Grimm version of Emperor's New Clothes story, the mutual knowledge is transformed into social (and common) knowledge by the child who shouts out that the emperor has no clothes.

or rules (which may be explicit or implicit) regarding their purpose, how they are managed, how they reach their conclusions, and so forth. Members are aware of and are committed to those norms, at least in their role as members. Gilbert (2004: 97–8) discusses a union, a court, and a poetry-reading group; in these, she claims, there are processes or mechanisms (formal and informal) by which the group belief is formed. These are not merely summative. The individual members of the court may have strong personal opinions, but the court itself has no opinion until the case is concluded, discussed, and voted upon. Hence the sum of individual concurring beliefs is not sufficient for the group belief. Nor is it necessary, since the mechanisms of group belief formation need not reflect the individuals' beliefs—one does not have to vote as one believes. These groups are like epistemic versions of a sporting team—a tug-of-war team might be the best analogy—with each participant committed to their collective goal and to an agreed, commonly understood mechanism for reaching that goal. (The justification for engaging in this form of group cognition is the thought that two or more heads are better than one.)[12]

Committees, courts, and unions are examples of what Gilbert (2004: 96) calls *established groups*, which contrast with sets of people not bound by an established structure, e.g. the set of people wearing striped socks today. Commitment is crucial to established groups. If I just ask a number of individuals their opinion on a matter and aggregate those opinions in order to reach what I hope to be a well-informed conclusion, those individuals do not form an established group. They do not have a conception of the group and *a fortiori* they are not committed to it. This idea is captured in Gilbert's:

> (G) A group *G* believes that *p* just in case the members of *G* jointly accept that *p*, where the latter happens just in case each member has openly expressed a willingness to let *p* stand as the view of *G*, or openly expressed a commitment jointly to accept that *p*, conditional on a like open expression of commitment by other members of *G*. (Gilbert 1989: 306; see also Gilbert 1987; Gilbert 2004)

(G), endorsed by Schmitt (1994: 262), exemplifies the *commitment* model of group epistemic and doxastic states. Raimo Tuomela (1992, 2004) proposes a very similar version of the commitment model, while Philip Pettit (2003) regards satisfaction of the commitment model as a necessary but not sufficient condition on the existence of a 'social integrate'.[13] Gilbert (2004) describes the kind of commitment involved in

[12] How far this is in fact true is another matter. See Kanai and Banissy (2010).

[13] Deborah Tollefsen (2002: 29) makes the very good point that it is odd to think that a group intentional state should be be given an analysis in terms of the intentional states of the individuals in the group, when we can ascribe the group state without knowledge of the states of the individuals. Nor do we ascribe an intentional state to individuals on the basis of knowing the brain states underlying that individual state (Dennett 1987: 48). Her comments thereby reject the commitment model and the supervenience approach. Tollefsen adopts interpretationism as her preferred means of understanding group states, rejecting the functionalism I promote in this chapter. According to interpretationism, if we can interpret an agent in a suitable way, then the agent is an intentional agent (Dennett 1971, 1987). I think we should see interpretationism as falling within the analogy approach, noting, as I point out on p.95, that the analogy in the analogy approach does not need to be a mere analogy.

(G), whereby several individuals come together and commit themselves as a body to something, as *joint commitment*. So (G) says that a group believes that *p* when its members are jointly committed to believing as a body that *p*. Assuming that knowing entails believing, then (G) places necessary conditions on group knowledge.[14]

As a consequence of the shared commitments entailed by (G), individuals making up such an epistemic social subject will typically be conscious of and involved in the epistemic work of the committee.[15] If indeed public commitment to the conclusion is required, then we would expect that individuals, in order to be comfortable with such a commitment, will want to know how the conclusion is reached and to be able to influence that conclusion.[16] Even without the requirement of commitment to the conclusion, merely being associated with that conclusion will often be enough to provide the motivation for understanding and engaging with the decision-making process. Some authors argue that such consciousness is among the norms governing the social epistemic subject.[17]

While (G) requires individual commitment to the group's duly reached conclusions, it does not require that each individual personally believes the conclusion. So group belief does not require mutual belief; hence (B-SUM←) is false. And since mutual belief does not involve joint commitment, (B-SUM→) is false. Hence (G) provides us with grounds for rejecting (B-SUM) and (K-SUM) for established groups. Schmitt (1994: 261) tells us that the Audubon Society does not believe that the sky is blue, while all its members do, even if the latter fact is also known to those members. Schmitt's reason is that the collective commitment to a proposition required by (G) is lacking in the case of 'the sky is blue'.

This, however, leads to problems for the commitment model as presented. Very frequently propositions that are common knowledge, like the proposition that the sky is blue, play a role in group deliberation.[18] Imagine that some committee reaches a resolution in a manner conforming to (G). For example, the committee agrees that it believes that *p* (≡ the department should hire a logician). The discussion leading to that decision depended essentially on certain propositions that were common knowledge in the group (e.g. *q* ≡ philosophy students need to be taught logic and *r* ≡ the department's only logician is about to retire). It would be absurd to say that the

[14] Gilbert ((1987: 192); (2004: 99)) regards the commitment in (G) as strong, imposing an obligation on group members not to disavow the group belief *p*, such that any disavowal may be rebuked by other group members. I am unsure that this is an inevitable feature of the groups that (G) seeks to describe, and I suspect that it varies from case to case. Perhaps group members may still affirm that *p* is the view *of the group* even if not of all of them individually. For example a court may come to an opinion with which some of the judges openly disagree in a minority opinion.

[15] Since the group's members will be aware of one another as members of the group and aware of the group's modus operandi, one might also call this the *mutual awareness* model, following Pettit (2003).

[16] This point will have greater force where the commitment has the strength claimed by Gilbert, as mentioned in footnote 14.

[17] I note though that Gilbert does not require that all members of the group always know what views the group collectively espouses. The group may come to a conclusion in the absence of some members, who are nonetheless committed to the group view. They may then ask, 'What do we think about X?'.

[18] This claim is consistent with the Gosplan example which shows only that it is possible, in unusual cases, for common knowledge to fail to play a role in group deliberation.

committee believed *p* but deny that it believed *q* or *r*, simply because it had failed actively and collectively to endorse the latter propositions. It did not need to endorse those propositions, precisely because they are common knowledge. Moreover, without ascribing group belief in *q* or *r* we cannot give a social–social explanation of the decision it came to: it would be natural to say that the committee decided on *p* because it believed that *q* and *r*. But that group-level explanation is ruled out by (G).

The same problem afflicts other propositions that may be used in group deliberation without having been the subject of group commitment. Just as (G) denies that the Audubon Society believes that the sky is blue, because the Society has not collectively endorsed that proposition, (G) likewise has the consequence that an established group may not be ascribed belief even in an obvious logical consequence of the propositions it does believe, since it may not have gone through the process of getting its members collectively to endorse the logical consequence. But that should not preclude group belief, since the logical consequence may play just as much a role in guiding the collective actions of the group as the propositions it has endorsed formally.

So the case for rejecting (K-SUM←) and (B-SUM←) on the basis of (G) is weak, since the same reasoning would deny that even common knowledge is socially known in normal cases. And that is unacceptable since it makes it impossible to explain reasoning within groups. In the next section we look at a different model of social knowing, which gives a different reason for denying (K-SUM←) and (B-SUM←).

4.2.2 *The distributed model*

Often we encounter circumstances such as the following. A complex task is information driven. But the information cannot be obtained and processed by any one individual. So several individuals are given roles in gathering different pieces of information while other individuals have the task of coordinating this information and using it to complete the task. A famous example from Hutchins's (1995*a*) *Cognition in the Wild* concerns the process of bringing a large ship, the USS *Palau* safely into port. Several crew members are given different landmarks whose bearings they are required to record and to communicate to a plotter who determines the ship's position and course. Another of Hutchins's (1995*b*) examples concerns the distribution of tasks between the two pilots (the Pilot Flying, PF, and the Pilot Not Flying, PNF) on a commercial aeroplane, in order to determine the correct airspeeds at the various points in the plane's descent when coming into land.

The key feature of such examples of the production of knowledge by a group is that the task is broken down into components, which are given to different members of the group. Membership of the group in this case is a matter of having a particular function within the overall system. This kind of group knowledge production is known, following Hutchins, as *distributed cognition*. The distributed cognition approach can be characterized in a number of ways; for current purposes, I propose the following. Distributed cognition identifies a *system* for producing knowledge, and studies how the various components, usually performing distinct sub-tasks, contribute to that overall task.

In typical cases of distributed cognition the components are often individual human subjects. There is division of cognitive labour—the different human participants are given distinct cognitive sub-tasks. As Hutchins has emphasized from the outset, almost always in such systems there are also non-human vehicles for representing and processing various pieces of information; these must be taken into account in articulating how the system achieves its cognitive goals. For example, in the navigation task there are distinct cognitive tasks for the sailors involved, such as taking bearings and plotting the vessel's course. In carrying out those tasks an alidade is used to measure and record the bearings, while the plotter has specially constructed maps and instruments in order to use the bearings to calculate the correct course. In the aeroplane, the PF controls the plane while the PNF operates the aircraft systems and has responsibility for completing the checklists required for each phase of flight. In addition, the pilots are aided by a variety of instruments, not only for recording important information, such as airspeed and aircraft weight, but also for calculating the correct airspeeds for the changing wing configurations (the configurations of flaps and slats that change in order to maintain lift as the aircraft reduces its speed during descent). When the correct airspeeds for the different stages of the descent are determined, these are represented by 'speed bugs', moveable markers around the rim of the airspeed indicator; this is done in advance of the information being required, since the pilots will be too busy to calculate the correct airspeed in real time; hence the need for a system to represent or 'remember' its correct airspeeds (Hutchins's paper has the title 'How a cockpit remembers its speeds').

We can now see that the distributed model of group cognition rejects (K-SUM) and (B-SUM) in both directions of those biconditionals. What counts for a piece of information being believed or known by the cognitive system constituted by the group is that the information is 'in' the system, available to be used by it as required for its purposes. That information need not be known by all concerned. The radar operator on USS *Palau* is keeping track of the position of other ships in the vicinity. Most of the time only he knows where those other ships are, since the rest of the system does not need to use that information. But if that information is needed, to avoid a collision or to transfer crew between ships, then because the radar operator has that information, it can be transmitted to other crew-members to be used by them in navigating the ship. In such circumstances the *Palau* knows the position of the others ships, even though most of the crew do not have that information. So (K-SUM→) and (B-SUM→) are false. At the same time, not everything known to many or all crew members is known to the ship as a collective entity if that information is not available to the ship's navigational processes. In the notorious collision between HMS *Camperdown* and HMS *Victoria* in 1893, several officers knew that the ships were going to collide but were unable or unwilling to use that knowledge to prevent the collision because of their deference to the authority of the admiral, Sir George Tryon (who went down with 357 other crew of the *Victoria* when it sank). In this case, with some degree of analogy to The Emperor's New Clothes, it would not be correct to say that *Victoria* knew that she would collide with *Camperdown.*[19] For the information held in the minds of those who individually knew this fact was not available to the ship and its navigational processes, e.g. to prevent the collision, even though those individuals were part of and even responsible for those processes. This remains the case whatever proportion of the crew had that individual knowledge. Hence the distributed model also rejects (K-SUM←) and (B-SUM←).

4.2.3 *Some differences between the two models*

Groups engaged in distributed cognition typically do not satisfy the commitment model. It may be correct to say that USS *Palau* knows (and so believes) that it is heading NNW at 14.2 knots, but that is not something which all the crew members have expressed a joint commitment to accept, not even all those involved in navigation. A midshipman may be on deck taking bearings and contributing to the generating of that knowledge, but that knowledge is not something he is aware of, let alone committed to.

The commitment model leads to questions and puzzles such as Pettit's discursive dilemma (2003). Commitment does not guarantee group rationality even if every individual is rational. This is because the opinions of the group's members might be aggregated in a way that leads to a paradoxical outcome. Imagine three subjects, A, B, and C, considering three propositions, p, q, and $p \wedge q$. A holds p to be true but q to be false, and so holds $p \wedge q$ to be false. B holds p to be false but q to be true, and so also holds $p \wedge q$ to be false. C holds all three of p and q and $p \wedge q$ to be true. Now imagine that A, B, and C have committed themselves in conformity with (G) to deciding which propositions to regard as the beliefs of the group by majority vote. Accordingly they regard p as true (A and C vote in favour), and q as true (B and C vote in favour), and they also regard $p \wedge q$ as false (since A and B both vote for the falsity of $p \wedge q$). Consequently the group beliefs are inconsistent. According to Pettit (2003) a group only becomes an agent if it adopts a belief aggregation procedure that can avoid this discursive dilemma.[20,21] Because distributed cognitive systems employ the division

[19] This was plausibly another case of common knowledge among the officers, maybe including even Tryon himself, that was not social knowing.

[20] The emphasis that List and Pettit (2006, 2011) place on resolving the discursive dilemma does give the appearance of trying to accept some aspects of the analogy approach while still adhering to the supervenience claim: 'group agents are distinguished by their capacity to mimic the way in which individual agents act and that this capacity must "supervene" on the group members' contributions' (List and Pettit 2006: 85). While not an incoherent combination, I do think it is an uncomfortable one.

[21] Standard examples of the discursive dilemma make it look as if all subjects form their attitudes to one of the propositions as a logical consequence of their attitudes to the others, in which case a sensible aggregation procedure would be to vote on the premise propositions and to commit to the group view of the conclusion proposition being whatever follows from the premise propositions agreed by that vote. This is 'the premise-based procedure' (List and Pettit 2006: 96). For example, A, B, and C vote on p and q and they are taken to agree on $p \wedge q$ because this follows from their agreement on p and on q. Not all cases, however, can be dealt with that way—not even this one. C might not hold $p \wedge q$ because she antecedently believed p and believed q. She might believe the conjunction first because she believes that it is the truth of p and q together that best explains some phenomenon, and so then infers the truth of p and q individually from her belief in the conjunction. So it might not be possible to identify a set of propositions that all

of cognitive labour, rather than aggregation of beliefs, the discursive dilemma is not a problem for distributed cognitive systems *per se.*

We may note, however, that a committee, the paradigm of the commitment model, may sometimes show some features of distributed cognition. For example, the committee may form specialist subcommittees or may choose to defer to the expertise of particular members. (By doing so it may also reduce the chances of the discursive dilemma arising.) Even so, a committee will usually bring the distributed knowledge back together to the whole committee so that all can share in the any fruits of distributed cognition. In that way the committee would be atypical of distributed systems, which do not usually share the knowledge produced across the system. The process by which the group comes to know according to the commitment model is not a distributed process—it is a process that is the same for all members of the group, involving them all individually in the same process of cognition.

4.2.4 *The commitment model and science*

Research teams in science might sometimes satisfy the commitment model. The team discusses the results they have been getting and their significance, possible interpretations, problems with the experiment and so forth. They may reach a view which can be expressed as the conclusion of the group, which may be reported as 'We have discovered that *p*; we believe that this shows that *q*'. One would normally expect the team members to avow such conclusions as the view *of their group*, even though some members may have reservations. The larger the research teams the less well they fit the commitment model and the more clearly they exhibit the structure of distributed cognition. Karin Knorr Cetina, in her (1999) *Epistemic Cultures: How the sciences make knowledge*, describes the High Energy Physics (HEP) experiments at CERN (the predecessor experiments to those carried out with the Large Hadron Collider). Knorr Cetina points out that the size, complexity, and long duration of the HEP experiments means that expertise cannot be centralized—the research leaders, she says, are not at the top but in the middle. It is neither necessary nor possible for participants in a HEP experiment to share all that they know with other participants, let alone to come together to agree on a common view.

Gilbert (2000: 37–49) applies the commitment model to scientific groups in particular; she argues that scientists are committed as groups to certain sets of beliefs. I think that this is quite possibly true for collaborative research teams, but I do not think that larger groups in science show this kind of collective commitment. It is true that shared beliefs are important to a scientific community, but the importance arises less from commitment than from the fact that shared background beliefs (among other things) are a prerequisite for cooperation and mutual understanding, as explained by Kuhn (1962). Those who reject beliefs core to a paradigm will not have reneged on a commitment; but they may find themselves excluded from the community because the basis for effective cooperation no longer exists. Gilbert proposes that her shared commitment explains why old hands are reluctant to buck the consensus:

subjects agree are the premises. A modus ponens version of the dilemma will show this clearly, as will many other instances.

neophytes and outsiders are helpful to scientific change because they are not bound by their joint commitment to that consensus. There are other explanations. Kuhn (1962: 62–5) argues that training with exemplars builds certain cognitive habits that will make alternative views invisible to those entrenched within a scientific community. Furthermore, established scientists may have more invested in a certain way of doing things; threats to the consensus are threats to their expertise and to the basis of their prestige.

We may think of scientific knowledge as being both produced and possessed by groups larger even than the teams at CERN. There are non-established research communities of specialists to which we can ascribe intentional states and actions ('palaeoanthropology tells us that *Homo sapiens* originated in Africa'). And there is a social, non-personal sense of 'knows' that is employed in saying that it is known that copper expands when heated (to use Tuomela's example). In such cases the group in question is not the small, organized research team, nor even a specialist research community, but the much larger collection of people constituting wider science. Neither members of the specialist community, such as semi-conductor physicists, nor participants in wider science think of themselves as a constituting group of which every member has expressed a willingness to let some *p* stand as the view of the group; nor have they openly expressed a joint commitment to accept *p* conditional on a like open expression of commitment by all other members.

Supporters of the commitment model may try to defend their model's claim to be a general model of group knowledge against the criticism that it does not accommodate the wider enterprise of science. They might argue that there is no group knowledge beyond the defined research team, in either specialist communities or in wider science. On the other hand, some proponents of the commitment model, such as Tuomela (2004), do include wider science within the remit of group knowledge. Given that the commitment model fails for wider science, what are we to make of truths such as (I)–(V)? Next, I propose that the distributed model does accommodate wider science. And in Section 4.4 I show in detail why we ought to regard wider science as involving a social epistemic subject.

4.2.5 *The distributed model and science*

Prima facie, wider science seems to fit the distributed model rather better than the commitment model. Science does have the key feature of distributed cognition, division of cognitive labour. The HEP experiments exhibit this very clearly; the participating scientists have their own areas of expertise and communicate with others on the basis of what needs to be shared for the local coordination of their parts of the project, what Knorr Cetina calls 'management by content'. While she says that this is something *like* distributed cognition, Ron Giere (2002*a*) argues that this *is* distributed cognition.[22] Giere points to the similarities between the HEP case and the USS *Palau*

[22] Recently Orestis Palermos (2020) has argued that research groups can be possessors of knowledge by exhibiting distributed cognition. I note that Palermos, in analyzing group knowledge, appeals to virtue reliabilism. This indicates that Palermos's view also exhibits the analogy approach—if one has a preferred analysis of knowledge for individuals, then one ought to think that the analysis holds also for groups (see

(while recognising that there are also important structural and cultural differences between the two cases).[23]

Scientists build on the work of other scientists. In modern science a scientist is almost never able to reproduce all the experimental results upon which her work depends for its epistemic justification. Trust in the scientific work of others is a ubiquitous and nowadays inevitable feature of science (Hardwig 1991). This goes also for theoretical results as well as for the outcomes of experiments. For although it is in principle easier for a scientists to follow and so internalize the reasoning of another scientist, in practice this too is often a matter of trust. Often those who use a scientific result are not themselves working in the field that produced that result. As described in Chapter 1, palaeontologists will often need to date a fossil from the age of the rocks in which it lies. That dating will be supplied by radiometry which depends on theories of radioactive decay. The typical palaeontologist will understand the basic ideas of that theory, but will have to accept from the physicists that the detailed reasoning involved is cogent and well-supported.

Science also displays another feature of distributed cognition, the role played by non-human vehicles of cognition. Hutchins focuses on physical means of representing information—alidades and sextants, cockpit dials and speed cards. These are important mediators between components of the distributed cognitive system. These are present in science also—printed and now online journal articles, reference resources, and data sets are the obvious example for wider science, for these are the principal means by which scientific information is communicated between scientists; we may include informal means of communication, letters, emails, and blogs, as well as educational resources, television broadcasts, podcasts, and the like. The means of representation also include physical specimens. When Darwin and other naturalists shared botanical and zoological samples and botanical and zoological gardens collected them, these samples were not just objects of study but also means of representing the species and other taxa to which they belong. We ought also to include non-human means of generating and representing information, which may include experimental equipment, satellites, computers running data analysis software, and even robot scientist—witness the automation of meteorology described in Chapter 1. Giere (2002*b*: 293) argues that visual representations in science can be thought of not merely as aids to cognition but as parts of systems engaged in cogni-

Corlett 1996 as another example of this). I, however, do not regard knowledge as analyzable and so, also in accord with the analogy approach, do not offer an analysis of group knowledge.

[23] Giere says, rightly in my view, that Knorr Cetina's description of an intended contrast to the HEP experiments, a molecular biology laboratory with centralized management, is also a case of distributed cognition. Its structure has some of the management structure of the *Palau* case. In his (2002*b*) Giere argues persuasively for the conclusion of this section, that science should be seen an instance of distributed cognition. Brown (2009) usefully supplements the case, in particular relating it to approaches in the sociology of science and learning theory.

tion.[24] In the next section I argue that the supervenience base for social knowing in science includes the material mean of representation.

There are however differences between wider science and Hutchins's examples of distributed cognition. The structures of the distributed cognitive systems in his cases have been designed to perform specific cognitive functions. The structures of wider science as a whole have not been designed (although some of their components have been). For example, the system of peer-reviewed journals is one that has evolved. That system may in due course be superseded—this is an issue of current debate. If it is superseded that will most likely be determined by what scientists find useful; scientific organizations may play a role in that debate, but there is no central world organization that could decide what will happen. In Hutchins's example of the cockpit of a commercial airliner, the system is designed to meet a specific goal—knowing at each point during the plane's descent what the correct speed and wing configuration should be. While local science may have well-defined goals around which the systems are designed, wider science is not like this.

Do these differences between science and the paradigm examples of distributed cognition show that the distributed model does not apply to science after all? One might reply that whether a cognitive system is designed or has come about by some other process is not relevant to how it functions. But that would be too quick: how a system came into being may well determine what functions it and its components have—whether something is a bug or a feature depends on design intentions. In order to think of a structure as a cognitive system whose parts have functions we need to assign a (cognitive) goal to science. Given that science is not a designed system, can we do this? P. D. Magnus (2007) holds that we cannot find a goal for science, and so, in his view, the characterization of science as distributed cognition is problematic.[25]

In Chapter 2 I argued that science is an institution. An institution can have a goal or function even if not designed to have that or any function. In this respect institutions can be like evolved organisms or parts of organisms—the latter can have goals and functions without being designed. In particular, I argued that science is an institution with a cognitive goal. Given that science does have a cognitive goal, we can think of it as a system whose components contribute to that goal, as in the biological analogue. (I expand on these points in more detail shortly.) Consequently science meets the characterization of distributed cognition given above.

[24] Giere (2002*b*: 293–4) also argues that scientific theories and models may be too complex to be fully realized as mental models, and so the models are reconstructed as external representations when needed. A slightly different approach, though one that Giere does not endorse, is to think of the mental as extending beyond the skin, so that external representations can be part of the individual's mind, as proposed by Andy Clark (1997). I shall not pursue this debate here, since I am concerned with the social subjects of cognition rather than individuals. But it is worth noting that Giere and Clark agree that there is a distributed cognitive system of which both the biological animal and the external device are parts.

[25] See Brown (2009) for a response to Magnus.

4.3 The failure of supervenience

In this section I argue that we should reject the supervenience approach generally and the commitment model in particular because they propose that the cognitive states of a group supervene on the mental states of the individuals composing the group. And the latter is false.[26]

(K-SUM) proposed that social knowing is a matter of being known by all or many; let us now consider a claim at the other end of the spectrum:

(K∃) It is socially known that *p* iff some person individually knows that *p*.

Since being known by all does not suffice for social knowing, *a fortiori* being known by some is not sufficient either. Discoveries kept secret by individuals or small groups of individuals and not used in a wider social context are not socially known by the larger social groups to which they belong.

Just as (K∃←) is obviously false, it may seem that (K∃→) is obviously true. If it is socially known that *p* (e.g. it is part of scientific knowledge that *p*) then surely someone must individually know that *p*.[27] This view is expressed by Tuomela (2004) '... a group cannot know unless at least some of its members know the item in question. The general ground for this assumption is that group properties supervene on their members' relevant properties.'[28] In addition to asserting (K∃→), Tuomela (2004: 112) commits himself to the seemingly obvious:

(SUP) Social knowing supervenes on the mental states of individuals.[29]

Although my concern is with veritistic approaches to social knowing, it may be noted that commitment to (SUP) is very wide.[30] It includes, for example, Martin Kusch's (2002) communitarian epistemology. As the quotation from Tuomela makes clear, this supervenience claim about knowledge is a special case of a more general supervenience claim, that the social supervenes on the individual, endorsed, for example, by Philip Pettit (2003: 184) and Christian List (List and Pettit 2006). Frederick Schmitt (2003: 2) says that it is 'nearly uncontroversial' that social relations and collectivities supervene on the non-social properties of individuals. Gilbert (1989: 417) also endorses a general supervenience claim: 'Viable sociological collectivity concepts will entail that facts about human collectivities, in particular about their actions, are

[26] Brian Epstein in his very interesting book *The Ant Trap* (2015) has also rejected the general supervenience claim. Hugh Mellor (1982) has also written things which directly imply the falsity of the supervenience thesis. (Although Epstein lists Mellor among the supporters of the supervenience claim, Mellor has confirmed to me that this is not the case.)

[27] Gilbert (1989) calls this view (as applied to belief) 'correlativism'.

[28] It is unclear whether Tuomela is referring to group knowledge as restricted to established groups and the like, or generally to any kind of group knowledge. (G) above clearly entails (SUP) as applied to established groups; correspondingly I do not take the failure of the supervenience claim as applied to the case of non-established social knowing to refute (G) regarded as a claim about established groups.

[29] I take mental states to include factive epistemic states such as knowledge. So (SUP) should not be understood as saying that social epistemic states supervene on the *intrinsic* states of individuals. Clearly social knowing depends on what is true, but so then does individual knowing.

[30] *Veritism* takes knowledge to entail truth, in a realist sense of 'truth'.

constituted by facts about the ideas and acts of will of human beings.' In what follows we shall see that (SUP) is false.[31]

Consider a process whereby some discovery becomes a contribution to scientific knowledge in the social sense. A scientist carries out some research which produces an important, novel result. The method employed is entirely reliable, and so the scientist has produced some individual knowledge for herself. She sends her work to a leading journal in her specialist field, the *Journal of X-ology*. This journal, although read by a relatively small number of experts, is taken by every library of a serious university or research institute, its articles are indexed and abstracted by the leading indexing and abstracting services in that area of science and can be readily found by online searches. The journal's editor sends the submitted article to referees, who seeing the quality and significance of the work recommend publication. They too have individual knowledge, as may also the editor. The paper is published. A small number of experts read the paper straight away, gaining more individual knowledge, and in due course the paper is referred to and cited in other papers and there is further increase in individual knowledge. At some point in this process of growing individual knowledge the discovery becomes socially known—it adds to the stock of scientific knowledge. What point is that? Could it be (a) when the scientist first makes the discovery? (b) when the discovery is published? (c) when a sufficiently large number of scientists have individual knowledge of the discovery? Answer (a) is not right: if the scientist were a hermit scientist, refusing to reveal the discovery, or had died very shortly after making the discovery, that discovery could have been made in the same way but without being socially known. The contribution to scientific knowledge in the social sense comes after the original discovery. Nor does (c) appear to be correct. There are some concepts for which the analogue to (c) would be a correct answer, such as the concept *crowd*. A central factor determining whether a collection of individuals forms a crowd is the number of individuals involved. But that is not the case for social knowing, for which weight of numbers does not seem relevant.[32] A solid but specialized piece of scientific (social) knowledge might be individually known by only a few experts, while The Emperor's New Clothes shows that something may be very widely known without being socially known. The explanation of the latter was that the knowledge in question was not socially accessible and could serve no social function. That answer also provides the reply we seek to our current question. What makes the difference between being individually known by some (or many) people

[31] Gilbert (2000) also says, 'According to the conception of collective belief I put forward, collective beliefs are a matter of how it is with the individual members of the group in question'; as she goes on to emphasize, the 'how it is' is not just a matter of the corresponding beliefs of the individuals. Other features of individuals are relevant according to her view, for example, as has been noted above, commitments by the individuals.

[32] A further piece of evidence in favour of this remark is that the collective term 'crowd' can be modified to reflect the fact that it lies on a continuum, e.g. 'large crowd', which is akin to the modification of 'tall' by the intensifier 'very'. But there is no modifier for 'scientific knowledge' that serves the function of indicating that the case in question falls further along the continuum. An expression such as 'widely known scientific knowledge' is not an intensification, but rather indicates a conjunction: the proposition that is scientific knowledge is also known by many individuals.

and being socially known, in a case of this sort, is the accessibility of the knowledge in question. Thus we should take publication, being placed in the public domain, as (in this case) the point at which the knowledge becomes socially known in addition to being individually known.

(We should note that it is *often* true that public accessibility of a piece of information is required for it to be able to perform a social function. And good science in particular typically requires a high degree of accessibility—there are obvious epistemic reasons to favour open data in science. But accessibility is not a necessary condition on any proposition being socially known in any context. The distributed model requires only that a social cognitive institution supplies the content of what is known to the appropriate institutional user of that information. A government may know something in virtue of its espionage agencies or because of a secret research programme. The known proposition may be accessible only to very few individuals at all.)

Now we are in a position to see why (SUP) is false and that being individually known is not even necessary for social knowing.

> **Case of Dr N** Dr N is working in mainstream science, but in a field that currently attracts only a little interest. He makes a discovery, writes it up and sends his paper to the *Journal of X-ology*, which publishes the paper after the normal peer-review process. A few years later, at time *t*, Dr N has died. All the referees of the paper for the journal and its editor have also died or forgotten all about the paper. The same is true of the small handful of people who read the paper when it appeared. A few years later yet, Professor O is engaged in research that needs to draw on results in Dr N's field. She carries out a search in the indexes and comes across Dr N's discovery in the *Journal of X-ology*. She cites Dr N's work in her own widely read research and because of its importance to the new field, Dr N's paper is now read and cited by many more scientists.

Was Dr N's discovery part of scientific knowledge? I argue that it was so *throughout* the period following publication. There is no doubt that it was at the end and also at the beginning. By publishing in a well-known, indexed journal, Dr N added to the corpus of scientific knowledge in the way that many hundreds of scientists do each month. Now consider the intermediate time *t*. As regards its status as a contribution to scientific knowledge, it seems irrelevant that Dr N and others who had read the original paper had died or forgotten about it. What is relevant is that the discovery was in the public domain, available, through the normal channels, to anyone, such as Professor O, who needed it. The position at *t* provides a counterexample to (K$\exists\rightarrow$)

Contrast that case with the following:

> **Case of Dr Q** Dr Q carried out his research in secret. He typed up his results and locked them in a drawer. He died; only ten years later did someone, by chance, come across his typescript in an attic. The results eventually reached the hands of Professor R, who arranged for their pub-

lication and cited them in her work, as a result of which they were read and discussed widely.

In this case, until it was published by Professor R, it is not correct to say that Dr Q's work was an advance in scientific knowledge.

Case of Dr N shows that a discovery can be a part of scientific knowledge even though no one has individual knowledge of it. When combined with Case of Dr Q we see that social knowing does not supervene on individual knowing. Consider the intervening period in the two cases. The mental states of all individuals in the two cases are the same. In neither case does any individual know anything or have any mental state relevant to the discoveries. The two cases, in the middle phase, are identical with respect to individual mental states. But in the case of Dr N, his discovery was part of (social) scientific knowledge during that period, whereas in the case of Dr Q it was not. Hence one can have a difference in social knowing without a difference in individual knowing or any other mental states of individuals. Hence (SUP) is false.

What is clearly significant in Case of Dr N, which makes it differ from Case of Dr Q, it that his results were available to anyone who needed them. His paper was available to be read in any university library, it was indexed in the appropriate indexing services, and could be found by an online search. As soon as Professor O needed information in the area of Dr N's expertise she found his paper immediately.[33] In Dr Q's case no one had any way of accessing his results until they fortuitously turned up on Professor R's desk. I shall consider the ramifications of this in a little more detail below. But for now it is worth pointing out that those sympathetic to taking the claim, for example, that 'libraries are repositories of knowledge', to be literally true should reject (SUP) straight away, if we add that the knowledge reposing there is social knowing. (Consider a library containing unique copies of certain books. The library burns down and the knowledge those books contained is lost. If this event has no witness, then it is a loss of social knowing without any individual mental state undergoing a change.)

[33] Accessibility comes in degrees. During the early Middle Ages the knowledge gained by the ancient Greeks was preserved by Islamic scholars. To the extent that the Christian scholars of western Europe were disconnected from these Islamic scholars, what was (socially) known to the latter was not knowledge among the former. It is not, however, that contact between these communities was impossible, and indeed there were exchanges of ideas and knowledge in Spain and in Sicily. For example, it was Fibonacci who introduced the Arabic numeral system to western Europe in 1202 (Germano 2013). Scientific contacts were fewer in the earlier medieval period but not non-existent. So Islamic knowledge could be accessed by Christian scholars, but only with some difficulty. Furthermore, the ancient texts were not entirely absent from the libraries of monasteries—rather their existence and significance were not properly appreciated. Again, the knowledge they contained was in principle accessible, but only with difficulty. Should we say that the Christian scholars did (socially) know the content of these works after all? And is that a problem for my view? The analogy approach regards such cases as unsurprising, since they arise also in the case of individual knowing. Memories that we do not use become weaker and increasingly difficult to retrieve over time. Sometimes we find we cannot retrieve a memory when we want to, but that it can come to mind with the right sort of prompt. Is this memory part of the individual's knowledge? It is a borderline case. *Knowledge*, like most concepts, is vague and we should expect borderline cases in both individual and social contexts. The fact that borderline cases arise in both contexts for the same reason, that accessibility comes in degrees, supports the analogy approach I advocate.

Note that the falsity of the supervenience claim does not show that there are mysterious group-level entities that act independently of other things. Rather, it indicates that the true supervenience basis for facts about a social epistemic subject extends beyond the mental sates of individuals: the supervenience basis includes non-human entities as well as human ones and also facts about the relationships between individuals (which need not themselves supervene on individual states).[34] Secondly, and as a consequence of the preceding point, we cannot reduce states of social knowing to states of individuals; so positing such states cannot be regarded as *merely* inflationary. That is, the additional ontology brings with explanatory benefits. Reference to social entities and their states of knowing enables sociological explanations of the kind offered by Durkheim, Parsons, and others. I have referred to this in Chapter 2, and will expand on the point later in this chapter.

So far I have taken (K∃→) and (SUP) to be tensed claims. That is, they claim that something is socially known at a time only if some individual knows it at that time, and that social knowing at a time supervenes on the mental states of individuals at that time. In Case of Dr N the proposition that is socially known at *t* is not individually known at *t*. Nonetheless, the proposition in question was known by someone at some earlier time and indeed at some later time. Can the rejection of (K∃→) and (SUP) be strengthened, so as to deny not only the synchronic dependence of social knowing on individual knowing or other mental states of individuals, but also any diachronic dependence? For example, can we find cases where what is socially known is *never* individually known?

I believe we can indeed describe counterexamples to:

> ($K\exists_{dia}\rightarrow$) If it is socially known that *p* at *t* then some person individually knows that *p* at some time or other.

Consider the Manhattan Project to build the first atomic bomb. This, like many other scientific-technological projects, involved division of cognitive labour. While individual team members knew how their bit of the project worked, they did not know the details of other parts. The project leaders knew how the various parts fitted together, but did not have detailed knowledge of each and every one of those parts. So the team as a whole knew how to build an atomic bomb, but no individual member of the project had this knowledge. (Cognitive division of labour will be a recurrent theme in what follows. In Chapter 1 we came across the example of the Large Hadron Collider (LHC) at CERN. The construction of the LHC and the ATLAS experiments conducted

[34] Does this mean that the knowing subject is an entity that has both human and non-human parts? Imagine a research group whose members communicate by sending hand-written messages to one another carried by human messengers (maybe they are afraid of email hacking). The messengers are part of the supervenience basis for the group's social knowing. Are the messengers members of the research groups that knows? That would not seem right. One possible response is to draw a parallel with the extended mind hypothesis for individuals. According to Clark and Chalmers (1998) what Otto knows includes what is in his notebook—the supervenience basis includes the notebook. But we do not therefore need to say that the notebook is part of Otto—this is another externalist thesis. Likewise means of information storage and exchange—libraries, the internet, messengers—can be part of the supervenience basis for a group's knowing without being part of the group that knows. My thanks to Mona Simion for pressing me to think about this point.

using it are a much bigger programme than the Manhattan Project and involves division of labour on a much larger scale.) So here we have an example of social know-*how* where no individual has that knowledge. Of course, a natural response is that this is not propositional knowledge-*that*. Nonetheless, if we accept the argument of Stanley and Williamson (2001) that all knowledge-how implies knowledge-that, then we may conclude that this case does involve social knowing-that without any corresponding individual knowledge-that.

A rather different case also provides an example of social knowing that *q* where no individual need know that *q* at that time or at any earlier time.

> **Interdisciplinary collaboration** Dr X, a physicist, and Dr Y, a mathematician, are collaborating on a project to demonstrate the truth of the interesting and important claim that *q*. They realise that the project can be broken down into three parts. Part one is a problem in physics, the problem of showing that *p*, which will be the work of Dr X alone. Part two is a problem in pure mathematics, that of proving that if *p* then *q*, for which Dr Y takes sole responsibility. Part three is easy, an application of modus ponens to the results of parts one and two. Accordingly they draft the outline paper with three parts: part one is blank except the conclusion, that *p*; part two is also blank except the conclusion, that if *p* then *q*; part three is complete, being the aforementioned application of modus ponens. The draft paper is left with a secretarial assistant with instructions to complete part one with the results of Dr X's work and to complete part two with the results of Dr Y's work, and then send it off to a leading journal, one that is read widely by the scientists in the field. In due course Dr X completes his experiments and physical reasoning for the conclusion that *p* and sends the completed part one of the paper to the secretarial assistant. At the same time Dr Y finalizes her proof that if *p* then *q*; she immediately sends the proof to the assistant as part two of the paper. Their work complete, the two scientists suddenly die, before they have communicated with one another about their success. The assistant mechanically completes the assigned task, without giving any thought to the content and sends the completed paper to the journal. The second part of the tale is a mirror image of the first. The journal's editor is too busy to evaluate the paper herself in any detail, but noticing the logical structure of the paper, she divides it up, sending the first part to a leading physicist and the second part to a leading mathematician. She leaves instructions for the editorial assistant (who has no knowledge of the paper's contents) that if both referees confirm that the arguments of their respective parts are sound, providing knowledge of the conclusions of those parts, then the assistant should go ahead and publish the whole paper. That indeed occurs and the paper is published.

Thus we have the following position: the published paper contains the conclusion that *q*, and the process leading to publication guarantees that it is indeed true that *q* and that any individual competent to understand the whole paper would come to know that *q* on reading it (as indeed might well happen after publication). Consequently, on publication the proposition that *q* becomes a contribution to scientific knowledge. In this case no individual knows that *q* at any time until the published article is first read in its entirety. Dr X knew that *p* and Dr Y knew that if *p* then *q*, but neither knew that *q* since neither knew what the other knew before they died. Neither their assistants nor the journal's editorial assistant understood the content of the paper. The editor knows the form of the argument, but, according to this story, does not know what the referees have reported (she leaves the matter entirely to the editorial assistant, or quits the editorship before the reports are in, or dies, etc.). So on publication the proposition that *q* becomes socially known, even though at no time has any individual known that proposition. Of course that proposition may become individually known once scientists read the article. But it need not. Clearly any *subsequent* individual knowledge is irrelevant to the possibility of social knowing that *q* without any individual ever knowing that *q*.

The case just described may be somewhat contrived, but the conclusion that (social) scientific knowledge can be generated without any corresponding individual knowledge is increasingly a reality. The collection and statistical analysis of scientific data is frequently automated. And when the resulting observations are also published automatically, on a reliable website for example, these can be a contribution to scientific knowledge in which no human had any part. As we have seen (p. 7), the publication of meteorological data is a good example of this. Automatic Weather Stations (AWS) collect data and transmit it to a computer that can be configured to publish that data on a live weather internet site, all without human intervention.[35] In a paper published in *Nature*, Ross King (King *et al.* 2004) describes how a functioning laboratory robot can formulate hypotheses, design experiments, and interpret results to determine the function of specific genes in yeast. While in that first phase of the work the hypothesis formulated and verified was one already known, King's robot has since formulated twenty original hypotheses concerning gene coding for enzymes and has confirmed twelve of them (King *et al.* 2009). Increasingly, AI and related computational methods are used to generate hypotheses from the data as well as test them (Sozou *et al.* 2017). It requires no great leap of the imagination to see such robotic science becoming sufficiently reliable and routine that it is produced, published, and even consumed with minimal human intervention.[36]

Now we can see the falsity of a diachronic version of (SUP):

[35] AWS and software for automatic analysis and internet publication are commercially available. Where data integrity is especially important, e.g. measuring pressure at airfields, manual checks are made on instruments. However, the instrument makes its own internal checks and AWS software also checks data for exceeding limits or showing step changes (which might indicate an engineering fault). In addition to web publication, the data may be assimilated into numerical forecasting model, which will itself include data checks. So even much of the process of checking the data quality is automated (Mike Molyneux, Met Office, personal communication). (For similar developments in quite a different field, see automated proof checking and theorem proving in mathematics.)

[36] For several years this field had its own journal, *Automated Experimentation*.

> (SUP_{dia}) Social knowing supervenes on the total history of the mental states of individuals (i.e. two possible worlds whose total histories are exactly alike with respect to the mental states of individuals will also be exactly alike at all times in the distribution of social knowing).

In w_1 information about the weather at L, a remote uninhabited spot, is published on the internet as the product of an automated system whereby raw data are collected from satellites and weather stations and processed by computer. The published information, the proposition that *p*, constitutes social knowing that *p*. World w_2 is like w_1 except that (a) the weather at L is somewhat different (so it is false in w_2 that *p*), but because of the remoteness of L, this affects no individual's mental states; (b) there is an unlikely accidental fault in the automatic data collection, processing, and publishing system, leading to the published output being the proposition that *p*. Since the published output in w_2 is false, it does not constitute social knowing. Worlds w_1 and w_2 are alike with respect to the mental states of individuals, at the time of the time of the publication of the weather report and at all earlier times (and, quite conceivably, at all later times also). But they differ with respect to the social knowing in those worlds. In one it is socially known that *p* but not in the other.

This section has tested the relationship between social states of knowing or belief and individual mental states. While a simple, summative account of the relationship has been rejected on all sides, those who concentrate on established groups favour a reduction of the relevant social states to individual states. On the face of it (G) is committed to such a reduction. It might be argued that what counts as 'open expression' depends on features of the world that go beyond the mental states of the individuals concerned. If so it might be that there are some instances that satisfy (G) yet do not strictly satisfy (SUP). Even so a version of (SUP) could be retained that broadened the supervenience base to include the actions of individuals as well as their mental states. In any case, it is clear that (G) envisages a close connection between group and individual mental states. The cases described in the section that refute (SUP) clearly refute (G) also.

Earlier we saw why it is that the distributed model rejects (K-SUM→)/(B-SUM→) and (K-SUM←)/(B-SUM←). This was a result of the fact that in distributed cognition cognitive tasks are broken down into parts and that the results of the sub-tasks are brought together in performing the collective task. Group knowledge on the distributed model does not depend on whether something is known by an individual or on how many know it or whether they are committed to it. Rather what is important is whether the information in question can play a role in achieving the cognitive goals of the group or system. That is why I have emphasized the importance of the accessibility of information. However, the accessibility of the information is not necessarily a matter of accessibility to all or most of the group. Rather it is a matter of accessibility to the parts of the network that may need that information. Science is a system that depends on transparency and internal criticism for its reliability, and which thrives on the availability of information, evidence, and ideas to those who cannot be predicted to need it. Hence in science the right kind of accessibility is accessibility to all. In other cognitive systems, such as a network of spies, accessibility will be constrained. In both cases, it is the conduits of information that determine what is known by the system and what is not known. Once one has this perspective, one can see that the network need not be constituted solely by conscious individuals. Non-mental forms of representation may be important sources of information or stores or conduits of that information in a distributed cognitive system. So the supervenience basis for group knowledge, according to the distributed model, need not be just the mental states of individuals. Consequently, we should expect (SUP) to be false.

4.4 The analogy approach and the social epistemic subject

Why is it that the group states we are discussing are appropriately called 'belief' and 'knowledge'? The summative approach has a clear answer: it is because the group state is a summation (or, better, aggregation) of the corresponding individual states. The supervenience approach aims to retain a close connection between the group state and the individual states, but when it comes to spelling out in detail what that connection is, the rationale for the group states being *belief* and *knowledge* are lost. According to (G) the individuals in the group do not need to believe that *p* for the group to believe that *p*. Nor is there an alternative explanation of why this state is a state of *belief*. Indeed Wray (2001) argues that the commitment model is an account not of collective belief but of collective *acceptance*.

The analogy approach takes a different view of why these group states are states of belief and knowledge. The correct reason is that the groups states function within the social entity in a manner *analogous* to the functioning of belief and knowledge in the individual. Consider the distributed model, an instance of the analogy approach. It tells us that the social entity knows as a result of information being gathered and processed by distinct parts of the entity in a manner that is analogous to the operation of the senses and the rational powers of persons. Just as these cognitive faculties of the individual person link to their actions, so the states of belief and knowledge of the social entity connect to the behaviour of the social entity.

This section expands on these claims for the analogy approach. It draws on a Durkheimian view of social metaphysics. According to the latter, we should understand social institutions in terms of their functioning within a social whole. Durkheim is explicit in seeing an analogy between society and an organism. I propose that we should see science as performing the same function in society as the cognitive faculties perform for an individual.

In my view, the analogy approach does not imply that the relationship between social knowing and individual knowing is *mere* analogy. Rather, focusing on the analogy allows us to see what social and individual knowing have in common such that both are subtypes of a more general kind.[37]

[37] My use of 'analogy' I take to be similar to the idea that traits in two organisms can be analogous even though not homologous. Analogous traits do not have the same origin, but they do perform the same functional role in the two organisms. For example the vertebrate eye and the octopus eye developed independently, but are both of the same functional type, eye. This is no 'mere' analogy, but a substantive similarity, involving a similar response to the same evolutionary pressures.

4.4.1 *Social analogues of individual cognitive states*

To make good on the analogy we need to look for social structures that are the analogues of individual cognitive faculties. Such structures, social cognitive structures, should have the following properties:

(i) They have characteristic outputs that are propositional in nature (*propositionality*).
(ii) They have characteristic mechanisms whose function is to ensure or promote the chances that the outputs in (i) are true (*truth-filtering*).
(iii) The outputs in (i) are the inputs for (a) social actions or for (b) social cognitive structures (including the very same structure) (*function of outputs*).

Most societies do contain social structures of this sort; in advanced societies the structures of science occupy a central position among them. Let us take as our example the group that comprizes semiconductor physicists. We will shortly see in more detail what the 'structure' here is. For the moment we can take it to be the organized community of these physicists.

(i) *Propositionality* The characteristic output of the community of semiconductor physicists is indeed propositional. As researchers the characteristic outcome of their work, whether experimental or theoretical or some combination thereof, is the journal article. The leading journals for these physicists include *Physical Review B*, *Journal of Physics: Condensed Matter*, *Electronics Letters*. Other characteristic outputs are also propositional, in the form of a conference presentation or poster, or a student textbook.

Not *all* the outputs of scientific activity are propositional. Scientists also develop new techniques, apparatus, and personal skills (although, if techniques and apparatus are also to be contributions to the field as a whole, they will usually be describable propositionally in order to be shared). Rather, the claim is that science does have structures and mechanisms whose characteristic outputs are propositional—which remains the case whatever else science does.

(ii) *Truth-filtering* There are processes within science designed to ensure that the outputs are true. The most obvious of these is the peer-review of journal submissions. While the journal referee has several duties not directly related to the truth of a submission (e.g. confirming that the work is of interest and value), a central task of a referee is to eliminate error. He or she has to decide whether the reasoning is sound, whether the theoretical proposals are consistent with known results, whether the methodology is appropriate, and so forth. Papers will be rejected that do not fulfil these requirements. These processes filter out those papers likely to be false and thus make it more likely that those that are published are true.

It is useful to make a rough distinction between experimental results and theoretical proposals. As regards the former the truth filters will be pretty robust. The referees will check that the methodology is one that is reliable and that any background theory employed is well-grounded rather than speculative or unsupported. This means that in good journals the results given will be ones that have a pretty high probability of being true and hence are such that other scientists may rely upon them. Where the results are particularly important other scientists will attempt to reproduce them. It is notable that credit will be given for so doing. The editorial guidelines of *Physical Review B* state, 'Papers must contain new results in physics. Confirmation of previously published results of unusual importance can be considered as new.'[38]

As regards theoretical proposals, the filters may be intentionally less robust. It may be necessary for several well-worked out proposals to be aired and to compete in order for scientists to be able to test and criticize them and for, eventually, evidence to mount up in favour of one over its competitors. Thus the ultimate goal of truth will be served by permitting the publication of theoretical proposals that may well turn out to be false. Even so filters are still applied: theoretical proposals will be published only if they 'do not contradict established experimental results' as the editorial policy of *Physical Review B* states. Proposals will have to meet minimum standards of theoretical plausibility and preferably have some empirical support. Thus proposals that have a low likeliness of truth are excluded.

Not all (or maybe even most) that gets published in scientific journals counts as a contribution to scientific knowledge; certainly, not all that gets published is of equal value or reliability. Usually there are further filters through which research must pass before it may be accorded the status of scientific knowledge. Results in a prestigious journal are more likely to be taken at face value than those in a lower-ranking journal, partly because it is believed that the better journals operate a more stringent filtering process. Experimental results that are not particularly dramatic, which are produced using standard techniques, and which are published in a reputable journal may be accepted as a contribution to knowledge without further scrutiny. But if experimental results are contentious, for example because they are in tension with earlier experimental results or because they appear to refute a well-confirmed and widely accepted hypothesis, then the results will not achieve the status of knowledge until they are suitably replicated. This is why, as we have seen, leading journals are willing to publish confirmations of previous important findings.[39] At the other end of the spectrum, a novel hypothesis will rarely become a contribution to knowledge on publication, but will typically require widespread independent confirmation and probably further theoretical discussion and debate before it achieves the status of knowledge.[40]

[38] http://prb.aps.org/info/polprocb.html, accessed 30 June 2007.

[39] But not always. In social psychology some journals are unwilling to publish replications, positive or negative, of earlier results. Such a policy is not truth-conducive, and so it is perhaps not surprising that social psychology suffers from a crisis of failed replications (Bird 2021).

[40] Do these remarks stand in tension with the case of Dr N discussed above, which assumes that mere publication can amount to a contribution to knowledge? Such cases are rare, since the filtering and assessment processes typically do not stop there; but they do exist. Experimental results, achieved with a non-contentious methodology, that are themselves not too strongly contentious, and which are carried out by reputable researchers and published in leading journals may well not need replication before

Truth-filters of the kind already discussed recur throughout science, from the criteria used to decide whether a student gets her PhD to the assessment of proposals for research funding. The student's training will include the acquisition of standard techniques that have a proven track-record of yielding results that are likely to be true. More generally young scientists are taught a respect for the truth; many scientists are driven by curiosity, which is the desire to know and to understand.[41] And one cannot know what is false and one cannot understand something if one's beliefs about it are erroneous. Curiosity involves inherently a preference for truth over falsity. Secondly, given the prevalence of truth filters in science, it is instrumentally good for a young scientist to acquire a respect for the truth (cf. Papineau 1988).

(iii) *Function of outputs* The best outputs of science are employed as inputs into (a) 'social actions' and (b) systems of the very kind that we have been discussing. The research results of semiconductor physics will often have direct applications in technology. For example, a discovery may lead directly to the design of a new kind of computer chip, the construction of a prototype, testing the latter, and eventually mass manufacture of the chip and marketing it. Such actions are themselves social in nature, since every stage involves the collaboration of teams of individuals. In many cases the interface between science and the production of a new piece of physical technology will involve large-scale division of labour. An extreme example of this is the design and construction of the new Large Hadron Collider at CERN. This involves enormous scientific and technical resources to overcome entirely novel challenges in three areas: the physics of the particle beams, the magnets controlling the beams, and the cryogenics needed to supercool the magnets. The project is managed by six departments (two of which are themselves purely administrative) and five scientific committees. The departments and committees oversee ten sub-projects and twenty-two working groups. Additionally the project depends on collaboration with non-European agencies as well as non-CERN projects in member states. It is the working groups that have to deal with the nitty-gritty of getting their sub-part of the project to work whereas the supervising committees have the responsibility of coordinating this work. Clearly no one individual can have all the expertise required by the project. There is considerable intellectual and managerial division of labour. The working physical object is the product of the large team at CERN and its external collaborators in the United States, Canada, India, and elsewhere, and depends on the scientific results generated both by the team and by its predecessors and other researchers in particle physics, cryogenics, magnetism, and so on. Scientific outputs, generated

becoming part of the body of knowledge. Another case may be formal results in theoretical research. Mathematical results showing that a certain theory has a given mathematical consequence may stand without further work since there may be no question concerning their correctness—which is not to say that formal work cannot be contentious, only that it is sometimes not contentious. Thus some pieces of scientific work may be suitable as the contents of Dr N's research.

[41] Williamson (2000: 31) asserts that curiosity is the desire to know. Whitcomb (2010) argues at length and convincingly for this claim. My addition 'and to understand' is not an additional component of curiosity but a special case of the more general desire to know—the desire to know how or why. See Section 3.5.1.

both within and outside the project, are in this case the inputs into a social activity of creating this remarkable machine.

Technology of this sort as well as of a more commercial kind is a consequence of 'social action' whose inputs are scientific results. This is the analogue to the use of individual knowledge in practical reasoning that leads to individual action. As individuals we may engage in reasoning whose outcome is not an action or intention to act but is further knowledge, which we may enjoy for its own sake, store in our memories, or use immediately in some further cognitive process. The same is true of the social knowing produced by science. Pure research is valuable for its own sake, satisfying a deep desire to understand our world. But much scientific research does not answer directly the questions we have but provides evidence that may help us decide between rival hypotheses. So the role of that knowledge is to be the input into a process of social reasoning which in due course may provide the answers we are looking for. Here 'social reasoning' means the process of debating rival hypotheses, submitting them to scrutiny, further development and investigation, a process which may take place in a variety of social fora—in the pages of journals, in conferences and workshops, as well as in discussions between individual scientists. Again we see a large-scale division of labour, this time in the production of scientific knowledge rather than a physical item such as a computer chip or the Large Hadron Collider. This division of labour may be deliberately organized, as in the Human Genome Project, but the division of labour in science is usually closer to that of market economies, developing organically rather than in a managed fashion. Projects are pursued because of the intrinsic interest of their results and because those results may be useful in assisting other scientists, whose identities may not be known, by providing data useful to their projects (and professional credit is given for so doing).

Not all the structures that yield social knowing must have analogues to the cognitive capacities of human individuals; we do not think that all knowing individuals including those of other species should have the same cognitive capacities as a normal human adult. Nonetheless, it is worth remarking that there are broad analogues. The process whereby experimental data become social knowing has a structural analogy to individual perceptual knowledge (which is *not* to say that individual experimental knowledge is merely perceptual knowledge); discussion of theoretical hypotheses parallels individual theoretical reasoning; libraries and databases are analogous to memory.

As regards the last of these, in fully understanding a society we must understand the material entities that are employed in human interactions—art, trade in goods, modes of transport, and the like. Modern society could not exist in anything like its current form if it did not possess the material means of rapid medium- and long-distance communication. Given that material entities can make societies what they are, it should not be surprising that the cognitive structures of society should depend on and in important respects be constituted by material things, such as libraries and their contents. Scientific knowledge is typically stored in an accessible fashion, often in print form in libraries, but now more often in one electronic form or another, available over the internet. Indeed, in order to play the role of output of and input into processes of social reasoning on the scale of modern science, social knowing *must* be stored in some physical (i.e. non-mental) form. Even if early science did depend on personal communications between scientists, modern science could not function that way. The existence of material structures such as books and journals or the internet is not an accidental feature of modern science, but is essential to it. The social cognitive structures that mirror individual cognitive structures inevitably have material components. For this reason social knowing does not supervene on states of individuals alone. This does not itself imply that social knowing could occur entirely without individuals. It is not *merely* the material inscriptions in libraries that constitute knowledge. Just as what makes something a human faculty of memory is its role within an individual's larger cognitive psychology, so what makes a library a repository of social knowing is its role within a larger set of institutions.

4.4.2 *The social epistemic subject*

We have just seen that there is a plausible analogy between cognition in social groups and cognition in individual persons. That analogy may suffice to explain the phenomena discussed so far: why we apply cognitive terms to social groups, as in (I)–(V); why in so doing propositions such as (K-SUM), (B-SUM), and (SUP) fail; and why, in particular the commitment model does not give a good account of (I)–(V).

So it is clear that the distributed model provides a better model for understanding group knowledge attributions in wider science than does the commitment model. That is no surprise. But then the two models were developed with different aims. The commitment model aims at articulating what genuine group doxastic and epistemic states are. The distributed model was developed with the aim of understanding how systems with differentiated components can achieve cognitive goals. Since the conditions that the commitment model lays down are so stringent, it is plausible that if there is any group knowledge at all, then the commitment model describes one way it can come about. The explicit commitment of individuals to the group does as much as could possibly be done to ensure that it is a unity. The distributed model does not have this ontological focus. It is true that Hutchins does *talk* of his systems (rather than just the individuals involved) as having cognitive states—recall that one paper is entitled 'How a Cockpit Remembers Its Speeds'. But he does not pretend to give an argument for this ontological claim. Hence the defender of the commitment model may respond to the foregoing as I indicated above:

> The commitment model provides our best account of how there can be a social epistemic subject. Insofar as science or other examples of the distributed model fail to fit the commitment model, we have no reason to suppose that they are cases involving genuine social epistemic subjects (so this challenge asserts). We should therefore see the distributed model as a model of the distributed *production* of knowledge, not the distributed *possession* of knowledge.

The challenge is made sharper by the fact that within the (designed) systems that Hutchins discusses, there is typically a node, a human individual, who does possess the crucial information that the system produces and is able to use it for the purposes of the system—the Pilot Flying in the case of the aeroplane cockpit, and the plotter in the case of the ship. So it might well appear to the social epistemologist that distributed cognition really concerns what I earlier called 'individual–social' epistemology, *viz.* epistemology that is concerned with the epistemic states of individuals as they relate to their social context, rather than with the epistemic states of social epistemic subjects ('social–social' epistemology).

What considerations are relevant to judging that there is an epistemic social subject in the case of distributed cognition, as found in science? In what follows I argue that the hypothesized epistemic social subject has sufficient *unity* to be considered to be a genuine entity; this is because division of labour can provide the principle of composition for some social entities. I then argue on functionalist grounds that this entity can have intentional, including epistemic states. I then turn to the objection that the knowledge in question is subjectless: we should accept my argument that there is social knowing but reject my argument that there is a social knower—this is a case of knowing without a knower.

4.4.2.1 *The unity of the subject* For any entity that has parts there must be something that unites those parts. In so saying, I am assuming that composition is restricted. (If it is not, then it is trivial that the components of a putative social epistemic subject form a whole.) For most physical objects that unity is mechanical—the parts will be physically joined to one another. For social entities, the principle of composition will not be like this. We ought to bear in mind that there might be different ways in which individuals can form social entities and that the resulting different kinds of entity may differ in their existence conditions. A set of individuals may form a social entity of ontological kind G but not of some other kind G′. So we may wish to consider that the principle of composition for a social *epistemic* subject ought to be one that plays some part in explaining how that resulting entity can be the possessor of epistemic states.

One concern that we might have in holding that there are any social entities at all, is that the individuality and free-will of human beings is a barrier to their being unified into a supra-individual entity—have not totalitarian leaders promoting social unity sought to suppress individuality and freedom? Some biologists emphasize the unity of a hive of bees or a colony of ants to the extent of arguing that these constitute 'superorganisms', which indeed may be regarded as more like true organisms than the individual bees and ants.[42] While these insect groups do show distributed cognition, one might think that it is the 'mechanical', genetically hard-wired nature of their co-operation that allows us to think of them as forming a single entity. But the cognitive cooperation of human individuals, especially in wider science, is not like that, being subject to the will.

The fact that a social entity's putative existence may be subject to the will of individuals is not sufficient to deny that that such existence is genuine. After all, the continued existence of my body is subject to my will (as is the existence of this mug,

[42] Hölldobler and Wilson (2009). See Gardner and Grafen (2009), Kümmerli (2009), and Canciani *et al.* (2019) for discussion.

that cake, etc.). If there is sufficient strength of will directed towards the unity of the social entity, that ought to allow for this objection to be overcome. So it is perhaps no surprise that the commitment model emphasizes commitment so centrally, since that is a demonstration that the individual members all will that their social entity should exist. And this might also explain the significance for Gilbert and Tuomela of the standing to rebuke, for the existence of that standing will reinforce the commitment of the individuals to the group's existence as an epistemic subject.

Individual commitment to the group is one way that the wills of individuals can support the existence of the group. But it is not the only way. The sociologist Émile Durkheim was concerned with the question: what holds a society together? The ways in which social cohesion is achieved he called *solidarity*. The question of the sources of solidarity asks: how do societies hold together and function effectively rather than break down or fragment? While, for Durkheim, this was a sociological question, responses to that question may also be seen as answers to the composition question for social groups. Durkheim (1893) distinguished between *mechanical* and *organic* solidarity as means by which society can be held together. In societies characterized by mechanical solidarity there is unification through the shared beliefs (experiences, obligations, etc.) of individuals. Cohesion is achieved by the individuals in the group valuing their similarities of belief and value, their kinship, and their cooperative practices. In societies characterized by organic solidarity, however, cohesion is achieved not by similarity among individuals but instead by difference. Unification is generated by mutual dependence arising from, above all, the division of labour. Where an economic system operates by division of labour that system's existence does not require commitment of the individual to the system; commitment by the individual to perform their function within the system will suffice. The fruit-canner does not need to have any conception of the economic system of which he is a part in order for him to contribute to its existence and operation, so long as he is committed to the grocer to supply his needs and to the grower to take his produce—commitments that economic self-interest will generate.

So individuals can compose a social unity when they cohere because of the mutual interdependence that arises from the division of labour. While Durkheim thought principally in terms of economic division of labour, solidarity can also arise from from cognitive division of labour.[43] And indeed, that is exactly what distributed cognition gives us.[44] Hutchins's examples of a landing a plane and navigating a ship show division of labour among two or more individuals (and also non-human de-

[43] I do acknowledge that there is an element of mechanical solidarity in many social entities even when organic solidarity predominates, including communities of scientists. The latter may well see themselves as a community in virtue of shared values, goals, and interests.

[44] I argue that distributed cognition is one way of implementing organic solidarity (in epistemic groups). Is there an analogous relationship between entities satisfying the commitment model and mechanical solidarity? Collective commitment to shared beliefs does look very much like a form of mechanical solidarity. There may be mechanical solidarity that does not involve commitment, so collective commitment is just one realization of mechanical solidarity. That said, it should be recognized that mere similarity of belief does not on its own constitute mechanical solidarity—it is also required that this fact be mutually recognized and engender a sense of belonging together. Although this falls short of

vices). Those individuals are trained for their particular sub-tasks, and various social, professional norms (and self-interest) ensure that they are reliably performed. Just as there is division of cognitive labour in Hutchins's examples, there is division of cognitive labour in the local science of an experiment or research project. It is true also of wider science, although in this case the division of labour has evolved without a plan being imposed upon the participants. That itself does not detract from unity, for the economic division of labour, generating Durkheim's organic solidarity is likewise an un-designed evolution. No modern science depends for its conclusions just on the intellect and the evidence of the senses of a single scientist nor even of a local team of scientists. The work of other scientists will be drawn upon in the design of equipment used, in auxiliary hypotheses and background knowledge, and in the statistical methods and computer software employed to analyse data. The scientist's conclusions may be worthwhile in themselves but they will frequently be used as evidence or background knowledge in some other scientist's reasoning. So wider science exhibits the division of labour that gives rise to organic solidarity. This is not restricted to fields or sub-fields, for the interdependence of science crosses such boundaries. Wider science is, for the most part, a single entity.

4.4.2.2 *The epistemic subject* I have just argued that there is a principle of composition (other than commitment) that can bind individuals together to form a social entity. This can apply to groups involved in distributed cognition and to science in particular. Given that there is a social entity composed of scientists, we need further to show that such an entity can be the subject of intentional states, in particular epistemic states.

The argument that entities engaged in distributed cognition have epistemic states is a functionalist one. We can identify particular behaviours and states of the system that are best understood as actions and the intentional states that explain those actions. In Hutchins's cases, those systems are designed to fulfil a function—controlling an aircraft's descent carefully, navigating a ship. So these systems have goals and engage in actions to achieve them. The successful execution of those actions to achieve those goals depends on the systems generating and being able to access certain crucial pieces of information. And as described by distributed cognition, the systems are designed precisely to produce that information and deliver it to those parts of the system that need it. The system's goals and actions are thus supported by an information processing function of the system. If a system has goals and performs actions to achieve them, and has an aspect or subsystem whose function is to provide information that guides those actions, then that function is a cognitive one. The states that the system is in as a result of performing that cognitive function are correspondingly cognitive/epistemic ones. For example, the *Palau* needs to enter Hampton Roads at Naval Station Norfolk (goal). So the ship changes course to 265°. Why? Because the *Palau* has taken its bearings and knows that Hampton Roads lies in a direction 5° south of due west.

collective commitment, it points in that direction in that both require groups members to have a sense of the group as a group and to know that certain beliefs or intentions are shared by its members.

Can the functionalist argument be extended from Hutchins's central examples to wider science? Wider science is not a system that has been designed (though it does have designed components). Nonetheless, as has been discussed, the lack of a design does not preclude possession of a function. Functional explanations are common in biology without that being any concession to the intelligent design view; rather, evolutionary accounts of function are required. Above I briefly argued that science has a cognitive goal. This claim is reinforced by the Durkheimian perspective, which sees various social institutions, especially those in societies characterized by organic solidarity, as fulfilling various functions, in a manner analogous to the functions of the various organs in a living being. Societies can have collective goals and engage in collective actions (such as those coordinated by governments); there are institutions that provide the information that allows those goals to be pursued and those actions to succeed; science is one of these institutions and an increasingly important one as the goals and actions become increasingly technologically sophisticated.

To the claim that the group can be an epistemic subject, a critic might respond that this is unnecessarily inflationary given that we can ascribe knowledge to some or all of the individuals in the group. For example, in Hutchins's examples we can identify a particular central individual who receives and acts on the information supplied by the other individuals (and objects) in the system. The response to the very last point is that while in some systems it may be effective to have a single, central individual in such a role, that is certainly not required, and many distributed cognitive systems lack such an individual—air traffic control is an important example. In large-scale local science, while the Principal Investigator or Programme Director may have an overview of the functioning of the project, she will not necessarily be a cognitive central node in the system; conversely, as Knorr Cetina describes the HEP experiments, those with responsibility for scientific outcomes of sub-projects and their communication to other parts of the experiment find themselves in the middle of the management structure. Such projects may produce papers with dozens of authors, where the content of the paper cannot be said to be fully known (in an individual way) by any of them (papers with hundreds of authors are increasingly common).[45]

4.4.2.3 *Subjectless knowing?* I have just argued first that we have reason to believe that there are social entities and then that we can sometimes attribute epistemic states to them. This approach assumes that if there is to be knowing at all, there must be a subject who is the knower, whether individual or social. Some philosophers of science are reluctant to ascribe a subject to this social knowing; they will agree with the claim that there is social knowing, but deny that we have to find an entity to which we can ascribe this state.[46] Popper (1979: 109) famously held that 'knowl-

[45] Usually those listed are not all 'authors' in the normal sense—they are participants in the project; even so many papers are collaboratively written with different authors contributing different sections according to their expertise.

[46] Tollefsen (2006), drawing on the work of Clark and Chalmers (1998) on the extended mind, shows that the lack of an identifiable brain is no reason to deny that groups can possess intentional states.

edge in the objective sense is knowledge without a knower: it is knowledge without a knowing subject'. Giere comments on Knorr Cetina's suggestion that in a large-scale experiment, it is the experiment that knows (Knorr Cetina 1999). Giere (2002*a*) prefers the idea that the knowledge produced by the experiment does not belong to any epistemic subject. The foregoing arguments suggest that we do not need to avoid identifying a knowing subject. On this view Knorr Cetina is right, when we take 'the experiment' to be the total system of the team working on the experiment and its analysis, plus the equipment and infrastructure that enables them to reach their results. We might prefer the terms 'project' or 'programme'. It does not sound at all odd to assign epistemic properties to such entities, saying, for example, 'the Manhattan project discovered how to harness the power of nuclear fission'.

A further reason for rejecting the subjectless knower view is that it is difficult to reconcile with the best argument that there is social knowing at all. The best argument is the functionalist one articulated above. That argument requires, in outline, that we regard social states as having functions and where the functions can be regarded as cognitive, the states are epistemic. But states of *what*? One cannot identify a state as having a function without there being a reasonably determinate thing that has that function. One cannot identify functions independently of the things they are functions of. Hence our best argument for social knowing requires that there is a social entity that does the knowing.

4.5 Conclusion

Science is a social and an epistemic institution. This book concerns the epistemology of science, but its social character cannot be ignored since it bears on its epistemic character. That is most emphatically the case when social groups—research teams, the scientific community that constitutes a particular field, and wider science—can be regarded as epistemic subjects. This chapter has argued for that claim. Other social epistemologists have also argued that social groups can be epistemic subjects, and have overwhelmingly endorsed the commitment model of social knowing. That model might describe a small or medium-sized research team. But it cannot describe a large research project, such as the HEP or ATLAS experimental teams, let alone whole scientific communities or wider science.

These larger groups are not 'established'; their nature is more organic and diffuse. If they are genuine epistemic subjects, then they refute the claim that a group's epistemic states supervene on the mental states of the individuals in the group. Their nature is better captured by Hutchins's idea of distributed cognition. This idea emphasizes two points: the division of cognitive labour often holds; social structures often include material, representational entities as well as human beings.

Although Hutchins asserts that the structures exhibiting distributed cognition are genuine epistemic subjects, this claim needs an argument. The later part of this chapter supplies that argument, by answering two questions: When is it that a social epistemic subject exists? And of what is the social epistemic subject composed?

A social subject exists when there is sufficient social glue (cohesion, solidarity) to join the components together. I have argued that the glue need not be joint commitment. Division of labour is another source of solidarity. Put less abstractly, the fact that certain people have mutually interacting jobs or roles may be sufficient to bind them together in the relevant way, as for example it is on a naval ship. That division of labour, those different job-roles, determine the distinct individual functions that contribute to the various functions possessed by the social whole thus bound together. If the latter include cognitive functions, then the social whole is an epistemic social subject. (While it may be vague when there is enough of the right kind of cohesion and distinctness that a social entity exists, that is no objection to the claims being made. It is often vague when a physical object has enough cohesion to be regarded as an entity, e.g. a cloud, a tmelting iceberg, etc.)

Of what is this social subject composed? Which people and objects are its parts? At least the individuals performing cognitive roles contributing to the function of the whole.[47] The lesson of distributed cognition and the anti-supervenience arguments is that we must include relevant non-human entities also. Describing the function of the navigation team on the *Palau* makes no sense without including the crew's navigational equipment and indeed the ship herself. While that answer gives us the correct supervenience basis for the epistemic state, that does not mean that these are the only components of the social epistemic subject. Jane knows that Edinburgh is the capital of Scotland but John does not; Jane's hand is not part of the supervenience basis for this knowledge. Still, that hand is part of the knowing subject (Jane) whereas John's hand is not. That is because Jane is not just an epistemic subject; we identify her as having cognitive functions as part and parcel of identifying other functional states. If she knows which direction is North, she may demonstrate this by pointing with her hand. Turning now to the *Palau*, the whole of interest extends beyond just the navigation team. For the cognitive function of knowing the bearing of Hampton Roads is related to the locomotive function of steering the vessel into Hampton Roads. And there will be crew members performing that function who have no role in the cognitive function. Given the connections between the various functions, those functions must be attributed to an entity comprising the whole crew (and the ship).

What are the consequences of this approach for wider science? Let us start with not-so-wide science, the science in the North Korean nuclear weapons programme. That scientific programme is functioning as part of the North Korean state, with a view to enabling that state to fulfil its goals. However, the social entity in question goes beyond just the officials of the state; it will include the citizens of North Korea also, since they are functionally connected to the weapons programme in a number of ways: their labour (even if coerced) pays for the programme; they have other obligations to the state and military (e.g. military service); they live within the boundaries of the country that the weapons programme aims to protect. So those citizens have a functional relationship to the knowledge generated by the weapons programme that the citizens of other nations do not. This explains why it is appropriate to say

[47] See footnote 34 on p. 91, which discusses whether every individual that plays any contributory function is part of the social subject. We should distinguish being part of the subject and being part of the supervenience base for its cognitive states.

'North Korea knows how to build a nuclear weapon, but Burma does not' (or 'the North Koreans know how to build a nuclear weapon, but the Burmese do not').

Now we can think about wider science, as practiced in universities and publicly funded research institutes across the world. Scientific collaboration, the division of cognitive labour, and the dissemination of results are international. Those results are typically accessible to anyone who has sufficient background knowledge to understand them. Yet others will be indirect consumers of the knowledge, for example, through the technology it creates. Consequently, membership of the social group of wider science may be considered to be very extensive, potentially encompassing much of the world's population. Such a conclusion is neither trivial nor absurd. For if we are all part of a global village, participating, willingly or otherwise, in a globalized economy, it is not absurd to think of others, even if geographically remote from us, as part of some of the same social entities.[48] Science in particular has been international from an early date, with scientists in different European countries as well as America communicating with one another regularly from the seventeenth century forward.

Accepting the analogy approach has the consequence that we can infer that results that apply to individual epistemology apply, *mutatis mutandis*, to social–social epistemology also. Because, for example, Gettier cases fail to be knowledge for individuals, they will exist and fail to be knowledge for groups also.[49] In the next chapter I consider *evidence* in science and beyond. It simplifies discussion to talk of the evidence possessed by an individual. The analogy approach allows us to infer that the conclusion—that any known proposition is evidence—applies to the knowledge and evidence of groups and institutions as well as to the knowledge and evidence of individuals.

[48] This point has an ethical dimension. For example, even if those of us in the West are under an obligation to help relieve poverty wherever it exists, that obligation seems all the stronger when the poor in question are those making the cheap goods we buy.

[49] Jennifer Lackey (2014) criticizes my view precisely on the ground, so she argues, that social knowing, as I describe it, fails to show some of the epistemological properties that we expect to find in individual knowing.

5

EVIDENCE

In this chapter I articulate and defend a functional characterization of the concept of evidence: evidence is that which allows us to make inferences that extend our knowledge. I show that this entails that all and only knowledge is evidence, as argued for by Timothy Williamson. I then examine a counterproposal, that evidence is non-inferential knowledge. The latter fails because it cannot account for the fact that in some cases we can lose initial evidence (e.g. by forgetting it) yet retain as known and well-evidenced those propositions inferred from that initial but now lost evidence. The empiricist identification of evidence with experience or observation (understood as perceptual), common in philosophy of science, exemplifies this mistaken non-inferentialist conception of evidence. It conflicts with scientists' use of 'evidence' which allows for evidence that is highly theory-laden or is theoretical in nature.

5.1 Introduction

Consider propositions of the form:

S has evidence *e* that *h*.

We may raise two questions about such propositions:

1. Which propositions may be substituted for *e*? (When is something S's *evidence*?)
2. When is *h* supported by *e*? (If something is S's evidence, when is it evidence *for* some proposition, hypothesis, or theory?)

This chapter concerns the first of these questions. The second is the question most usually addressed by philosophers of science and epistemologists (and I shall do the same in Chapters 7 and 8). But the first is crucial also, even if it is ignored by most philosophers of science.[1] It is all very well knowing what it would take for *e* to be evidence for *h*. But that doesn't help us understand whether *h* is in fact well supported by the evidence unless we also understand which propositions can legitimately play the role of *e*. Clearly it makes a great difference to S's hypothesis being well-evidenced whether all of S's relevant beliefs count of as evidence or just those beliefs that reach some higher epistemic standard, such as knowledge. For example, as we saw in Chapter 3, René Blondlot took the results of his experiments to be evidence for his theory of N-rays. No doubt his *beliefs* about those experiments

[1] For example, Peter Achinstein's (2001) *The Book of Evidence* is a magisterial answer to question 2, but says next to nothing regarding question 1.

Knowing Science. Alexander Bird, Oxford University Press. © Alexander Bird 2022.
DOI: 10.1093/oso/9780199606658.003.0005

stood in a supportive evidential relation to his theory. But those beliefs about the experiments were false and were the product of self-deception. In holding those beliefs, did Blondlot thereby have evidence for the theory of N-rays? To answer that question we need to understand not only the *evidence for* relation but also what it is to *be* evidence.

In characterizing evidence philosophers of science have generally not focused on epistemic strength. Rather they held that to be an evidence proposition a proposition must have a certain content, character, or origin. It is standard in philosophy of science to hold that evidence is always about what is observed. Furthermore, it is also typically held that to observe is essentially to perceive. (The issue of epistemic strength has not been entirely absent. As we shall see in the next chapter, one motivation for the standard view is that being observational/perceptual ensures that evidence does have a high degree of certainty.) These widespread empiricist assumptions also mesh with a common view among epistemologists that evidence propositions are only those that have not been inferred from other propositions. While this looks like a foundationalist conception of evidence, it is also important to those coherentists who give a special place to what BonJour (1985) calls 'cognitively spontaneous beliefs', those beliefs that do not arise from inference from other beliefs in the subject's system of beliefs. I shall be arguing that evidence propositions must reach a high degree of epistemic strength, namely knowledge. But they do not need to have any specific kinds of content, character, or origin—any known proposition is evidence, whether inferred or uninferred, whether perceptual or highly theoretical.

Although the questions 1 and 2 are connected, they are independent in that we can answer the first without answering the second. Detectives may gather evidence without knowing in advance which hypotheses they may use it as evidence for. Some philosophers of science seem to think that scientists always gather evidence in the light of a hypothesis they are testing, but this is clearly false. Naturalists and geologists of the eighteenth and nineteenth centuries typically gathered samples, made drawings, collected data and so forth with no hypothesis in mind. This was evidence which may later and often did become evidence for some particular hypothesis. And even when data is collected in order to test hypothesis X it might subsequently become evidence that is used in the assessment of hypothesis Y. The status of the data as evidence is independent of whether it is evidence for X or evidence for Y.

In order for something to be evidence for some belief, claim, or hypothesis, that evidence must be propositional.[2] Yet, what is referred to as evidence is often not

[2] Can experience itself be evidence? That depends on whether experience has conceptual content, which is a contentious matter. If experiences can have conceptual and so propositional content (Peacocke 1983; Brewer 2005), then we might be in a position to argue that experiences can be evidence, so long as we can also argue that experiences that are hooked up to the world in the right way can be or can contain states of knowledge. If experiences cannot have conceptual content (Dretske 1981; Raffman 1995; Peacocke 2001), then something else, related to experience, will have to play the evidence role. That might be the belief brought about by experience, or a belief about the character or non-conceptual content of experience. The latter would mirror the approach to non-propositional evidence taken in this paragraph. Since I take a strongly anti-empiricist view of evidence in this chapter and the next, the exact relationship of experience to evidence is not a burning issue for me.

propositional. The bloody dagger is held to be evidence of the defendant's guilt; likewise the fossil in a museum is evidence for the theory of evolution. This is a legitimate use of the term 'evidence'. It is not the focus of our current concern with propositional evidence; but it is nonetheless connected with it. For in our reasoning about the defendant's guilt the dagger appears as a component of relevant *facts*. It is the *fact* that the blood on the dagger was the victim's blood and the *fact* that the dagger was found in the defendant's possession that tell against the defendant and contribute to the jury's reasoning. Likewise, in using the fossil to support the theory of evolution, we point to the fact that it was found in sedimentary rock 80 million years old, the fact that it is a jaw bone, including some teeth, the fact that it is not the jawbone of any current animal, and so forth. So when we point to some *object* as evidence we are pointing to it as a source of *propositional* evidence.

Timothy Williamson (1997, 2000) argues that the answer to question 1 is that the propositions in S's evidence are precisely the propositions that S knows:

(E↔K) *p* is in S's evidence if and only if S knows *p* ('E=K').

Note that Williamson argues that we should not expect our concept of evidence to be such that:

If *e* is in S's evidence, then S is in a position to know that *e* is in S's evidence

i.e. the claim that evidence is luminous (or some similar requirement that a proposition's status as evidence should be apparent to S). Williamson argues that nothing meets this condition of luminosity. Of course, this does not mean that we never know that some proposition is evidence. A difficult investigation might require that we review our evidence. Perhaps something we took to be evidence is not evidence at all. In such cases it may be important to know exactly what our evidence is. In other cases it need not be.

While we may regard (E↔K) as a substantive and necessarily true identity claim, akin to 'water is H_2O', we should note that Williamson's argument is an apriori one. If it is apriori that evidence and knowledge are one and the same, why do we have a concept of evidence at all?

I propose that we answer this question and question 1 by posing another. What is the purpose of evidence? Whereas knowledge may be intrinsically valuable, we do not seek evidence for its own sake. Rather we want evidence because evidence will allow us to draw conclusions in the course of inquiry. It is evidence that allows us to infer that a theory is true or to know that a defendant is guilty. So evidence is sought and is used for the purpose of making inferences. Indeed, most of the philosophical discussion of evidence concerns this feature: above all we ask when will evidence support an inference to a conclusion; as recognized by question 2, the term 'evidence' is most naturally used in the context of a proposition's being 'evidence *for*' some proposition (or sometimes 'evidence against' and other cognates). The significance of evidence for the task of inference suggests an answer to our question, why do we have a distinct concept of evidence: we have a concept of evidence in order to characterize certain propositions *functionally*. Put in general terms, the concept of evidence serves to characterize propositions in terms of their role in inference.

This chapter looks at the relationship between evidence and inference in detail in order to better understand what evidence is. I have two objectives:

(a) To articulate a functional characterization of the concept of evidence: evidence is that which allows us to make inferences that extend our knowledge.
(b) To provide an indirect argument for my characterization by showing that it entails Williamson's claim that all and only knowledge is evidence, for which he provides independent argument.

The question of what evidence is, is primarily a question for general epistemology. And so the approach I take in this chapter does not focus on scientific evidence in particular. But the conclusion that all and only evidence is knowledge is one that has significant consequences for philosophy of science. Most importantly, when a scientist considers the evidence regarding some hypothesis she entirely properly may take into account theoretical propositions that are not observational let alone perceptual in character. For example, it is evidence for a theory in quantum gravity that it predicts a value for the fine-structure constant that is accurate. That claim is not undermined by the fact that the existence and value of the fine-structure constant are highly theoretical. That is so because the relevant theories are sufficiently well established that we *know* that the value of the fine-structure constant is approximately 1/137. This contrasts with the view that if a scientist reasons that way, then the claim about the value of the fine-structure constant is really a proxy for the observational/perceptual evidence that stands behind that claim and with the view that really what the scientist should be understood as doing is attempting to predict the total pattern of her sensory stimulations. I look at these consequences of (E↔K) towards the end of the chapter and then again in the next chapter.

5.2 The concept of evidence characterized

In Chapter 2 I argued that the aim of belief is knowledge; correspondingly, the aim of science is the acquisition of (scientific) knowledge. When we make an inference from what we already believe, we add to our beliefs. The aim of such an inference, in science and elsewhere, should therefore be the production and so expansion of our knowledge. That aim can only be achieved if the premises of such an inference are of the right sort. If my inference starts from propositions I believe only because of wishful thinking or self-deception (such as Blondlot's observational beliefs in his N-ray experiments), then they will not achieve this goal. So we need to characterize the propositions from which an inference can proceed that will lead to a successful (knowledge-producing) inference. This is the role of the concept of evidence.

The characterization of the concept of evidence I therefore propose is this:

(EC) *p* is in S's evidence if and only if S can gain knowledge by inference from *p*.

(EC) captures the intuition that evidence is what we infer from in order to get knowledge. According to (EC) evidence is, so to speak, the raw material in the process of achieving the aim of belief and of science. It is because the aim of belief and science is knowledge, that we want a concept that fulfils the role accorded to evidence in (EC).

(EC) acknowledges an asymmetry between the evidence and what that evidence is evidence for. Yet the asymmetry afforded by (EC) is local not global. Given some particular inference, it is the 'what is inferred from' that is the evidence. (EC) is consistent with its being the case that when we make a successful inference from some evidence, the inferred proposition, being knowledge, itself thereby becomes evidence, evidence that can be used as the premise of some further inference.

(EC) is the conjunction of the two implications:

(EC→) If *p* is in S's evidence then S can gain knowledge by inference from *p*;

and:

(EC←) If S can gain knowledge by inference from *p* then *p* is in S's evidence.

I shall go on to give reasons why (EC→) and (EC←) provide accurate characterizations of the concept of evidence. But first some clarifications.

In the above *p* must play a *non-redundant* role in the relevant inferences. Take any knowledge-extending deductive inference: we can include *p* as a premise in that argument, and indeed we can construct an argument in which *p* might appear to play a role (e.g. using and-introduction followed by and-elimination). In such cases *p* would be redundant, and (EC←) would not license inferring that *p* is in S's evidence. The concept of redundancy also extends to the components of *p* if *p* is complex, e.g. a conjunction: the proposition $(r \wedge s)$ would not be part of one's evidence just because one can gain knowledge by inference from *r*.[3]

We may consider a simple token inference to be composed of three elements: (i) a set of premises, (ii) an inference procedure, and (iii) a conclusion. Extended inferences are concatenations of simple inferences. What we may call the *epistemic quality* of a subject's belief in a proposition *p* concerns whether subject knows *p*, or has a justified belief in *p* that falls short of knowledge, or unjustifiably believes *p*. Let us take the case where a subject's belief in the conclusion comes about solely as a result of the inference. In that case the epistemic quality of the belief will depend on epistemic characteristics of the premises and of the inference procedure. The relationship between the epistemic characteristics of the premises and the epistemic quality of the conclusion will be the focus of this chapter and will be discussed at length. Here I shall consider the epistemic characteristics of the inference procedure so that we can set it aside.

The epistemic characteristics of the inference procedure will depend on a number of factors. Consider, for example, a subject carrying out a deductive inference. The quality of the procedure will depend on formal features of the deduction—e.g. whether it is valid. It will also depend on factors concerning the subject, e.g. whether

[3] I do not think that the notion of redundancy can be formalized. For example one might think that *p* is a redundant premise in the set $\Gamma \cup \{p\}$ in a derivation of *q* if *q* can be derived from Γ alone (and $\Gamma \nvdash p$). But it might be that any derivation from Γ alone is very different from the former (all such proofs might be much longer). In such a case *p* would not be redundant in the sense relevant to the above principles, and should be included amongst S's evidence for *q*.

she is intellectually capable of entertaining the premises and conclusion, and that she does not make errors in implementing the inference rule. If the inference procedure is non-deductive, its quality will also depend on relevant features of the world.[4] Let us call an inference procedure *faultless* iff it suffers from no shortcomings in its epistemic characteristics. A simple deductive procedure carried out carefully by a competent logician will typically be faultless. Likewise, an electrician who infers from the movement of the needle in a properly functioning ammeter that the battery to which it is connected is not dead will typically be employing a faultless inference procedure. A faultless inference procedure will be one that leads from true premises to true conclusions. If one uses the procedure and reaches a false conclusion from true premises because of the procedure, then the procedure suffers from an epistemic shortcoming.[5] This means that the properties of being faulty and of being faultless are strongly externalist epistemic properties: a procedure might be faulty simply in virtue of the way the world is: another procedure that is identical as regards the subject may be faultless. While I do not claim to have characterized 'faultless' with perfect precision, the notion does capture the intuitive idea that when an inference leads us to a false conclusion, we may sometimes regard the inference procedure, construed broadly so as to include external features of the world, to be at fault. In this chapter, we shall ignore such possibilities of epistemic fault in the inference procedure, in order to focus attention on the epistemic quality of the premises in an inference.

While the epistemic characteristics of the premises and of the inference procedure will be important in determining whether an inference gives the subject new knowledge, they are not the only relevant factors. S might already know the conclusion to be true, so S cannot gain any knowledge by the inference. Or S might have strong evidence against *q*, so that it would be irrational to believe *q* despite this new inference. We may regard a subject as *receptive* if she has no thoughts relevant to *q* other than those employed in this inference.

We can now explicate the proposition (C) 'S can gain knowledge by inference from *p*' as it is used in (EC), (EC→), and (EC←). (C) is true precisely when it is possible for a receptive subject, S, as a consequence of some faultless inference procedure from *p*, a non-redundant premise, to come to know the proposition that is the conclusion of the inference.

5.3 All evidence supports knowledge-producing inferences

In this section I provide the argument for:

(EC→) If *p* is in S's evidence then S can gain knowledge by inference from *p*.

[4] See e.g. Norton (2003).

[5] Must every inference from true premises to a false conclusion be regarded as faulty—for example, inferring that one's lottery ticket will lose in a million ticket fair lottery, when in fact one will be the lucky winner? I discuss that specific question below. However, note that I am defining 'faultless' in order to spell out the content of the phrase 'can gain knowledge by inference' more precisely, not in order to characterize the intuitive notion of 'faultless'.

The idea behind (EC→) is that evidence has a sufficiently high-grade epistemic quality that a subject is capable of gaining knowledge by inference from it. (EC→) contrasts with the claim that evidence might be epistemically low-grade and might be incapable of supporting knowledge-producing inferences. Hence the clarification in Section 5.2 excluded purely accidental reasons why a subject might fail to produce knowledge from her evidence. A counterexample to (EC→) must be one where a failure to produce knowledge must be attributable to the epistemic weakness of the evidence.

Thus (EC→) would be false if either of the following were true:

(A) A receptive subject S infers *q* from evidence *p* by a faultless procedure, but S's resulting belief, *q*, is not justified. *Or*

(B) A receptive subject S infers *q* from evidence *p* by a faultless procedure, but S's resulting belief, *q*, is not true.

That (A) is false is clear. It is at the very least the function of evidence to justify our hypotheses. If a subject uses a faultless inference procedure to gain a new belief from her evidence, then the new belief is justified.

The falsity of (B), although less immediately apparent, follows straightforwardly if all evidence is true, since all propositions inferred by a faultless procedure from true premises will be true, as was discussed above. One might wonder, however, whether this claim about faultlessness is too strong. Might it not be sufficient for faultlessness that it justify beliefs, without the additional requirement that the procedure is perfectly reliable? For example, the inference from 'Jane has only one ticket in a million ticket fair lottery' to 'Jane will not win the lottery' might be thought to employ a procedure that is not perfectly reliable but nonetheless justifies its conclusion. Should not such a procedure count as faultless in our sense? In the case where Jane does win the lottery, may not the falsity of the conclusion be regarded as epistemic bad luck external to the inference as opposed to a result of an epistemic fault of inference? It is my view (Bird 2007*b*) as well as that of Cohen (1998), Nelkin (2000), Sutton (2007), and others that in lottery cases, a belief (on this evidence) that Jane will not win is not justified. This is an issue that has been much discussed and cannot be settled here. Nonetheless, these are some of the relevant considerations. First, the justification claim regarding Jane's ticket leads to the lottery paradox: one is then justified in believing of every ticket that it will not win, and so justified in believing that none of the tickets will win, while also being justified in believing that one of the tickets will win (one knows that in this fair lottery one winning ticket will be drawn). Denying justification regarding any particular ticket is the easiest way to avoid the paradox. Of course, one is justified in believing that Jane's ticket is very unlikely to win (which does not lead to a paradox). But the inference 'it is very likely (unlikely) that *p*' to 'it is true (false) that *p*' is not a justification preserving inference—such an inference is faulty because its procedure is imperfectly reliable.[6] Secondly,

[6] For this reason I do not endorse Williamson's specific argument for the conclusion that a proposition that is a justified true belief but not knowledge fails to be evidence (even though I accept that conclusion). Williamson (2000: 200) considers a proposition that, he says, is justified (in this case by inductive

one may subscribe to some version of the safety principle concerning justification, e.g. that one cannot be justified in believing that *p* if in some close possible world one falsely believes that *p*.[7] For the worlds in which Jane's ticket wins and in which her ticket loses are very much alike. Thirdly, as I discussed on p.32, Jane is justified in thinking 'my ticket might win', but she is not justified in thinking 'my ticket might win but it will not win'; so she cannot be justified in believing the second conjunct of the latter proposition. Fourthly, as I also remarked there, if Jane were justified in believing that her ticket will not win, she would be justified in throwing the ticket away and would never be justified in buying a ticket, even if the lottery is organized by a philanthropist so that each ticket has a positive expected value. Fifthly, if one accepts, as I have argued in Chapter 2, belief aims at knowledge, there is something wrong about a belief that one knows cannot achieve this aim (assuming the widely held view that there is no knowledge in lottery cases).

So if the evidence propositions are true, a faultless inference procedure will lead to a true conclusion. But are all evidence propositions true? A central function of evidence is the testing of hypotheses. In particular it is a requirement on any hypothesis that to be credible it must be consistent with the evidence. Dick Heuer (1999: 47), an intelligence analyst, states 'A hypothesis may be disproved ... by citing a single item of evidence that is incompatible with it.' A century and a half earlier, the legal theorist Thomas Starkie (1834: 444) tells us 'It is essential *that all the facts should be consistent with the hypothesis*. For as all things which have happened were necessarily congruous and consistent, it follows, that if any one established fact be wholly irreconcilable with the hypothesis, the latter cannot be true. Such an incongruity and inconsistency is sufficient to negative the hypothesis, even although it coincide and agree with all the other facts and circumstances of the case to the minutest extent' (emphasis as in the original). Although Starkie uses the term 'facts', it is clear from the context that he is talking about evidence; indeed this passage comes from a section entitled 'Circumstantial evidence' in his book *A Practical Treatise on the Law of Evidence*. (Indeed the way in which Starkie moves between 'evidence' and 'facts' is itself evidence of the factiveness of evidence.)

That all evidence is true is the best explanation of the requirement on any hypothesis that to be credible it must be consistent with the evidence. If evidence must be true, then this requirement excludes a set of propositions all of which must be false. The requirement that hypotheses be consistent with the evidence would lose its force if false propositions were among our evidence, for then perfectly true propositions would be excluded as being inconsistent with our evidence.

A response to this argument suggests that the injunction that a hypothesis must be consistent with the evidence is derived from the prohibition on believing contradictions. If one permitted hypotheses inconsistent with the evidence one would be

reasoning) but which is not knowledge. There is nothing Gettier-like or otherwise untoward about the case, as Hughes (2014) points out. So, in my view, either the proposition is both justified and known, or it is neither.

[7] One might prefer the stronger safety principle, which would apply in this case also: S cannot be justified in believing that *p* if S knows that in some close possible world S falsely believes that *p*.

permitting contradictions. If one believes one of a pair of inconsistent propositions, one must reject the other. But this response is inadequate, since the requirement of consistency with evidence is stronger than the prohibition on contradictions. If one is inclined to believe both of two propositions that contradict one another, the prohibition on contradictions means that one cannot fully believe both. In order fully to believe either of the propositions, one will have to reject the other. The prohibition on contradictions does not say which to reject. It may be that further evidence is required in order to decide between them. The requirement of consistency with the evidence is stronger since it does tell us which of the two contradictory propositions to reject. If a hypothesis and the evidence contradict one another, then the hypothesis is to be rejected. No further evidence is required. Jon Williamson (2015: 37) suggests that what explains the preference for rejecting a hypothesis in such cases is that the evidence makes the hypothesis *unlikely* to be true. That explanation permits the possibility of false evidence. For example, one might think that justification is sufficient for evidence. A justified belief might be false, but in general conflict with a justified belief will make a hypothesis unlikely. Williamson's explanation fails to recognize the nature of the prohibition on inconsistency between evidence and hypothesis, which requires rejecting the hypothesis outright, not simply regarding it as unlikely (as the quotations from Heuer and Starkie make clear). Of course, one may not always know what one's evidence is. In which case one might be in the position of doubting that one's evidence is indeed evidence when it conflicts with a favourite, well-confirmed hypothesis. But it remains the case that if one did know in such a case that one's supposed evidence is indeed one's evidence, then one would be able to reject the hypothesis as false without further ado. This makes sense only if one's evidence is true. If one's evidence is true, then what is reliably inferred from it is also true. Hence (B) is false.

Since (A) and (B) are false, we may conclude that evidence must be capable of supporting inferences that yield beliefs that are both true and justified. We know that not all true justified beliefs are knowledge. But it would be very odd if evidence could support inferences to true justified beliefs but not to knowledge. Furthermore, Gettier cases typically arise because of reliance on justified but false lemmas. But the falsity of (B) means that we can rule out false lemmas. So there is no reason to suppose that evidence is capable of supporting inferences producing justified true beliefs without being able to support knowledge-producing inferences. We may conclude that evidence is indeed capable of supporting knowledge-producing inferences.

5.4 Evidence is anything supporting knowledge-producing inferences

In this section I discuss:

(EC←) If S can gain knowledge by inference from *p* then *p* is in S's evidence.

While (EC←) is a natural counterpart of (EC→) neither implies the other. (EC←) can be true without requiring (EC→). One could deny that evidence must be capable of supporting knowledge-producing inferences, yet agree that if some inference does produce knowledge, its premises must be among one's evidence.

(EC←) asserts the epistemological significance of evidence. If (EC←) were false, then there are propositions that are not evidence but from which one can nonetheless make knowledge-producing inferences. Thus to deny (EC←) would be to deny the place given to evidence in epistemology. If (EC←) were false, it would be possible to gain knowledge by inferring from propositions not among our evidence, in which case there would be no reason for us to care about which propositions are among our evidence. A criticism of S's claim to know *q* may be made thus, 'Proposition *q* is not supported by your evidence!' The falsity of (EC←) would permit S to reply, 'True, *q* is not supported by my evidence. But I did infer it from non-evidence propositions that support knowledge-generating inferences.' If (EC←) is false, then it becomes unclear why we care about evidence. If something else other than evidence, call it 'schmevidence', could lead to knowledge, then it might be perfectly alright to ignore the evidence so long as one had enough schmevidence, since that could be sufficient for us to expand our knowledge. But we don't think that. The rhetorical force of terms such as 'Evidence-Based Medicine' and 'Evidence-Based Policy' is precisely that it is absurd to suggest that it is satisfactory to form inferred beliefs on the basis of anything other than evidence.[8]

There is one response to my defence of (EC←) that deserves serious consideration. We might care about evidence even if (EC←) is false, if evidence is the *ultimate* source of premises for our inferences. That is to say, we might take evidence to consist of propositions that are themselves not inferred from any other propositions. Let S make an inference from premises *a*, *b*, and *c*. From these S infers propositions *i* and *j*. And from *i* and *j* S infers the conclusion proposition *p*. And let's say that S thereby comes to know that *p* is true. (EC←) claims that all of *a*, *b*, *c*, *i*, and *j* are in S's evidence. The counter-proposal under consideration claims that *i* and *j* are not evidence propositions. It is only *a*, *b*, and *c* that are evidence, assuming that they have not themselves been inferred from other propositions. This proposal fits well with the empiricist conception of scientific evidence as always observational/perceptual in nature. Perception delivers experiences that are at the basis of the inferences we make in science. It is reports of these perceptual experiences, gained in observation, that are correctly called evidence in science, not anything inferred from them.

Because this is an important idea I discuss it in detail in Section 5.6. In order that we can progress to the next section with (EC←) presumed to have been established, I shall briefly preview one of the central reasons for rejecting this counterproposal. The counterproposal ignores the fact that we generally do not remember the non-inferential propositions from which our inferred knowledge ultimately derives. Nor is that failure to remember to be regarded as a mere wrinkle, a departure from the ideal depicted by the theory. For our memories are designed *not* to store such information. Psychologists distinguish three types of memory: perceptual memory, short-term memory, and long-term memory. What we learn non-inferentially, e.g. in perception, will enter the first and often the second of these kinds of memory

[8] Which is not to say that the claims of, for example, the Evidence-Based Medicine movement are trivial, because their claims also involve contentious assertions about what counts as good evidence in medicine.

(whose duration is short) but enter the third only if particularly significant. Long-term memory will store information we are likely to use again, typically information we infer from what we perceive (or from other knowledge) rather than the perceptual experiences themselves. Given the way that memory works, evidence could not fulfil its epistemic role if it were restricted to non-inferential knowledge—we just do not retain enough non-inferential knowledge to support, even in an ancestral way, the knowledge-generating inferences we make. As I shall explain in Section 5.7, scientists do not use 'evidence' to refer only to non-inferential beliefs or knowledge; for example, proponents of Evidence-Based Medicine are not asking us to base clinical decisions on what we know through perception but rather on the results of properly conducted randomized trials or meta-analyses of such studies—such evidence is the outcome of sophisticated statistical inference, doubly so in the cases of meta-analysis.[9]

5.5 Evidence and knowledge

The preceding sections articulated and defended the view, (EC), that *p* is in S's evidence if and only if S can gain knowledge by inference from *p*. That completes my first task, to provide a functional articulation of the concept of evidence. In the light of that account, we may ask: what must something be like in order to fulfil that function? I shall defend the claim that all and only knowledge can fulfil this function. So we may conclude that all and only knowledge is evidence. That is a conclusion already reached by Williamson (1997).[10] Since Williamson presented independent arguments for his conclusion, we may regard his arguments as providing indirect support for (EC).

I claim that if (EC→) is true, then evidence itself cannot be less than knowledge:

(E→K) If *p* is in S's evidence then S knows *p*.

The principle 'No False Lemmas' states that if an inference depends non-redundantly on a false premise then it cannot generate knowledge (Harman 1965: 92). The principle can be extended. If an inference depends non-redundantly on a premise that is unjustified, then the conclusion of the inference will fail to be knowledge. (This will be so whether justification is conceived of externally, for example in terms of reliability, or internally, for example in terms of rules of rational belief-formation.) When an inference depends on several premises non-redundantly, then a subject is justified in believing the conclusion of the inference only if she is justified in believing all the premises (assuming that the subject has no other reason for believing the conclusion). So in addition to the No False Lemmas principle we may state a No Unjustified Lemmas principle. Consequently, for an inference to generate knowledge, its essential premises must be both true and justified. If the old JTB account of knowledge were correct, then we could simply infer from the conjunction of the

[9] A meta-analysis provides a statistical analysis of a set of several first-order studies, thereby aiming for a more robust result than any individual study provides.

[10] For further arguments in support of the equation of evidence and knowledge, see Bird (2004, 2018) and Littlejohn (2011, 2013). And for arguments against, see Goldman (2009) and J. Williamson (2015).

two principles (No False Lemmas and No Unjustified Lemmas), a third principle, No Unknown Lemmas:

(NUL) S can gain knowledge by non-redundant inference from *p* only if S knows *p*.

As it happens a Gettier belief (one that is true and justified, but does not amount to knowledge) could satisfy No False Lemmas and No Unjustified Lemmas without satisfying No Unknown Lemmas. Nonetheless, the best explanation of why No False Lemmas and No Unjustified Lemmas are correct principles is that they follow from the more general principle, No Unknown Lemmas. There is nothing special about true justified beliefs. If *p* is a Gettier belief, then we should not expect an inference that depends on *p* to be knowledge generating. Typically *p* is a proposition that is accidentally true, e.g. $p = q \vee r$, where *q* is false but justified but *r* is true but unjustified. But if *p* is accidentally true its truth cannot make any difference to the epistemic quality of what is inferred from *p*. Since propositions inferred from false lemmas are not knowledge, neither then are propositions inferred from accidentally true propositions. In the kind of Gettier case just mentioned, an inference from *p* that yields a justified conclusion will depend essentially on the contribution of the disjunct *q* of *p* (since only the disjunct *q* is justified). But in that case the inference depends on a false lemma, and the conclusion is not knowledge. Hence an inference from a premise which is a Gettier case of a justified true belief that is not knowledge will not yield knowledge. Nothing short of a known premise will yield knowledge.[11]

Having argued for No Unknown Lemmas, (NUL), we now note that this principle plus (EC→) gives us the conclusion that all evidence is knowledge, (E→K).

Hence, the left to right implication of (EC), (EC→), entails that all evidence is knowledge. Correspondingly the right to left implication of (EC), (EC←), entails that all knowledge is evidence, when conjoined with the highly plausible principle that inference from knowledge is knowledge generating:

(KI) If S knows *p* then S can gain knowledge by inference from *p*;
(EC←) If S can gain knowledge by inference from *p* then *p* is in S's evidence;
therefore
(E←K) If S knows *p* then *p* is in S's evidence.

(KI) is simply the idea that inference from what we already know has the power to extend our knowledge. The phrase 'can gain knowledge by inference from *p* is still interpreted as discussed in Section 5.2, i.e. where the subject is receptive to *p* and the inference procedure is faultless; so (KI) is not refuted by cases where the subject

[11] There are some who think that there are unusual cases where one can gain knowledge by inference from a true and justified proposition that fails to be knowledge—see Luzzi (2010)—or even from false propositions—see Hilpinen (1988), Warfield (2005), Arnold (2013). Hilpinen and Coffman (2008) maintain that the right response is that in successful cases of the latter sort there are always true propositions in the vicinity that do the relevant work (see also Littlejohn (2013) for arguments that evidence propositions must always be true). I extend this to claim that there are always *known* propositions doing the work. It is important to note that while it is useful to represent an inference as proceeding from a specific set of premises to a conclusion, actual inferences are rather messier, with multiple related propositions playing a role in supporting an inference. See Bird (2018) for details.

has other beliefs relevant to the conclusion of the inference or where the inference is epistemically unsatisfactory. Under the intended interpretation, a counterexample to (KI), would be a case where S comes to believe *q* as a result of an inference from *p* where S knows *p*, yet S fails to come to know *q* because of some inadequacy in the epistemic quality of *p* (and not because of some other facts about S or facts about the inference procedure).

An aside concerning the relationship of (KI) to closure It is difficult to see how (KI) could be false: what more could be required of the premise of an inference that aims at extending knowledge than that the premise itself be known? However, the fact that (KI) looks sufficiently similar to the closure principle for knowledge might make one think that someone such as Nozick (1981) who rejects closure should reject (KI) as well. Alleged counterexamples to closure claim that (known) valid inferences from known premises can sometimes lead to conclusions that cannot be known—for example where the premise is a known proposition about everyday circumstances ('S is typing at a computer in a university office') and the conclusion is a consequences of that proposition that denies some relevant sceptical scenario ('S is not a brain-in-a-vat on a planet orbiting Gamma Cephei being deceived into believing that she is typing at a computer in a university office'). Are such cases, if we accept them, counterexamples to (KI)? It is difficult to say with confidence that they are, precisely because it is difficult to say what has gone wrong for the subject in such cases. For, prima facie, what could be epistemically faulty about a valid inference that is furthermore known to be valid? And the subject in such a case may unproblematically be assumed to be receptive. That said, I think we can be confident that the fault cannot be laid at the door of the *premise* of the subject's inference. The counterexample posits that S *knows* that she is typing as a computer in a university office. Could S's epistemic relation to the proposition that S is typing at a computer in a university office be of any higher quality, such that a higher quality epistemic attitude to that proposition *would* permit inferred knowledge that S is not a brain-in-a-vat on a planet orbiting Gamma Cephei etc.? Clearly not; this is not analogous to the case making an inference from a premise that is false or unjustified; it is not that the premise is epistemically not up to the task of supporting a knowledge-generating inference. For that reason, I believe that a denier of closure must say that in some cases a known valid inference can be epistemically faulty. The oddity in so saying is just why many find it difficult to deny closure; but it is at this locus, on the inference procedure, that the debate is centred. If so, then such counterexamples to closure, even if accepted, need not be taken as counterexamples to (KI). For example, one might argue, as some epistemic contextualists do, that closure is a correct principle when restricted to a given context (Cohen 2000): what goes wrong in the counterexamples is that there is a cross-context application of closure; such applications constitute faulty inference procedures. In any case, the worst damage that a denial of closure could do to my argument is to require a modification to (KI) and the other principles under discussion. For the closure denier does regard many inferences as potentially knowledge extending, and so needs to supply some replacement for closure. Whatever that replacement is, it will suggest an appropriate replacement for (KI). *Mutatis mutandis* the arguments of this chapter will stand. For example, one modification would be to say that say that the 'can gain knowledge' in all the principles applies only to cases where the conclusion, when true, is knowable at all to S. The corresponding alteration to the central claim of this chapter would be that the function of evidence is to provide the premises of inferences that can extend knowledge *where knowledge is possible at all.* Likewise, if one prefers a contextualist approach, then the changes will require restriction to a given context, and the corresponding conclusion will give a contextualist account of evidence (which is just what one would expect a contextualist to endorse). (I note that the argument for (E←K) stands even if we have to have modify preceding arguments to accommodate the denial of closure. If we adjust the interpretation of 'can gain knowledge' that applies equally to (KI) and to (EC←), and so (E←K) still follows. If we take a contextualist approach, then the same contextualist restrictions will apply to what counts as evidence in (E←K) as applies to what counts as knowledge.)

In this section I have argued that the following are both correct:

(E→K) If *p* is in S's evidence then S knows *p*.
(E←K) If S knows *p* then *p* is in S's evidence.

which together entail:

(E↔K) *p* is in S's evidence if and only if S knows *p*.

If we combine (KI) with (NUL) we get the following extended 'knowledge by inference' principle:

(KI+) S can gain knowledge by inference from *p* if and only if S knows *p*.

Recalling:

(EC) *p* is in S's evidence if and only if S can gain knowledge by inference from *p*,

(KI+) and (EC) directly give us:

(E↔K) *p* is in S's evidence if and only if S knows *p*.

In summary (where '⇒' denotes entailment):

$$\begin{array}{ccccc} \text{(NUL)} & \wedge & \text{(EC}\rightarrow\text{)} & \Rightarrow & \text{(E}\rightarrow\text{K)} \\ \wedge & & \wedge & & \wedge \\ \text{(KI)} & \wedge & \text{(EC}\leftarrow\text{)} & \Rightarrow & \text{(E}\leftarrow\text{K)} \\ \Downarrow & & \Downarrow & & \Downarrow \\ \text{(KI+)} & \wedge & \text{(EC)} & \Rightarrow & \text{(E}\leftrightarrow\text{K)} \end{array}$$

The fact that (EC), along with the plausible (KI+), entails (E↔K) is indirect evidence is favour of (EC). For Timothy Williamson (1997, 2000) has argued persuasively for E=K, *viz.* (E↔K), on independent grounds. The grounds are independent since Williamson's arguments proceed for the most part on the basis of whether we would regard certain propositions as part of a subject's evidence, whereas the arguments above concern principles regarding the function of evidence. In so doing those arguments illuminate the concept of evidence in a way that Williamson's arguments do not (and are not intended to). There is no reason to suppose that 'evidence' and 'knowledge' are really the same concept. If they were, one might ask, why do we have a concept of evidence? Why don't we make do just with the concept of knowledge? We talk of 'evidence for' but not of 'knowledge for'. We have a concept of evidence distinct from that of knowledge to mark a special role for evidence. Evidence is characterized functionally as that which supports knowledge-yielding inferences. (KI+) tells us that it is knowledge that supports knowledge-yielding inferences, which explains why Williamson's equation is right.

5.6 Is evidence non-inferential knowledge?

In Section 5.4 I argued for:

(EC←) If S can gain knowledge by inference from *p* then *p* is in S's evidence.

I raised a counterproposal to (EC←). If (EC←) is false, then the case for (EC), (E→K), and (E↔K) is undermined. So in this section, I examine the counterproposal in detail and show that it is not correct.

Unless you are a sceptic, you will think that new knowledge can be gained by chains of inferences. (EC←) says that intermediate propositions in the chain of inferences have the status of evidence propositions. The picture that (EC←) licenses is this. A scientist make an inference that generates a conclusion. Then that scientist or another scientist uses that first conclusion to make another inference to some second, further conclusion. (EC←) says that if the inference to the second conclusion is knowledge-generating then the first conclusion is evidence. For example, Tycho Brahe made observations of the planets. From this evidence, Kepler inferred that planets travel in ellipses (and his other laws). Starting from Kepler's conclusions, Newton inferred a further conclusion, that the planets are subject to an inverse square central force law. (EC←) claims that Kepler's law that planets travel in ellipses is evidence—assuming that Newton came to know the force law as a result of his inference.

The counterproposal, which I articulate and discuss in detail below, limits evidence propositions to non-inferential knowledge. It rejects (EC←) because it denies that the inferred propositions, such as Kepler's laws, are evidence. This counterproposal still gives evidence a key epistemological role. This role is foundational. In the chain of inferences just considered, it is only the initial propositions that are the evidence propositions. The intermediate propositions do not count as evidence, according to the counterproposal. Evidence propositions are those propositions that form a basis of our inferences and are not themselves inferred from anything else. So in our example, only Tycho's observations are evidence, while Kepler's elliptical law is not. Alvin Goldman (2009: 86), for example, proposes:

(E=NPJ) The proposition *p* is among S's evidence if and only if *p* is non-inferentially, propositionally justified for S.

The non-inferentialist view is also implicit, for example, in the assumption common among empiricists (and many non-empiricists) that evidence is observational.[12] It is explicit in Patrick Maher's (1996: 158) account of evidence:

> (E↔EK) The proposition *p* is among S's evidence if and only if S knows *p* directly by experience.

Common to Goldman and Maher is the claim that a proposition is evidence only if it is not inferred:

> (E→NI) The proposition *p* is among S's evidence only if S does not know or believe *p* on the basis of an inference.

Williamson's argument for (E→K) proceeds by refuting what he takes to be the main alternative to (K→E), *viz.* that only knowledge that is certain is evidence. But this is not the only alternative. Another proposes that evidence is non-inferential knowledge. Since a proposition might be non-inferential knowledge without being certain, Williamson's argument for (K→E) is incomplete. It needs supplementation by an argument against (E→NI). One might indeed think that the main point of identifying evidence with non-inferential beliefs or knowledge is the conviction held by some that such beliefs and knowledge will be certain. One might also think nonetheless that it is theoretically significant to identify the foundations upon which the structure of knowledge rests, and that the concept of evidence identifies these foundations, marking the division between what one has inferred and what one has ultimately inferred it from.

Maher (1996: 158) writes:

> Even if a proposition is known to be true, if this knowledge [E] is not directly based on experience then E is not evidence and hence not evidence for anything. For example, let E denote that a substance taken from a certain jar dissolved when placed in water, and suppose we know E directly from experience. Let H be the proposition that the substance that was in the jar is soluble and suppose we know H by inferring it from E. Then we know both H and E from experience, but E is evidence for H and not vice versa since only E is known directly by experience.[13]

If one accepts the argument from above for (E→K), while adhering to (E→NI), then one holds:

> (E→NIK) If *p* is among S's evidence, then S knows *p* non-inferentially.

[12] See Rowbottom (2019: 136) for a related view that takes observation to be non-inferential.

[13] Goldman (2009: 200) discusses an example from Williamson in which I watch a film of balls being drawn from a bag, with replacement. I see that each of draws 1 to n reveals a red ball. I infer probabilistically the proposition *p* that draw $n+1$ is red too. Williamson says that *p* is not in my evidence because *p* is not known. Goldman says 'An alternative explanation is readily offered by E=NPJ [the proposition *p* is among S's evidence if and only if *p* is non-inferentially, propositionally justified for S]. My evidence does not include the proposition that draw $n+1$ was red because, although that proposition is justified, it isn't *non-inferentially* justified. On the contrary, it is justified by inductive inference. So the unsatisfied necessary condition for evidence, according to E=NPJ, is that evidence confers non-inferential justification' (emphasis in original).

This position might produce the following account of evidence:

> (E↔NIK) If S knows *q* by inference, then S's evidence for *q* consists in that non-inferential knowledge from which *q* is (ultimately) inferred.

(E↔NIK) proposes that evidence is to be equated with the propositions that are the ultimate source of one's knowledge, the propositions from which one infers one's knowledge but which are themselves not inferred. Maher's account of evidence as knowledge given directly by experience, (E↔EK), entails not only (E→NI) but also (E→NIK). It is consequently very similar to (E↔NIK). The principal difference between (E↔EK) and (E↔NIK) is that the latter would permit non-inferential apriori and innate knowledge to count as evidence. In this section I:

(i) Show that (E→NIK) is false.
(ii) Refute on this ground Maher's account of evidence (E↔EK), as well as the related account (E↔NIK).
(iii) Plug the gap in Williamson's argument for (E→K) and so shore up his grounds for asserting (E↔K).
(iv) Refute an alternative but related view of evidence as causally (rather than inferentially) foundational.

5.6.1 *Loss of evidence: forgetting*

Imagine that a subject S makes a chain of inferences starting from the proposition *e* and leading to the proposition *p*. Let it be that these inferences add to S's knowledge. We may take it that S does have evidence for the proposition *p*, for S knows *p* on the basis of inference. Let us assume that everything relevant to *p* that S knows is contained in the chain of propositions constituting the inferences leading to *p*. Therefore S's evidence for *p* lies somewhere in this chain of propositions. (E→NIK) requires that it is the knowledge at the starting point of this chain that is evidence, *viz.* the proposition *e*. The intermediate propositions, although known, are not evidence.

It follows that in order for S to retain knowledge of *p* without inferring anew from different evidence, S must retain knowledge of *e* so long as S has any evidence for *p*. In general then, (E→NIK) requires that in order to possess inferential knowledge, one must retain, as knowledge, all the evidence from which it was originally inferred. But that is implausible, since one can lose one's 'ultimate' evidence without also losing knowledge of what is inferred from it. For example, one can forget one's evidence without ceasing to know the things inferred from that evidence. Suppose S knows *p*, having inferred it from *e*; S then, having forgotten *e*, goes on to infer *q* from *p*. According to (E→NIK), S's evidence for *q* is not *p*, since *p* is inferential knowledge; S's evidence for *q* ought to be *e*, since *e* consists of propositions not inferred by S but from which S has, in a transitive historical sense, inferred *q*. But *e* cannot be part of S's evidence for *q*, since at the time of coming to know *q*, S no longer knows *e*. By (E→K), *e* is no longer part of S's evidence. Hence (E→NIK) and also (E↔NIK) and (E↔EK) are false.

In Maher's example quoted above, it might be important for S's purposes to know whether the substance in the jar is soluble. As in the example, S comes to know this by inferring it from the fact that the sample dissolved, as he saw. In due course S, while still remembering the important fact that the substance is soluble, forgets how he came to know this. For all S can recall, it might equally have been thanks to someone's testimony. At this point S comes to learn that the substance in the jar is glucose; and so S infers that glucose is soluble. What evidence does S have for the belief that glucose is soluble? According to (E→NIK), none. For the proposition 'The substance in the jar is soluble' is not evidence (it is known by inference); and the proposition 'The substance in the jar did actually dissolve when placed in water' is not known at all, having been forgotten. But it is absurd to suggest that S has no evidence for the belief that glucose is soluble.

This argument may be thought to trade too much on human failing. An ideal epistemic subject does not forget. (E→NIK) may be supposed to apply only to such subjects. I shall consider such an idealization shortly. But first let me remark that failure to remember should not be regarded as a wrinkle, a departure from the ideal depicted by the theory. Thomas Young (of Young's slits and Young's modulus, etc.) expressed the value of forgetting this way (Robinson 2006: 19):

> It seems to be by a wise and benevolent, though by no means an obvious arrangement of a Creative Providence, that a certain degree of oblivion becomes a most useful instrument in the advancement of human knowledge, enabling us readily to look back on the prominent features only of various objects and occurrences, and so class them and reason upon them, by help of this involuntary kind of abstraction and generalisation, with incomparably greater facility than we could do, if we retained the whole detail of what had once but slightly impressed on our minds.

Young is right—as I noted above, our memories are designed *not* to store all information that impinges upon our minds. While what we learn, for example, in experience remains in perceptual and short-term memory only for a brief period, what is retained in long-term memory is only what is significant and so likely to be used again. And what is retained is typically only the information abstracted or inferred from the original experience, not the content of the experience itself. I can remember the melting point of lead (327.5°C), but not the visual or auditory experience by which I learned this fact. As Young says, this is good engineering (by evolution, which may be regarded as a 'Creative Providence'). The processes of perception and memory work rather like the detectors in the ATLAS experiment described in Chapter 1, which process and then eliminate in real time the large majority of the data generated by its collisions, storing only summary data and information about significant events.

5.6.2 *Loss of evidence: undermining*

We will now consider an idealized subject who does not forget what she has once learned. Even a rational memory-perfect individual can lose evidence by *undermining*. New evidence may misleadingly cast sufficient (rational) doubt on some previous piece of evidence. As a result the proposition in question is no longer knowledge and so is not evidence any more. In Williamson's example a subject sees a red ball and a black ball enter an empty bag. A ball is withdrawn and replaced 10,000 times.

Each time the ball drawn is black. The initial knowledge that there is a red ball in the bag is now undermined. The subject should not now rationally believe that there is a red ball in the bag—at most he should believe that it seemed to him that a red ball was placed in the bag. This does not show that the original perceptual knowledge was not after all non-inferential, that it was inferred from the evidence that it seemed that a red ball was placed in the bag. Apart from the fact that this simply fails to do justice to the phenomenology of perceptual knowledge, this response, to be perfectly general, requires that non-inferential knowledge must be certain in the sense of not being subject to any possible undermining. But there is no reason to suppose that non-inferential knowledge has this feature, or that any knowledge has. Undermining of this sort can happen to all sorts of knowledge, not just perceptual knowledge. Testimony, even if reliable, is often open to undermining by counter-testimony. A proposition which is known non-inferentially may also be knowable inferentially; it may thus also be undermined inferentially.

I now consider a case where non-inferential knowledge is undermined by misleading additional evidence. By 'misleading evidence' I mean a proposition that is genuine evidence (the proposition is true and is known to be true) but which, in the circumstances in question, points *away* from the truth about some pertinent hypothesis. For example, I am investigating the hypothesis that this is a fair coin. The hypothesis is true. I test the hypothesis by tossing the coin ten times and each time the coin lands showing heads. That test outcome is evidence against the correct hypothesis that the coin is fair and so is misleading evidence. In Williamson's example, the 10,000 draws of a black ball is misleading evidence relative to the proposition that one of the balls is red. Misleading new evidence can point away from the truth regarding some proposition that was previously known and so was itself, then, evidence. Sufficient misleading evidence can lead to the previously known proposition no longer being known. And since knowledge is necessary for evidence, the proposition in question loses its status as evidence.

At time t there is just enough evidence e, so that when S infers p from e, S knows p. However, at t^*, some portion of e, the set of propositions u, is undermined by new information. Normally that would deprive S of the knowledge of p. However, let it be that between t and t^* S has also acquired further additional evidence e^* in favour of p such that the total current evidence $e–u+e^*$ is sufficient to support knowledge of p. At t^* S can know p even though S did not infer p from the evidence S has at t^* for p, so long as S is still appropriately sensitive to that evidence. This, along with forgetting, shows that S can still know a proposition p even though p was inferred from propositions that are no longer part of S's evidence. We might stretch the interpretation of (E→NIK) to permit this, since p was inferred from what was S's evidence at the time. Nonetheless, (E→NIK) is still in trouble. Let S infer q from p at t^*. Then the non-inferential propositions from which q is ultimately inferred are the propositions in e, including u. But S's evidence at t^* does not include u—the propositions in u were undermined, and so are not known by S, who for that reason may not even believe them any more. So S's evidence cannot be identified with S's non-inferential knowledge, and hence (E→NIK) is false.

	t	t^*
non-inferential knowledge	e	$e{-}u{+}e^*$
	↓	
inferred knowledge	p	p
		↓
inferred knowledge		q

In this table '↓' represents S's inferences. At t^* S does not have all the evidence, e, from which she earlier, at t, inferred p, because by t^* some significant part of that evidence, u, has been undermined. Nonetheless, S continues to know p, because S has acquired a total set of evidence, $e{-}u{+}e^*$, that does support p. S has not inferred p anew from this total evidence, but is cognitively sensitive to it, allowing for knowledge of p. S infers q from p and comes to know q. What is S's evidence for q? (E↔NIK) has no good answer, whereas (E↔K) says that S's evidence is p.

To illustrate such a case consider this example:

Nipper of the Yard Detective Superintendent Leonard 'Nipper of the Yard' Read, sees—and so comes to know the proposition u—that Reggie was in Una's Cafe near King's Cross Station at 11:00 p.m. At 11:20 p.m. a mail robbery is committed at the station. Nipper's (perceptual, non-inferential) knowledge u, when combined with other evidence forms an evidence set e that allows him to come to know the proposition p, that Reggie committed the robbery. Although he himself knows with full confidence that Reggie committed the crime, in order to ensure a secure conviction in court Nipper pulls Reggie in, extracts a confession and obtains damning forensic evidence. The confession and forensic evidence constitute the proposition e^*, now also known by Nipper. Let it also be the case that (previously unknown to Nipper) Reggie has an identical twin brother, Ronnie. At the time of the crime Ronnie is locked up in high-security Pentonville gaol, which is also near King's Cross. Shortly after obtaining the confession and forensic evidence Nipper is (correctly) informed by a reliable source that Reggie has an identical twin brother who was in London near King's Cross at 11:00 p.m.; but the source fails to add that this was because the brother was in the nearby prison. Thinking about Reggie's criminal record, Nipper reflects on the fact and comes to know the proposition q, that Reggie is the single most prolific criminal in London (having now exceeded Jack the Hat's record of twenty-three robberies and other crimes in the preceding year).

Note first that the mere fact that Reggie's identical twin, Ronnie, was somewhere in the vicinity of Una's cafe is not itself a defeater for Nipper's perceptual knowledge that Reggie was in Rosie's Cafe, for Nipper could not have seen Ronnie there, thanks to Ronnie's being locked up. (This is *not* a case like seeing the one real barn in fake barn country.) On the other hand, when Nipper receives the information that Reggie has an identical twin brother who was near King's Cross at 11:00, his knowledge that it was Reggie he saw in Una's cafe at that time is undermined—Nipper no longer knows *u*. Nonetheless Nipper still knows *p*, that Reggie committed the crime: he is in possession of the confession and compelling forensic evidence *e**. His evidence is now *e*–*u*+*e**, which suffices to support knowledge that *p*. Nipper's retention of the knowledge of the truth of *p* does not require that Nipper must infer the proposition *p* that Reggie committed the crime, for a second time, on this occasion inferring it from the confession and forensic evidence. A second inference is not needed so long as Nipper's belief in that proposition *p* is sensitive to this new evidence *e**. (I shall discuss causal and counterfactual sensitivity at greater length below.)

The theoretical role for evidence captured in (E→NIK) is that evidence should be the non-inferential knowledge from which inferred knowledge is inferred. So what is the foundational, non-inferential knowledge which, according to (E→NIK), is Nipper's evidence for *q*, the inferred proposition that Reggie is the most prolific criminal in London? It is not (i) Nipper's perceptual knowledge *u* that Reggie was in Una's cafe, because, thanks to undermining, this is no longer knowledge; it is not (ii) Nipper's knowledge *p* that Reggie committed this robbery, since this is inferred knowledge, not non-inferential knowledge; it is not (iii) his knowledge *e** of Reggie's confession and the damning forensic forensic evidence, for even if this is non-inferential knowledge, it is, *ex hypothesi*, not knowledge from which Nipper has made any inference. While it is likely that Nipper possessed (and perhaps even inferred from) a lot of relevant background knowledge, that background knowledge was not enough to let him know *q*, that Reggie is the most prolific criminal in town. None of the crucial propositions, *u*, *p*, or *e**, fulfils the role of being non-inferential knowledge from which this inferred knowledge that *q* is inferred.

As Nipper of the Yard suggests, knowledge of inferred propositions can survive the loss of the evidence from which they were inferred. Such inferred propositions can be the basis of further knowledge-producing inferences. Therefore we should conclude that it is not merely knowledge that gets transmitted by inference but also the status of being evidence. That is to say, in the above, *p*, although inferred, is the evidence on which knowledge of *q* is based.

The falsity of (E→NIK) has been demonstrated by pointing to the vulnerability to undermining of non-inferential propositions. Intermediate inferred propositions are also vulnerable to undermining and this also shows that (E→NIK) is false. Consider a case like the previous one where *p* is inferred from non-inferential evidence *e* via a chain of inferences. Proposition *i* is an intermediate proposition in this chain and so is inferential knowledge. (The inference goes from *e* to *i* to *p* and finally to *q*.) In the new case, it is *i* that is then undermined by misleading new evidence. However, the subject S has also acquired new, independent (non-inferential) evidence, *e**, that supports *p*. As before, S does not make any new inference, from *e** to *p*. But S is causally sensitive to *e**. For example, imagine she discovers the undermining of *i*; knowing *e**, she does not give up her belief in *p*, but she would have given up that belief had she not been in possession of *e**. This causal sensitivity is enough for S to retain knowledge of *p*. Under these circumstances, what is S's evidence for the final inferred proposition, *q*? It isn't *e* because although *e* is retained in this example, the chain of rational inferences from *e* to *p* is broken at the undermined proposition *i*. The evidence is not *e**, since S makes no inferences from *e**. So (E→NIK) cannot identify any propositions that are S's evidence. Yet S comes to know *q* by inference. Not only would be it absurd to say that S can come to know an aposteriori proposition by inference but without any evidence, to do so would empty the concept of evidence of its relevance. (E↔K), on the other hand, can identify S's evidence for *q*—it is the proposition *p*, a proposition that S knows and from which she inferred *q*.

5.6.3 *Causal sensitivity*

The use made of Nipper of the Yard depends on the idea that evidence may support knowledge without that knowledge being inferred from that evidence. Inferring a proposition from evidence is one way in which a belief in that proposition can be appropriately responsive to the evidence, but it is not the only way. Causal sensitivity and counterfactual dependence are more general than the relation of inference. It might be that if *p* were not among S's evidence, then S would not continue to believe *q*; or that if asked why one should believe *q*, S would cite *p*. The truth of these counterfactuals is consistent with S's having at no point inferred *q* from *p*. In the example discussed above, I said that Nipper could retain knowledge that Reggie committed the crime, despite the undermining of his original evidence. That retention does not require that Nipper must infer anew the proposition that Reggie committed the crime from the confession and forensic evidence. Perhaps Nipper does not give the matter a second thought. Why does he still know, despite not having drawn a fresh inference? The truth of the following counterfactuals would typically be sufficient to show that Nipper has sufficient sensitivity to the new evidence for knowledge even if he has not made an inference from it: were he asked why anyone should believe that Reggie committed the robbery, he would cite the confession and forensic discoveries; and had he not himself had that new evidence, he would have ceased having a high degree of belief that Reggie was near King's Cross on learning of the existence of Reggie's twin Ronnie.

I now consider whether we can find an alternative to (E↔NIK) that accommodates the fact that evidence can support knowledge through causal sensitivity while retaining the guiding idea that evidence stands at the far end of a chain of epistemic support (inferential or causal). (If available, this would solve the problem of undermining, but not of forgetting.) I argue that the most plausible replacement for (E↔NIK) does not do this, since causal sensitivity is not an antisymmetric relation.

(E→NIK) might reasonably be supposed to accommodate cases where an existing belief is inferred anew from a fresh set of evidence. But it is implausible to suppose that this is in fact what always occurs when new evidence is acquired. We are constantly acquiring new evidence that is relevant to a huge range of existing beliefs. We do not refresh those beliefs by repeatedly re-inferring them. But that does not make the beliefs we have irrational. The counterfactual causal sensitivity of our beliefs to one another can be enough to ensure this. The failure of causal, reliabilist, and counterfactual (e.g., tracking) analyses of knowledge notwithstanding, it is undoubtedly the case that it is typically the reliability (often causal) of the connection between the facts and a belief-like mental state that makes that mental state one of knowing. Let S know *p* and let it be that this knowledge is therefore reliably related to the fact *p*. Then S's belief in *q* might be reliably related to the fact *q* by virtue of a reliable connection between that belief and S's knowledge of *p* plus a nomic correlation between *p*-like and *q*-like facts. Hence S may know *q*. The reliability of the connection between S's belief in *q* and S's knowledge of *p* may require causal or counterfactual sensitivity of the former to the latter, but need not require that S has inferred *q* from *p*.

Those who would still wish to restrict evidence to a foundational subset of knowledge might take this on board by suggesting that (E↔NIK) should be replaced by an alternative, where the idea of being inferred from is replaced by that of being causally sensitive to. This move would also accommodate those who would hope to bypass the foregoing discussion by having a very weak account of what it is to infer one proposition from another, whereby causal sensitivity is sufficient for inference. Such a move would yield something like:

> (E↔CIK) If S knows *q*, then S's evidence for *q* consists in that knowledge to which S's belief in *q* is causally sensitive, but which is itself not causally sensitive to other knowledge.

The final clause, that evidence is knowledge which is itself not causally sensitive to other knowledge, is required to match the key element of (E↔NIK), that evidence is the root or foundation of knowledge. While (E↔K) says that all knowledge is evidence, (E↔NIK) makes an asymmetrical distinction between evidential and non-evidential knowledge. Non-evidential knowledge, on this view, depends on evidential knowledge in a way in which non-evidential knowledge does not depend on evidential knowledge. In (E↔NIK) this is assured by the historical dependence implicit in 'inferred from'. In (E↔CIK) it is the final clause that generates this asymmetry. Without the final clause we would have:

> (E↔CEK) If S knows *q*, then S's evidence for *q* consists in that knowledge to which S's belief in *q* is causally sensitive.

(E↔CEK) might well be a starting point for an account of 'evidence for'. However, it does not provide an alternative to (K→E), with which it is consistent. In particular (E↔CEK) does not provide the asymmetry between evidence and what it is evidence for that is sought by the objector. This is because two pieces of knowledge may be causally sensitive to each other. The thesis of the theory-dependence of observation states that one's observational knowledge may be causally sensitive to which theories one believes (and knows). At the same time, obviously, theoretical belief (and knowledge) is causally sensitive to the observations one makes. In a related kind of case, my knowledge of some arithmetical truth may be sensitive to my knowledge of the reliability of the calculator I use—it was malfunctioning, and I have had it repaired; my belief and knowledge of the arithmetical output are thus sensitive to my knowledge that it is now functioning reliably. At the same time, my knowledge of that reliability is also sensitive to my knowledge of the calculator's output. Should something cast doubt on the output, such as conflict with my mental arithmetic or with the results of another calculator, then my knowledge of the calculator's reliability might be undermined.

Since causal sensitivity is not asymmetrical, (E↔CEK) will not do as a replacement that seeks to retain the spirit of (E↔NIK) (*viz.* that evidence is the root or foundation of knowledge); the additional clause that gives (E↔CIK) is required. But now the problem arises that (E↔CIK) may prevent too much from being evidence. The discussion of the last paragraph raises the possibility that no knowledge is causally insensitive to other knowledge. According to (E↔CIK) there would then be no evidence. Even if that were not itself an absurd conclusion, it would be in conflict with the current argument for (E↔CIK), which is the thought that evidence is a special kind of foundational knowledge, whereby the possibility of non-evidential knowledge depends on its relation to the evidence.

5.7 Evidence and science

In this chapter I have argued for a functional conception of evidence:

> (EC) *p* is in S's evidence if and only if S can gain knowledge by inference from *p*.

which tells us that to achieve the status of evidence a proposition must be capable of supporting inferences that deliver knowledge; furthermore any proposition that is so capable counts as among the subject's evidence. The latter tells us that evidential propositions are not restricted by content nor by the means by which they come to be believed. The proposal thus rejects a conception of evidence that appeals to empiricists, whereby evidence propositions must concern what a subject perceives or observes. I mentioned and rejected another conception of evidence that would appeal to some empiricists (and perhaps some non-empiricist foundationalists), *viz.* that evidence is restricted to propositions that are not themselves inferred from other propositions. Instead *any* proposition, so long as it can support knowledge-producing inferences, can play the role of evidence, whatever the content of the proposition, whatever its relation to experience, and whatever the means by which it come to be believed and known. That includes propositions that may have been inferred from other propositions and propositions that may be very distantly removed from perception. In the light of (KI+), we see that the propositions that fulfil this role are precisely the propositions that constitute the subject's knowledge.

It is worth noting that the use of 'evidence' in science accords with the conception promoted here rather better than with the empiricist and non-inferentialist views of evidence, such as Maher's (E↔EK): the proposition *p* is among S's evidence if and only if S knows *p* directly by experience. Part of the evidence for Einstein's general theory of relativity is the anomalous precession of the perihelion of Mercury. The precession is *inferred* from many individual observations of Mercury and that inference involves considerable mathematical work as well as substantive auxiliary hypotheses.[14] Secondly, the precession *is not perceptible.* So this standard use of 'evidence'

[14] Maher (1996: 160–1), we should note, does recognize that we do call inferred propositions evidence, which conflicts with his proposal. He himself uses an example in which the advance of the perihelion of

in science does not accord with an empiricist view, but does with the functional (inferentialist) conception. Examples of this sort are readily gathered from the scientific literature. Consider, for example, a paper entitled 'Evidence for the spin-0 nature of the Higgs boson using ATLAS data' (ATLAS Collaboration 2013). What is the nature of this evidence? The evidence concerns facts determined by the ATLAS Collaboration concerning the decay of the Higgs particle into photons, in addition to evidence from the CMS Collaboration concerning the decay of the Higgs into Z bosons. Facts about the decay of the Higgs are clearly not perceptible facts. Moreover, they are inferred facts. These facts are determined by multiple processes of inference, many of which are built into the detectors and the computer programmes that process the information streams they produce.

It is nonetheless true that the facts presented as evidence (the precession of Mercury, the decay of the Higgs boson) are often referred to by scientists as 'observations'. This shows only that the scientists' conception of observation is one that is only remotely linked to perception (this is a point to which I shall return at length in the next chapter). Historically, observation has typically been heavily perceptual, but even then it has rarely been exclusively perceptual. An astronomical observation may involve looking at a planet through a telescope, but the key information in the recorded observation will include the elevation of the telescope, its geographical location, and the time at which the planet was seen; these are pieces of information which are not garnered by seeing through the telescope, and in the case of time, not perceptible at all (even if one has a watch). Likewise observations by nineteenth-century chemists and physicists recorded values of variables such as temperature and electric current, which cannot be perceived. Positivists may argue that the real observations are perceptions of the instruments. Even so, it remains true that the use of even perceptual reports needs to be given context, which is not obviously perceptual: is what the observer did to avoid a parallax error something that is itself perceptible by her? The empiricist proposal to limit 'observation' to perceptual encounters is a revision to our use of the term 'observation'. Observation in modern science is even less perceptual and more heavily theory-laden. The Higgs boson has been observed, but no one has perceived it; nor indeed has anyone seen its trail on a cloud chamber photograph. Rather, as I described in Chapter 1, a vast array of different kinds of detectors, all designed in a theory-intensive way, are connected to a computer which delivered its results in a statistical form. In less esoteric science, modern radio telescopy delivers nothing of significance that is perceptible, but rather the evidence is in the form of data collected in a computer, which, furthermore, has been processed by statistical software. The same goes for the evidence that is used in constructing weather models, which is collected and processed automatically. I will expand in detail on this argument against the empiricist conception of observation in the next chapter.

Mercury is evidence. In effect he permits inferred knowledge to be regarded as evidence when it stands in for or replaces the non-inferential knowledge from which it was inferred. Section 5.6 entails that this is not an adequate explanation of our use of 'evidence'.

Furthermore, while evidence in science is often observational, it is not always so. If the empiricist, non-inferentialist conception of evidence were correct, the phrase 'theoretical evidence' (denoting evidence for a hypothesis from theoretical knowledge) would be an oxymoron. Nevertheless many published papers in science do refer explicitly to theoretical evidence: 'Theoretical evidence for strong correlations and incoherent metallic state in FeSe' (Aichhorn *et al.* 2010), 'Experimental and theoretical evidence for hydrogen doping in polymer solution-processed indium gallium oxide' (Huang *et al.* 2020), 'Empirical and theoretical evidence for herd size as a risk factor for swine diseases' (Gardner *et al.* 2002), 'Theoretical evidence for the sensitivity of charge-rearrangement-enhanced x-ray ionization to molecular size' (Hao *et al.* 2019), and so on.

According to the conception of evidence I have promoted and defended in this chapter, we may gather evidence in order to establish the truth of some proposition *p*. Then, once enough evidence has been gathered such that the truth of *p* is established, and so known, the proposition *p* itself can be evidence for some further proposition *q*. Returning to the Higgs boson. Evidence for its existence was provided by the ATLAS and CMS experiments of 2012. Once physicists agreed that this evidence does show that the particle exists, then its existence becomes a fact that itself is evidence that can be cited in confirming the standard model of particle physics. This illustrates the import of (E←K) and of (EC←). What makes something evidence is not that it was produced in some specific way, but rather that it can be used for a certain purpose: making knowledge-producing inferences. Correspondingly what scientists care about when they consider whether a proposition (such as an assertion that the precession of the perihelion of Mercury is 5557 seconds of arc per century, or that the Higgs boson exists) counts as evidence is not what general *kind* of source (perception, observation, inference form theory) produced the proposition, but whether the particular process that produced it warrants that proposition sufficiently that it can be relied upon in making further inferences. I have argued that such warrant is sufficient if and only if it is enough to make the proposition in question known. Consequently, the functional view developed in this chapter can embrace theory-laden evidence and theoretical evidence with ease.

5.8 Conclusion

At the outset of this chapter I gave myself two objectives: (a) to articulate a functional characterization of the concept of evidence that relates to inference; (b) to relate my characterization to Timothy Williamson's claim that all and only knowledge is evidence, E=K.

The guiding idea behind (a) is that evidence is something we want for a purpose. We don't collect evidence for its own sake, but in order to make inferences from it (e.g. concerning the truth of a hypothesis). Of course we can make inferences from any old propositions and beliefs. But not just any old proposition or belief is evidence. We want evidence for epistemically worthwhile inferences, the ones that produce knowledge: evidence is that from which knowledge-producing inferences can be made.

This conception allows any proposition to be evidence, so long as one can infer knowledge from it. This stands in contrast to a conception of evidence as that from which inference ultimately starts: evidence must be non-inferential. The empiricist version of this non-inferentialist conception of evidence says that evidence is that which is known directly from experience, or, in a more scientific vein, evidence is always observational. This conflicts with scientists' use of 'evidence' which allows highly theoretical knowledge and knowledge that has been gained by multiple scientific inferences to be regarded as evidence. (We examine the notion of observation in the next chapter.) The inadequacies of the non-inferentialist conception of evidence were exposed by considering the possibilities of forgetting one's evidence or having it undermined by further evidence.

6

OBSERVATION

In this chapter I first characterize the concept of observation as it has been used in philosophy of science. There the concept combines a claim about the functional role of observation—observation supplies basic evidence in science—and a claim about its nature or identity—observation is perception. I argue that if one retains the second, empiricist claim, then anti-realist scepticism about science is warranted, as is shown by an inability to give reductive accounts of testimony and instruments in terms acceptable to the empiricist. Those terms include a commitment to epistemological internalism, since it is internalism that motivates the fundamental role given to perception by empiricism. We should, therefore, reject empiricism. The empiricist, perceptual conception of observation is in any case at odds with how scientists use the term 'observation'—scientists often claim to observe imperceptible entities such as quarks and magnetic fields. So I propose a non-empiricist account of observation that allows an observation to be the result of inferences from data. To support this, I draw a distinction between data and observation. And I consider whether science needs perception at all (it does not).

6.1 Introduction

Observation is widely held to be the epistemic bedrock of science. For many empiricists this has been the basis of some form of sceptical anti-realism. Such anti-realists I call the *sceptical empiricists.*[1] Even so, almost all realists accept the foundational role of observation, resisting anti-realism by arguing that we can make rational and reliable inferences regarding the truth or truthlikeness of our theories from premises entirely founded on observation. These realist empiricists I call *optimistic empiricists.* What makes these views *empiricist* is the widely held assumption:

(O-P) Observation is perceptual in nature.

Bas van Fraassen (2001: 154), today's arch anti-realist, writes, 'We can detect the presence of things and the occurrence of events by means of instruments. But in my book that does not generally count as observation. Observation is perception, and perception is something that is possible for us without instruments.' Likewise, the pragmatist Hasok Chang (2005: 880), says: 'A *quality* is *observable* if it can be registered by human sensation .... *Observation* is an act of gathering information by human sensation without optional interpretations' (emphasis in original). At the

[1] For a recent defence of an instrumentalist conception of science, see Rowbottom (2019).

Knowing Science. Alexander Bird, Oxford University Press.
DOI: 10.1093/oso/9780199606658.003.0006

same time, here is the realist Anjan Chakravartty (2007: 3–4): 'There are things that one can, under favourable circumstances, perceive with one's unaided senses. Let us call them "observables" ... Unobservables, then, are things one cannot perceive with one's unaided senses.'[2]

Consider this argument (where an 'observational proposition' is a proposition whose truth can, in principle, be determined directly by observation):[3]

(OBS) All evidence is observational;
(INF) From observational premises only observational conclusions may rationally be inferred;

therefore

(SCEP) Only observational propositions can be known.

(Sceptical Empiricism)

The sceptical empiricist will hold this argument (or something much like it) to be sound. The empiricist identification of observation with perception, (O-P), is important. Without (O-P) this argument would not present such a sceptical challenge. For if the 'observable' includes imperceptible items, such as quarks or magnetic fields, then this argument would leave very little in the supposedly unknowable category of 'the unobservable'. And since acceptance of (O-P) is so widespread, we can usefully combine the two ideas as one:

(OBS*) All evidence is observational, and observation is perceptual in nature.

The optimistic empiricist, a typical scientific realist, holds that science can generate the knowledge that (to most of us) it appears to, starting with a foundation of observation, accepting (OBS*). Since the optimistic empiricist accepts (OBS*), but rejects (SCEP), she therefore rejects (INF). She will argue that there are rational inference procedures that can take us from the observational evidence to conclusions about the unobservable.

So: the sceptical empiricist (an anti-realist) and the optimistic empiricist (a typical realist) agree regarding the truth of (OBS*). They differ then regarding (INF). A standard realist move rejects (INF) by appeal to the rationality of Inference to the Best Explanation (IBE). The anti-realist rejects that move, arguing that that IBE is not a rational inference procedure—because it is inconsistent with Bayesian conditionalization, for example. I shall not consider those debates in this chapter (I do look at IBE and Bayesianism in the remaining three chapters). My aim in this chapter is

[2] The identification of observation with perception is pervasive in philosophy of science. It is one reason why so much of the debate surrounding the 'theory-ladenness of observation' focuses on the theory-ladenness of perception (see e.g. Fodor 1984; Brewer and Lambert 2001).

[3] Determined 'directly by observation' means without being inferred from other propositions known by observation. (An observational proposition can be known directly by observation and without inference; that is consistent with its being possible to know that proposition by inference, for example by induction.) An 'observational proposition' might alternatively be defined as a proposition concerning solely what may be observed. (I'm not entirely confident that the latter can be articulated precisely.) The two ways of defining 'observational proposition' are not equivalent. Nonetheless, my discussion does not turn on anything that is sensitive to that difference.

to attack the shared empiricism of both the sceptical and the optimistic empiricist. (OBS*) is false.

The strategy of this chapter is as follows. I will argue that the optimistic empiricist is over-optimistic. My grounds are not a general concern with (INF). Rather my objection is that the realist has a problem in explaining the role of testimony and the use of instruments in science in a way that remains true to empiricism. The sceptical empiricist is right in this at least: if one is an empiricist one should also be a sceptic. An important component in this argument is the role played by epistemological internalism. The sceptical arguments assume internalism. That is a legitimate assumption in this context because empiricism is motivated by internalism. One could (and in my view should) avoid scepticism by rejecting this internalism. If one rejects internalism one has no reason to be an empiricist.

The reason for adopting this strategy is that I doubt whether the debate between realism and anti-realism at a global level is a fruitful one, one reason being that I do not think that there is a single, uniform inference rule that is 'Inference to the Best Explanation'. I explain all this in Chapter 9. Whereas I do think that there are rather more specific arguments regarding testimony and instruments that are more problematic for the optimistic empiricist, the empiricist who wants also to be a realist.

My discussion then focuses on the sceptical empiricist. I show that the argument *Sceptical Empiricism* is question-begging. The key idea here draws on the conclusion of Chapter 5 regarding evidence. And I also argue that the sceptical empiricist cannot reasonably hope to reinterpret the aim of science in empiricism-friendly, instrumentalist terms. In important areas of science the only perceptual experiences a scientist has when conducting her experiment are with a computer screen. But it cannot be the aim of science to predict our perceptions of computer screens.

Neither sceptical nor optimistic empiricism is tenable. I therefore reject (OBS*).[4] Having rejected (OBS*) and so (O-P) in particular, we need to rethink quite radically what we think observation is. I start by noting that scientists *do* say that quarks and magnetic fields are observable. I hold that they are right to do so. The correct understanding of observation is functional: observation provides a representation of an object of study that can serve as basic scientific evidence. Perception plays no role in our understanding of observation. Indeed it plays only a minor role in modern science at all.

To help make my case, I shall start by examining the development of the empiricist conception of observation from the scientific revolution onwards. At the end of the chapter I conclude that empiricism's emphasis on observation-as-perception is not a product of philosophical necessity; instead it is a product of a particular—albeit important—phase in the history of science.

6.2 Empiricism, observation, and perception

In this section I articulate the empiricist conception of observation and chart the historical development of that conception.

[4] Indeed, I shall also reject (OBS) even when 'observational' is interpreted correctly, as the scientists understand the term. See Section 6.6.3.

I take the core of the empiricist view of observation to be the following:

(O-E) The *functional role* of observation: observation supplies basic *evidence* in science.

(O-P) The *nature* or identity of observation: observation is *perception*.

Two further claims are implicated in the empiricist epistemology of observation. First:

(O-C) The epistemic *foundational strength* of observation: observation should be *certain* or very close to certain.

This motivates the empiricist adherence to (O-P). And, secondly, also firmly embedded in empiricist philosophy of science is evidentialism:

(EV) Belief in a scientific theory is justified only by its relationship to the evidence.[5]

In a later section, I will also emphasize empiricism's association with epistemological internalism through (O-C).

6.2.1 *Empricism's ideal scientist*

The motto of the Royal Society is 'Nullius in verba', which translates as 'Take nobody's word for it'. The phrase is an abbreviation of Horace's line 'nullius addictus iurare in verba magistri', which means 'bound to swear allegiance to no master', the context being Horace's decision to give up poetry and take up philosophy (Sutton 1994). Seventeenth-century fellows of the Royal Society would have known that Horace meant that he would not be bound to any philosopher's doctrines—likewise the Royal Society would not follow any established school of philosophy, in particular not Aristotelianism. The fellows would have also read the motto in the more general sense given in its usual translation.[6] Hunter (1989) describes how the ideas of taking nothing on authority and testing all claims were central to the search for a motto for the new society: they considered alternatives such as 'Omnia probate' ('test all things'). When the Society met for elections on St Andrew's Day, one fellow remarked, 'I had rather have had it been on St. Thomas's Day, for he would not beleeve till he had seen and putt his fingers into the holes; according to the motto, *Nullius in Verba*' (Stimson 1949: 64).

The ideal scientist is an individual who has performed all the relevant experiments or observations himself or herself, depending in no way on the authority or reports of others. Any important result achieved by another scientist must be replicated

[5] Evidentialism is usually articulated with respect to any proposition, not just scientific theories. In passing I note that I do not support unrestricted evidentialism, since some propositions can be known to be true without relying on evidence for them—ordinary perceptual propositions, for example. And knowledge entails justification. So some propositions can be justified without their being justified by evidence. But that quibble is not relevant to the justification of scientific theories.

[6] *Not* the common mistranslation 'there is nothing in words'—for example, 'As the founders of the Royal Society of London put it in 1663: *Nullius in verba*—there is nothing in words. It is facts we seek.' (Popper and Wächterschäuser 1990), quoted by Greg Dawes 'Nullius in verba, nihil in verbis, sapere aude' https://blogs.otago.ac.nz/emxphi/nullius-verba-nihil-verbis-sapere-aude/, last accessed 6 January 2022.

so that the scientist sees the evidence for herself. The evidence must be delivered by the scientist's own perceptual experiences. The scientist is therefore able rationally to assess the likely truth of a theory against this evidence, without depending on any other scientist. Science for the Royal Society was, nonetheless, a social activity as Bacon had portrayed in his *New Atlantis* (1627). But the sharing of ideas is not the sharing of evidence or reasons. To justify one's belief in a theory, one needs to see the evidence oneself. And so the meetings of the Society centred on the performing of experiments that all could observe and participate in.

This conception of the ideal finds its philosophical expression in empiricism. For the empiricist all our knowledge comes ultimately from sense-experience—hence Locke's denial that there is innate knowledge and Hume's denial that 'demonstration' (formal proof) can give knowledge of matters of fact. The function of observation (O-E) and its nature (O-P) are integrated: basic evidence is perceptual, all knowledge is founded on nothing other than perception. But what exactly is perception such that it suffices as a foundation for knowledge? For as Descartes maintained, and the empiricists, Berkeley especially, were aware, what is often held to be perception—experience of external objects—is liable to error, in which case it cannot function as a *certain* foundation, (O-C). To give observation/perception its foundational strength, perception is *not* held to be a relationship with some external object. Rather, for the early empiricists especially, an observation/perception is an awareness of the inner aspect of a sense experience (acquiring the name 'sense datum'). For the empiricist (and indeed for the rationalist) scientific cognition is a private process.

6.2.2 *The empiricist conception of observation*

Mill (1843) explicitly links observation in science to perception, but argues that what is typically called 'observation' is not strictly the observation at all. For our observation *reports* are usually inferences from perceptions, inferences that are 'performed with the speed and certainty of an instinct' so that we cannot distinguish them from the perceptions themselves. In truth, '[o]bservation extends only to the sensations by which we recognize objects; but the propositions which we make use of, either in science or in common life, relate mostly to the objects themselves' (Mill 1843: 643). Mill also remarks that the inferred description is liable to error: it is the uninferred basic perception that is certain. Although Mill did not give an explicit account of the details of such an inference, it is clear from his discussion that a model for many inferences would be:

It appears to my senses as if *p*;
My senses are reliable indicators of the way things are;
therefore
p.

(The phenomenalist inference)

In this inference '*p*' may be what the scientist reports as her observation, but in truth her observation is 'it appears to my senses as if *p*'.

Mill's distinction between the true observations (the sensations or perceptions) and the seeming observations (how we describe the things we experience after inference) is a major part of what was at issue in the 'protocol sentence' controversy of the Vienna Circle.[7] On the one hand the foundationalist Schlick (1934) promoted what Hempel deemed a phenomenalist conception of the protocol sentences, which report the subjective sense-experiences of the subject (à la Mill). Neurath (1932), on the other hand, developed a physicalist and inter-subjective conception whereby protocol sentences referred to physical objects and processes. Hempel himself favoured the inter-subjective approach: observations report directly perceptible facts, such as the clicking of a Geiger counter. Both Feyerabend and Quine took the inter-subjective conception further: observation sentences report facts that competent speakers of the language would unanimously agree on, using perception.

The debate therefore centred in large part on the significance of (O-C). Should the demand for certainty be paramount, so that what we regard as the basic statements of science are reports of subjective experience? Or should the demand for certainty be relaxed so that the basic statements can involve reference to objective, physical entities? It is important to note that those who favoured the latter, physicalist or inter-subjective approach did not relax the certainty condition far. An observation is still the closest one could get to the certainty of a report of subjective sense-experience while satisfying the need for inter-subjectivity—observations are still reports of what is perceived. The debate over protocol sentences concerned only the extent of the certainty condition on observation. All those engaged with the debate subscribed fully to (O-E) and (O-P). Quine (1990: 2), for example, writes 'It is common usage to say that the evidence for science is observation', *viz.* (O-E), although Quine himself prefers to discuss observation sentences to observation itself: the observation sentences are those (intersubjectively agreed) sentences prompted by sensory stimulation, *viz.* (O-P).

6.3 Challenges to optimistic empiricism

Much pre-modern science lacked the characteristics of ideal science, and so was poor science by empiricist lights. Observation was not held in medieval science to be especially important, whereas reason alone was thought often to be a sufficient basis for knowledge. And authority, especially the authority of ancient authors, was held in high regard; a scientist thereby relied on the cognitive processes of predecessors. Modern science, however, does not always appear to meet the ideal either. An optimistic empiricist is obliged to argue therefore that further reflection will show that, fundamentally, modern science does indeed have these features even if obscured by the complexity of actual science. So the empiricist would say that the social nature of the science practiced by the early Royal Society and scientists since must be distinguished from medieval or Renaissance deference to the authority of ancient authors. The open, collaborative nature of modern science enables other scientists to

[7] *Beobachtungsatz* (observation sentence—Schlick's preferred term) and *Protokollsatz* (protocol sentence—Neurath's term) were terms used interchangeably to denote evidence reports.

undergo the same process of cognition—that is why the repeatability of experiments is so important. By repeating experiments, other scientists can repeat the cognitive experience of the original scientist.

6.3.1 *Testimony and the social nature of science*

Although the empiricist ideal rejects dependence on authority, the social nature of science involves unavoidable dependence on testimony in some cases. For although experiments may be repeated, not all observations are experimental in this sense—where the observation concerns an unique event of scientific significance, for example. In 1769 James Cook undertook an expedition to Tahiti to observe the transit of Venus on behalf of the Royal Society. Not all astronomers could make those observations. However, according to Hume (1748: 111), we can accept the testimony of others when we have independent evidence of their reliability (Coady 1992).[8] On Cook's expedition the observations were made by Cook, Daniel Solander, and Charles Green, all of whom had proven experience as astronomers; Cook was an officer in the Royal Navy (and therefore a gentleman), Solander was a Fellow of the Royal Society (as Cook would become), and Green had been assistant to the Astronomer Royal. Since we know such people to be reliable, we can infer that what they say is true:

James Cook FRS says that *p*;
James Cook FRS is a reliable informant;

therefore

p.

(The testimonial inference)

This reductionist approach, which Adler (2015) calls the 'unadorned empiricist model', faces insurmountable problems. The core of the problem is the scale of the trust in others' testimony that is required by modern science (Hardwig 1991). Consider just the simple case of verifying that Cook was a reliable informant. Very few individuals among all those interested in astronomy would have known Cook well enough to judge his reliability for themselves. The remainder would have to use other sources to verify that premise in the above inference. That will be just further testimony. Perhaps for some individuals, there is a chain of reliable informants between them and Cook, such that each informant in the chain can vouch for the veracity of the next. That is, Z can confirm Cook's reliability, thanks to personal experience of Cook. Y can verify Z's reliability, and so can report that Z is reliable in providing information regarding Cook's reliability. And X can verify Y's reliability, and so report that Y is reliable in informing one of Z's reliability in informing one of Cook's reliability. And so on. Note that for this to satisfy the empiricist model, not only would there need to be a chain of reliable informants, but also each would have to pass on

[8] Hume seems more concerned with the justification of our crediting testimony in general, rooted in part in our knowledge of human nature, than in the justification of one's accepting the testimony of particular individuals. That reflects Hume's interest in a science of human nature (Faulkner 1998). But he is clear that any such general justification is modified by the facts of the particular case, e.g. the interests and capacities of the individual.

these increasingly nested reports of reliability. It is implausible that any such nested reliability reports are produced. And if they were, a rational individual receiving a multiply nested reliability report ought to be inclined to regard it as *un*reliable, given the high chance that one of the many links in the chain will turn out, after all, to be insufficiently reliable.

Most people will not even be connected to Cook by any chain of reliable informants. They might instead put their trust in institutions, such as the Royal Society and its journal, the *Philosophical Transactions*, in which Cook and Green published their observations (Green and Cook 1771). Again, however, it is improbable that most readers would be in a position to verify the reliability of the *Philosophical Transactions*. Perhaps the individual would be able to verify for themselves enough of the facts reported in that journal and so be able to gauge its reliability. But that is unlikely, since the articles published cover such a wide range of topics. And in most instances they would be reporting novel facts that the individual would not know or be able to check for herself. So individual verification is implausible. As a result, we would be back to others' testimony regarding the *Philosophical Transactions*, and we are little or no better off than we were. Likewise, one might trust the Royal Society only to elect reliable individuals as fellows and only to give reliable individuals important scientific tasks. Again, attempting to verify that trust, for most of us, would inevitably depend on testimony, testimony that itself cannot be verified.

The scale of modern science just amplifies this problem. John Hardwig (1991) emphasizes, as I have done, the very great extent of the division of labour in science. A scientist can have the expertise to understand the details of only a tiny proportion of the many fields of science. Yet, science is interconnected and increasingly so. One scientist will have to rely on the expertise of many others to draw the conclusions she wishes to. In Chapter 1, I explained how a palaeontologist will depend on radiometric dating to test hypotheses about the relationships of fossils and other remains. Although she will understand the broad principles of radiometric dating, she will not have any expertise in the precise details that are required to make the exact determinations of age that she will use in her palaeontological work, nor will she know all the relevant underlying atomic theory and the evidence that supports it. Modern interdisciplinary science will very often depend on the cooperation of scientists from a greater number of specialisms than in this relatively simple example. It is beyond the intellectual capacity of any individual to acquire a sufficient understanding of all the fields involved. And as Hardwig also points out, the quantity of work done in collaborations is vastly beyond what any individual could achieve in a lifetime. He reports on an experiment in particle physics that required 280 person-years of labour to complete (Hardwig 1985, 1991). The scale of the ATLAS experiments described in Chapter 1 is hugely greater even than this.

There is a long tradition of denying that testimony can deliver knowledge, from Plato to Locke (and beyond), precisely because testimony differs from perception or reasoning from one's own perception.[9] If that were correct, then none of us, the

[9] See Chakrabarti (1994) for details of the extent of this tradition.

most eminent scientists included, would have anything more than the most rudimentary scientific knowledge, such is the extent of our dependence on testimony in science. To avoid such scepticism, the empiricist must show that this reliance on testimony can be justified by means acceptable to the empiricist. As we have seen, this simply cannot be done. The scale of our reliance on testimony in science is so great, and our individual perceptual experience of the world and of other people in it is, by comparison, so limited, that no one could hope to confirm for themselves the reliability of more than the tiniest fraction of the scientific testimony on which they depend. Reductionism about testimony must be rejected, and as a result a non-sceptical empiricism faces an insurmountable obstacle. Likewise the possibility of an autonomous scientist, one who relies solely on her own cognitive capacities, must be rejected. It is important to appreciate the scale of the disparity between reality and the empiricist 'ideal' of autonomy. The latter cannot be regarded as an idealized model that provides an insightful approximation to reality. Rather, it is a radically mistaken model, and the empiricism that requires it is radically mistaken too.

Often those who accept the anti-reductionist points made above hold that there is something special about testimony—it is apriori that one is entitled to accept testimony as true. In effect, having failed to reduce testimony to perception, this treats testimony as like perception. As Burge (1993: 458) puts it, 'We are entitled to rely, other things equal, on perception, memory, deductive and inductive reasoning, and on … the word of others'. This too I regard as mistaken.[10] Both reductionism and apriorism exhibit epistemological internalism: because trust in testimony is founded, ultimately, on one's experiences or on apriori argument, the subject is able (so the arguments claim) to assess the reliability of that trust for herself. Whereas an externalist approach to testimony, such as Goldberg's (2007) reliabilist treatment, avoids these issues. According to the latter, the primary issue is whether the source of testimony is in fact reliable.[11] In the case of science, however, I suggest that a more fruitful approach is to think of the social epistemic subject (the relevant scientific community, for example) as knowing the relevant scientific facts (as discussed in Chapter 4), and the individual as knowing because they are connected to the community in the right way (e.g. by reading its leading journals). Either way, scepticism about the possibility of gaining knowledge in science without producing it all oneself can and should be avoided by rejecting empiricism and internalism.

[10] Burge maintains that we are apriori entitled to rely on propositions presented as true by a rational source, because rationality is a guide to truth. At best this shows that the rational source can know what the truth is. But it does not show that the source will choose to tell us the truth. Misconduct in scientific research (fraud, questionable research practices) shows that sometimes scientists do not so choose.

[11] Goldberg does require that for the hearer to know, the hearer must be able to recognize the content of the speaker's assertion and be able to discriminate between reliable and unreliable sources. The latter is not, however, a return to reductionism. It serves as an anti-luck requirement, ruling out knowledge by the gullible or those in fake-barn-like cases (i.e. where the hearer luckily happens to listen to the one reliable informant surrounded by many liars).

6.3.2 *Instruments*

Another departure from the empiricist ideal is the fact that scientists use instruments. Some instruments give one perceptual experiences more or less like those of unaided perception—a telescope or a low-powered optical microscope for example. Other instruments, such as a high-powered optical microscope or an electron microscope, give the viewer a perceptual experience rather less like any they would have without the instrument. In some cases the instrument creates an object with which one perceptually engages, such a photograph either of the entity under study (e.g. Eddington's photographs of the 1919 solar eclipse) or of its effects (a photograph of traces in a cloud chamber). Often the instrument measures quantities that themselves cannot be perceived (air pressure, electric current, magnetic field intensity). One can perceive the instrument in real time or one may perceive some record of measurements (such as the marks on a barograph). Increasingly, all there is to perceive when using an instrument is a computer screen. Although scientific realists sometimes talk of instruments *extending* human perception, it is clear that instruments increasingly *replace* perception.[12] And that is because our sense organs are not appropriate means to find out about the things in which modern science is interested.[13] While an optical microscope will allow one to see a flea's spiracles, reading the field strength from the display of a Gauss meter is not a means of seeing a magnetic field. Nor is performing a statistical analysis on a database from a particle accelerator a way to perceive subatomic particles.

Let us consider first the use of an instrument in the apparent extension of our senses (such as a telescope or microscope). For phenomenalists such as Mill and Schlick, the use of instruments introduces the need for a further inference to replace or to supplement the phenomenalist inference:

> Using the instrument, it appears to my senses as if *p*;
> Using the instrument, my senses are a reliable indicator of the way things are;
>
> *therefore*
>
> *p*.

Intersubjectivists, such as Neurath and Quine, might be willing to regard their intersubjectivity as extending to some cases involving the use of instruments such as the telescope. Hacking (1983) argues that scientists see with microscopes, while Toon (2014) argues, drawing on the extended mind thesis, that the instrument becomes part of the scientist's perceptual faculties. Such views (plausible in my opinion), while respecting the letter of the core empiricist claim (O-P), that observation is perceptual, do so at the cost of weakening or abandoning the commitment to claim (O-C), that observation should be certain, which motivates empiricism. Instruments and other

[12] For a very different view of the cognitive relationship between perception and instruments, see Chapter 6 of Rowbottom (2019).

[13] Our sense organs are the products of the accident of our species' specific place in the history of one particular planet. It is implausible that they should, even when 'extended', be useful for perceiving entities and properties that are remote in scale, location, and history from that local, historical, biological accident.

techniques involved in observation can introduce error and distortion. Often they produce results that do not reflect reality in every respect. Chromatic aberration in telescopes and microscopes is clearly an artefact of the instruments and not a record of any feature of the object under study. Other features of an image will be debatable. When Camillo Golgi (1898) described what he called an 'internal reticular apparatus', others rejected his discovery as an artefact of his staining technique.[14]

Clearly, however, even the extension of perception cannot avail in cases where what the scientist perceives is the dial of a Gauss meter (a magnetometer) or lines of data on a computer screen. The empiricist will need the following inference:

The instrument indicates that *p*;
The instrument is a reliable indicator of the way things are;
therefore
p.

(The instrumental inference)

The empiricist will want to be sure that the major premise can be known by means acceptable to empiricism. As already mentioned, in some cases that is easy, such as in the case of the terrestrial use of a telescope. Matters are more contentious when we establish the reliability of the instrument by checking that it delivers the correct results on objects whose natures are already known, and then infer that the instrument is reliable in telling us about entities that we could *not* know about with unaided senses. Hacking (1983) argues that when a microscope delivers images of a microscopic labelled grid manufactured by independent and well-understood techniques, we can be confident that the microscope is able to provide information about microscopic objects.

Nonetheless, given that we know that instruments can often be reliable in some respects but not in others, an empiricist has some reason to demur in at least some such cases. In yet other cases what the instrument is reporting on is not at all like anything we humans can perceive—a magnetic field, for example. The reliability of an instrument may be a consequence of the truth of the theory with which it is designed (plus facts about the construction of the instrument). For example, some magnetometers work by exploiting Faraday's law. But how do we know that such laws are correct? A sceptical empiricist will deny that we can know. The optimistic empiricist will be willing to employ Inference to the Best Explanation (IBE) to justify the claims of reliability in such cases. In my view the sceptical empiricist is correct.

The complaint is sometimes made against the sceptical empiricist that he is willing to use IBE in everyday circumstances, but not in scientific circumstances when the conclusion is some proposition concerning what cannot be perceived. This complaint would be justified if IBE were a single, unified inference procedure. If it were, then its being reliable in one set of circumstances would imply its being reliable in all,

[14] This was one of the first organelles (functional subunits of a cell) to be discovered, now known as the Golgi apparatus.

and it would not make sense to question its reliability concerning the imperceptible any more than it would to question whether a pocket calculator is reliable in Oxford, once its reliability had been established in Cambridge. But IBE is not like that. IBE covers a range of structurally similar inference procedures, some of which may be more reliable than others. One reason for this is that what counts as a 'good' explanation varies from one field of inquiry to another. A theory in fundamental physics, such as the standard model, may be a good explanation partly in virtue of specifying laws that display a certain mathematical elegance. Whereas the explanatory goodness of the Alvarez meteor hypothesis of the K-Pg extinction lies in its ability to account for a range of disparate phenomena. The standard model of particle physics is also a good explanation in that it can explain the data to a high degree of mathematical precision. The theory of evolution through natural selection lacks that good-making feature—it does not give quantified explanations. It shares with the Alvarez hypothesis the ability to explain diverse phenomena. But evolution offers a single explanatory framework (the same idea of the retention of favourable traits through natural selection explains both the existence of systems of sight and also the origin of cardiovascular structures). Whereas the Alvarez hypothesis links distinct explanations—it explains the iridium anomaly by the high incidence of iridium in meteors and it explains the demise of dinosaur species through the effects of a dust cloud caused by the meteor impact. There is not one IBE but many IBEs. (This is a point I shall return to in the next chapter and in the final chapter.) And some of these IBEs will deliver conclusions concerning perceptible entities and others will deliver conclusions that concern imperceptible entities.

Since there isn't a single, unitary inference procedure, IBE, of general reliability, we cannot infer from the confirmed reliability of our use of an IBE in perceptual cases to the reliability of an IBE in non-perceptual cases. The sceptical empiricist is entitled to complain that the optimistic empiricist has not empirically justified the use of instruments whose reliability is dependent on theories that go beyond the perceptible.

Before leaving the discussion of instruments, we should also note that many instruments are used by scientists whose knowledge of the laws and principles underlying their construction is limited or non-existent and who are not in a position to check the reliability of the instrument by other means. Their dependence on the instrument is yet another instance of the unavoidability of trust discussed in the preceding section.

6.3.3 *Conclusion*

Testimony and the use of instruments show that the core empiricist view outlined above cannot be correct. The central problem with both is that they bring about beliefs without the believer having perceptual contact with any aspect of the world that is responsible for the truth of beliefs. The proposition that *p* may be a crucial component of scientist A's reasoning regarding some theory. But A has no perceptual engagement with the fact that *p* or anything in the world intimately related to such a fact. In testimony, all A has is a perception of scientist B's words. In the case of the use of an instrument, A might have only a perception of a dial or a computer screen. The empiricist is forced to try reductionism, using the testimonial inference or the instrumental inference.

If reductionism were successful, the result would still do violence to our image of science. Imagine a scientist who has a justified belief in the standard model of particle physics, thanks to participation in the ATLAS experiments, in which both testimony and instruments play a very large part. (O-P) means that the observations she makes are not of subatomic particles. Rather they are of pages in journals and computer displays. And so (O-E) means that the scientist's basic evidence does not consist of newly discovered facts in physics, but is rather facts about computers and about social science. Yet, scientists do not call these things observations, nor do they cite them as evidence.

In any case, reductionism is false. There is just no reconstruction of a scientist's justified scientific beliefs from her perceptual beliefs alone via reasoning about the reliability of the relevant testimony or instruments. Optimistic empiricism is mistaken: its realist optimism about science is inconsistent with its empiricism.[15]

It is important to note that it is not simply empiricism on its own that leads to these sceptical conclusions. Scepticism is a consequence of empiricism plus epistemological internalism. This is a point on which I expand below, in Section 6.5. As I will argue, it is not a reasonable move to save empiricism by rejecting internalism and adopting externalism. For the very motivation for empiricism is internalism.

6.4 Empiricism and scepticism

If optimistic empiricism fails, then perhaps we should be sceptical empiricists. I now consider and reject that move. Empiricism in any of its guises is false.

6.4.1 *Sceptical empiricism*

Let us remind ourselves of the core commitments of empiricism:

(O-E) The *functional role* of observation: observation supplies basic *evidence* in science.
(O-P) The *nature* or identity of observation: observation is *perception.*

plus

(O-C) The epistemic *foundational strength* of observation: observation should be *certain* or very close to certain.
(EV) Belief in a scientific theory is justified only by its relationship to the evidence.

Thanks to (EV) and (O-E), if a scientific inference is to give us knowledge of the truth of some scientific theory, then it must start from premises that are observational in nature. (In the light of (O-P), an 'observational' proposition is one whose

[15] There are parallel metaphysical commitments of empiricism. Ladyman (2000, 2018) demolishes this in van Fraassen's philosophy of science. In my (2007*c*) I take on empiricist/Humean metaphysics more generally.

truth or falsity can be determined by perception.) Sceptical empiricism claims that premises that are purely observational cannot rationally decide among competing theories concerning the unobservable. Hence our evidence underdetermines theory. This is the sceptical argument we saw at the beginning of this chapter:

(OBS*) All evidence is observational (perceptual);
(INF) From observational premises only observational (perceptual) conclusions may rationally be inferred;
therefore
(SCEP) Only observational (perceptual) propositions can be known.

(Sceptical Empiricism)

(The component (OBS) of (OBS*) is not the same as (O-E). While (O-E) says our *basic* evidence is observational, (OBS) says that *all* our evidence is observational. Nonetheless, replacing (OBS) by (O-E) makes no difference to the conclusion. For if (O-E) and (INF) are correct, only observational conclusions can be inferred from basic evidence. Even if we allow those inferred observational propositions to count as evidence, (INF) will only allow inferences from that evidence to observational conclusions. Either way, the rationally inferable propositions will still only be the observational/perceptual ones.)

Standard realist responses to *Sceptical Empiricism* are, as mentioned above, optimistic empiricist responses that take issue with (INF) while accepting (OBS)/(OBS*). They propose that inference patterns such as Inference to the Best Explanation can allow us to move from observational premises to non-observational conclusions. Thus realists will often say that theories can have virtues (or lack them) such that while the logical relationship of two theories, A and B, to the evidence E may be the same, the support given by E to A may differ from the support E gives to B. I will look at IBE in detail in the next chapter. This chapter is concerned with arguing against (OBS)/(OBS*). I argue that the sceptic's argument cannot rely on (OBS) as a premise; nor, therefore, can it rely on (OBS*).

In the preceding chapter I argued for the equation of evidence and knowledge, E=K. According to E=K any fact that is knowable could be part of our evidence. The sceptical empiricist argues that what is knowable is limited. It will beg the question if the sceptic appeals to a limited conception of *evidence* in order to reach this conclusion. (OBS) says that our evidence is of a certain kind (observational). Given the equation E=K, the premise (OBS) is identical to:

(OBS-K) All knowledge is observational.

But the sceptic's conclusion is that we can have no knowledge of facts, and their constituent entities and properties, that are not observable. The premise (OBS)/(OBS-K) and the conclusion (SCEP) are identical. So the question is being begged. This does not show that (OBS) itself is false, just that the sceptical argument cannot employ it as a premise.

The import of this argument can be seen as follows. What evidence, one might ask, do we have for the standard model of particle physics? From amongst the mass of evidence, one might cite the proposition H, that the Higgs boson exists, since the Higgs's existence was predicted by the model (plus various other assumptions). The sceptical empiricist will reject this: H is not a observational proposition; it therefore cannot be evidence. (Although scientists *do* say that the Higgs is observable, we will put this aside for present—it isn't observable in the empiricists' sense of perceptually observable.) Why should we accept that the imperceptible/unobservable nature of the Higgs entails that the existence of the Higgs is not evidence? That can be only because the sceptical empiricist already holds that the existence of the Higgs cannot be known. But to hold already that facts like the existence of the Higgs cannot be known is just to be a sceptic. The sceptical argument only gets off the ground if one is already a sceptic about the imperceptible/unobservable.

Imagine that the empiricist were to deny that their argument assumes scepticism. It makes only an assumption about what evidence is (i.e. observational/perceptual); it makes no assumption about what can or cannot be known. Then the empiricist will thereby be committed to saying: the proposition H is not evidence; but it might well be known. Let's say it is known then. So now we cannot say that H is *evidence* for the standard model. But we can say: H is something we know to be true and which supports the truth of the standard model. In which case the status of H as 'evidence' now seems irrelevant—it is still the case that the standard model is supported by propositions we know. And that's good enough for one to have a justified belief in (and maybe also knowledge of) the correctness of the standard model.[16]

6.4.2 *Empiricism and the aim of science*

An alternative approach for the empiricist is to deny that the scientist (*qua* scientist, anyway) intends to know about gravitational waves, top quarks, or magnetic fields. Rather, the aim of science is to generate empirically adequate theories—theories whose observational/perceptual consequences are true.[17] This view, van Fraassen's Constructive Empiricism, satisfies the core claims of empiricism, in particular affirming that observation is perceptual. Criticisms of Constructive Empiricism tend to focus on whether science and scientists could rationally aim and expect to achieve more than empirical adequacy. Here I want to emphasize the sheer perversity of having empirical adequacy as an aim.[18]

Empirical adequacy is perhaps conceivable as an aim of science when a scientist's perceptions are of things of scientific interest, such as litmus tests, images in a microscope, or cloud chamber photographs. Being right in predicting such things could seem a worthwhile enterprise. But much of modern science simply is not like that.

[16] This does mean giving up on evidentialism, (EV). And so the argument thereby shows that evidentialism only makes sense in the light of E=K.

[17] This question is orthogonal to the question discussed in Chapter 2, about whether the aim of science is truth or knowledge. The constructive empiricist can accept much of that argument, and conclude that the aim of science is knowledge of empirical adequacy rather than empirical adequacy *per se*.

[18] This aim is not entirely exclusive to full-on anti-realists. See Hempel 1952: 653.

In many areas data are collected and processed automatically, and the resulting electronic database is interrogated by a scientist operating a computer terminal. Both the ATLAS experiments and the Japanese meteorological system AMeDAS, discussed in Chapter 1, are like this. The only perceptual interaction is with the computer screen. So what does empirical adequacy amount to in the case of a theory that can be tested only by such means, as is the case with the Higgs hypothesis? It looks as if it amounts to the truth of a set of predictions of the form 'if one builds a collider to these specifications, then operates the collider in this way, has one's computer linked to the collider's data storage system, runs this programme on the computer, and then types "ABC" on one's keyboard, then one will have a perception of one's computer screen displaying "XYZ".' It is hard to see why anyone should care about the truth of such statements *per se*, let alone be willing to pay $13.5 billion to find out whether they are true.[19]

It might be replied that the relevant perceptions are not those involving the computer screen. Rather, the data thereby collected can be used to make further predictions that do have perceptual consequences. It is the correctness of these perceptual predictions that are the concern of science. In the case of AMeDAS, it is not the data on the screens that are the observational consequences, but the weather predictions that we infer from them and which are often all too perceptible. Nonetheless, when it comes to high energy physics, it is not at all clear that there are any such predictions concerning perceptible entities. It is very difficult to devise *any* kind of test of novel hypotheses in particle physics. The ATLAS experiments are all that we have as regards the Higgs hypothesis. It is true that this hypothesis is itself a component of the standard model of particle physics. One could, perhaps, construct an argument that links the standard model to quantum mechanics more generally and then to experiments that can involve perception (maybe concerning the photoelectric effect or a quantum double slit experiment). But that does not really help the constructive empiricist. The chain of inferences would be very long and involve lots of auxiliary assumptions. More importantly it would proceed via theories that are independently very well confirmed anyway. So, even using this contrived route we do not get any *new* perceptual predictions from the Higgs hypothesis. It is of course revealing that the scientists themselves do not feel the need to make any link between the Higgs hypothesis and anything perceptual. For them the results of the computer analysis of the collider data are enough.

6.5 Empiricism and epistemological internalism

In rejecting empiricism I do not thereby endorse empiricism's traditional opponent, rationalism. Both empiricism and rationalism are motivated by concerns that are indicative of epistemological *internalism*. Empiricism and rationalism both seek foundations for knowledge. The empiricist identifies these with sense-experiences, what

[19] $13.5 billion is what Forbes magazine estimates the cost of finding the Higgs boson to be: https://www.forbes.com/sites/alexknapp/2012/07/05/how-much-does-it-cost-to-find-a-higgs-boson/#1ddd44ec3948, accessed 13 March 2019.

were called 'ideas' or 'impressions' in earlier parlance, later 'sense-data'. The rationalist denies that these are the sole foundations for knowledge and may also deny that they are (on their own) sufficient foundations for much knowledge at all. The rationalist claims that rational intuition provides an important basis for knowledge that cannot be provided by sense-experience. Descartes, for example, argues that sense-experience provides a basis for knowledge of the external world only when supported by an apriori argument that clear and distinct ideas can be taken to be veridical. The empiricist does not accept the need for any such a supplement (or alternative) to sense-experience. An optimistic empiricist, such as Locke, holds that sense-experience is not only necessary for knowledge, but can be sufficient for knowledge in that it does not need the kind of rationalist underpinning that Descartes sought to supply. A central worry for empiricists is that this optimism is unwarranted: from the individual's subjective sense-experiences can one legitimately infer how things are with material objects outside the mind? Berkeley emphasized this pessimistic, sceptical consequence of Locke's empiricism. He sought to avert the conclusion by adjusting his metaphysics, by denying that the world we are thinking about is a material world external to our minds. According to one reading of Hume, he accepts the sceptical conclusion—he is a sceptical (metaphysical) realist (Wright 1983).

What empiricism and rationalism share is the aim of providing a foundation for knowledge in propositions that is certain. Indeed, for Locke, knowledge *is* certainty (Locke 1690: 4.14.4). These foundations are taken to be certain in the sense that when we reflect on the basis for believing those propositions, we see that we cannot be wrong about them. If I look at a piece of cloth in unusual lighting I might be uncertain that it is in fact red, but (so it is claimed) I can be certain that it looks red to me—about its *appearance*, that I am having a subjective experience of redness, I see that I cannot be wrong.[20] This requirement on knowledge shows that Locke's conception of knowledge is an internalist one.[21] The rationalist prefers to found knowledge-claims on the deliverances of rational intuition, but for the same internalist reason: their self-evidence allows the thinker to see that they cannot be wrong.

Internalism is a family of views about the conditions for knowledge or justified belief, and includes the claim that justification depends only on states the subject can reflectively access (Ginet 1975: 34; BonJour 1980: 63) and that knowledge requires knowledge of all the preconditions of knowledge (Nozick 1981: 281). Internalist views include or are closely related to the KK principle, that a subject who knows that *p* also

[20] Whether there is any state that really is such that I cannot be wrong in believing that I am in that state may be questioned (Williamson 1996*a*). Nonetheless, the attractiveness for empiricists of founding knowledge on such states is that they are taken to be luminous. In Locke's words, 'It is the first act of the mind when it has any sentiments or ideas at all, to perceive its ideas; and so far as it perceives them, to know each what it is, and thereby also to perceive their difference, and that one is not another. This is so absolutely necessary, that without it there could be no knowledge, no reasoning, no imagination, no distinct thoughts, at all. By this the mind clearly and infallibly perceives each idea to agree with itself, and to be what it is ...' (Locke 1690: 4.1.4).

[21] For Locke's internalism see Nagel (2015).

knows or is in a position to know that they know that *p* (Prichard 1950: 94).[22] Some recent internalists are happy to ascribe the label to themselves (Wedgwood 2002*b*), whereas the internalism of other philosophers must be inferred; often internalism is an unstated motivation for their explicit claims.

The internalism of Locke's empiricism is clear. This internalism also plays out in our discussion of the obstacles to optimistic empiricism in the philosophy of science. The empiricist's problems arise from the obligation to justify the major premises in the testimonial inference (the premise stating that the informant is reliable) and the instrumental inference (the premise stating that the instrument is reliable). Accepting that such premises need in every case to be justified by the subject is to accept the internalism we have been discussing. It is to accept that for a subject to have knowledge it must be the case that everything relevant to the justification of the subject's belief, including its status as providing justification, is internally accessible to the subject.

The significance of internalism is best appreciated by understanding how matters look when we adopt some form of externalism. Simple reliabilism says that a subject is justified in believing that *p* precisely when that belief is brought about by a reliable method—a method that reliably brings about true beliefs. And she knows that *p* when, in addition, it is true that *p*.[23] What makes this externalist, failing to meet the requirements of internalism, is that there is no demand that the reliability of the method be apparent to the subject—she can know simply in virtue of the fact that the method just is reliable. Let us accept simple reliabilism for the sake of argument. Consider an astronomer who just trusts Captain Cook's reports or a scientist who just uses her Gauss meter and accepts its readings. According to the reliabilist those people can thereby gain knowledge, even though they have not considered let alone justified the major premises in question. If Cook, when reporting astronomical observations, is a reliable informant and if the Gauss meter is a reliable instrument then that is sufficient for those individuals to know.

There is a voluminous literature on the debates between internalism and externalism, and on how reliabilism should be formulated, and whether indeed reliabilism is the best form of externalism. Philosophers of science tend to be more sympathetic to internalism than the average epistemologist. That, I suspect, is because a philosopher of science thinks that a scientist cannot really be justified in reaching a conclusion if she has given no thought to the reliability of the methods she uses. Imagine a scientist who, before the Geiger counter is developed, wants to measure radioactivity. As a result of some accidental ingestion of a psychedelic mushroom, she puts together an object with parts randomly seized from her laboratory and is convinced that this will measure radioactivity. By a remarkable fluke she has con-

[22] For a survey of the views falling under the label 'internalism' see Pappas (2017). For the relationship of internalism to the KK principle see Okasha (2013) and Bird and Pettigrew (2021).

[23] The difference between justification and knowledge arises if it is allowed that a reliable method can on occasion lead to false beliefs, i.e. one can have false but justified beliefs. That however does mean that simple reliabilism is subject to the Gettier problem (Zagzebski 1994), which is one reason among others why simple reliabilism is not correct.

structed a reliable Geiger counter. Does its reliability mean that she now knows which objects are radioactive and which are not, by using her instrument? We are inclined to deny this. That shows that a simple reliabilism cannot be right. On the other hand, that does not mean that we should adopt some form of internalism. It is not the case that for an individual to gain knowledge by using an instrument, she must in every case have a detailed justification for a belief in its reliability. Nor do we expect that in science. Recall our palaeontologist who relies on radiometric dating. She has a broad understanding of the method she uses, but may well not know sufficient detail to be able to justify a claim about its reliability. In a laboratory one might use a standard, commercially produced Gauss meter. Usually these are highly reliable. But it is not unheard of for some instruments to have been manufactured with a fault or to develop one while in use. We do not regard it as a condition on using one that the scientist must on every occasion eliminate the possibility of a rare fault afflicting her device. Turning to testimony, our scientist reads a paper in a journal and believes its result. In so doing she may give some thought to the reputation of the authors, the rigour of the journal, and the prior plausibility of the result itself. But in order to gain knowledge, we do not expect her to have ruled out the possibility that this is one of those cases of fraud or of a 'questionable research practice' which still bedevil science. So the care that we expect scientists to show in forming and in arguing for their beliefs is not enough to satisfy the internalist.

I accept that a rational scientist should take care to understand her methods, but I have denied that this is either sufficient for internalism or motivated by internalism. What then is the motivation? We can explain why, in certain circumstances, a rational individual will take that kind of epistemic interest in her methods. Note first that there are some circumstances in which it is not rationally required that one understand one's methods or have any idea as to whether they are reliable. Let us imagine that one has the aim of achieving X. There are several methods, A–D, that one could use and which might deliver X. One does not value achieving X especially highly and it would be costly to find out which of these methods is reliable. On the other hand, applying the methods is cheap and success or failure in applying any of these methods to achieve X is immediately obvious. And, furthermore, one can keep trying until X is achieved. In these circumstances it would not be rational to launch an investigation in order to find out which of A–D will deliver X. Rather the rational thing to do would be to employ trial and error: try method A; if it doesn't succeed, then move on to method B and so on. Science, however, is not like this. We value the end of science highly. Often applying the methods of science is costly in intellectual and other resources. Furthermore, we typically do not know independently of the method in question whether the aim of science (whether that is taken to be knowledge or truth or even empirical adequacy) has in fact been achieved. (An analogy: if I use a medicinal cream to sooth a painful burn, I know independently whether it is working—by a change of sensation; however, if I get vaccinated in order to protect myself from a disease, I, a layperson, cannot know by independent means that this has been achieved—it is only by knowing that the vaccination is reliable that I know that I am immune. Science is like the latter. Only because the scientist knows that the method of science she uses is reliable does she know that it has achieved the aim of science.) Under these circumstances—we value the aim of science; the methods of science are costly to deploy; we do not know independently whether the aim of science has been achieved—it is rational to investigate the methods of science in order to know which ones work and which ones do not. So, to conclude, we do have an explanation of why a rational scientist should take an epistemic interest in her methods, but that explanation has nothing to do with internalism.

Let us say that some form of externalism is correct (maybe a reformed reliabilism, maybe not, maybe instead a knowledge-first externalism to which this book is sympathetic). What consequences does that have for empiricism? It might, at first sight, appear to aid the optimistic empiricist, since she is no longer faced with the unachievable task of providing an empiricist justification for the problematic major premises in the testimonial and instrumental inferences. It is not obviously inconsistent to be both an empiricist and an externalist, and so this might look like a happy combination.

That, however, is not the right way to view things. For it is internalism that motivates empiricism. Under internalism, one must ask, which mental states can justify a scientific belief? 'Perceptual states' is one answer, the empiricist answer. But under externalism, there is no general requirement that there is an internal state that justifies a belief. Frequently, it is the external reliability of a belief-forming procedure that justifies it or makes it knowledge. Why should a philosopher of science hold (O-E)&(O-P)&(O-C), the view that says observation is foundational evidence, is perceptual in nature, and is certain? If observation has these features, it can satisfy the internalist. But if one is not an internalist, one has no need to defend such claims. As an externalist, one might still think that observation plays an important role in science. But one will not hold that it must, in virtue of being perceptual, provide internalist certainty. The externalist is free to abandon the empiricist emphasis on perception. In the next section I develop a non-empiricist account of observation, one which accords much more closely with scientists' use of the term 'observation'.

I conclude by considering a rationale for empiricism that does not obviously draw on internalism. This rationale points out that science employs ampliative inferences, which will tend to lead to propositions that are less certain than those from which they are inferred. Our most interesting novel hypotheses are thus risky, but we want to maximize their chances of being true. So one might think that the proper caution of science would recommend starting with propositions in which we can have a high degree of confidence, which is to say reports of our perceptions. This rationale, however, makes two mistakes. First, it assumes that scientific reasoning leads from the more certain to the less certain. But that is far from the case. When we make measurements in science we will often take multiple readings, because their average value is likely to be more accurate than any single reading. Data can lead by inference to a result that has a high degree of certainty even if the basic data points, the results of direct perception, are themselves of low reliability. In such cases quantity can make up for lack of quality. Something similar happens at a higher level also. A hypothesis might be supported by several independent lines of experimentation or reasoning. The hypothesis might thus be better confirmed than any one of the results or inferences that support it. Furthermore that consilience allows confirmation to flow from the (now well-confirmed) hypothesis *to* the supporting claims. Sir Arthur Eddington's (only partially ironic) injunction, 'Never trust an experimental result until it has been confirmed by theory' captures the thought that in some cases our theoretical beliefs are more soundly based than than experimental ones, and that just as the latter can support the former, so can the former support the latter.

The second error in this rationale for empiricism is the idea that perception, the supposed starting point of our scientific reasoning is especially reliable. We all know that casual perception can be very unreliable. Careful perception can do better. But even that is limited by biology—evolution has given people, on average, fairly reliable senses, but there is no reason for it to have given us *very* reliable senses. We tend to think of instruments as being used to *extend* what is perceptible—the optical microscope is an example. Instruments do other things besides. Some *assist* our senses by making them more precise and reliable—judging lengths by eye using a ruler versus by eye alone or using an astrolabe as an inclinometer. Some instruments *replace* perception because they can be made to be more reliable than perception—using a thermometer or a clock instead of perception of temperature or duration. And some instruments *ignore* perception, in that their outputs have nothing to do with perceptible qualities at all. Instruments such as the Gauss meter often mentioned in this chapter are used to measure quantities or detect phenomena that we humans could never perceive.[24] The history of scientific instruments has been the history of assisting and extending perception in the first instance, then replacing perception, and finally ignoring perception altogether.

Empiricism neither entails nor is entailed by internalism. Internalism is a powerful *motivation* for historical empiricism, and it remains so for some contemporary empiricists, particularly those of a pessimistic, anti-realist stripe.[25] I suspect that empiricism, among optimistic empiricists especially, has become partly or fully detached from its internalist origins. It has simply become the default view among philosophers of science. Recall the argument *Sceptical Empiricism*:

(OBS) All evidence is observational;
(INF) From observational premises only observational conclusions may rationally be inferred;
therefore
(SCEP) Only observational propositions can be known.

[24] I say 'humans' because a number of animals do have magnetoception. Even a homing pigeon, if engaged in science, would prefer to use a Gauss meter than its own ability to perceive magnetic fields.

[25] Such as Darrell Rowbottom (personal communication). Rowbottom's *The Instrument of Science* (2019) is a recent and powerful exposition of instrumentalism .

The debate between realists and anti-realists has centred on (INF). The realists have accepted (OBS) and have understood (OBS) in the light of (O-P): observation is perceptual in nature. Whereas if we reject empiricism and so (O-P), the argument around *Sceptical Empiricism* looks very different. For, as I remarked on p. 136, this argument does not have sceptical consequences if observational propositions include propositions about magnetic fields and subatomic particles. As it turns out, scientists *do* say that such things are observable. So we now turn to the project of understanding what observation is, in a way that respects what scientists say about observation and which is not tied down by empiricist presuppositions.

6.6 The nature of observation

Empiricism is false. That is the conclusion of Sections 6.2–6.4. The empiricist promoted a perceptual conception of observation. Without empiricism, the motivation for that conception is gone. So what is the correct view of observation and its role in science? In this section I develop an alternative, functional account of observation. This allows us to make a useful distinction between data and observations.

Historically, evidence has been taken to have a functional role, that of providing basic evidence (O-E), and has been held to be perceptual in nature (O-P). While these coincided for much early science, modern science shows that they come apart. Only the functional role of observation was ever essential. The perceptual character of observation was only ever accidental. Dudley Shapere (1982: 507–8) expressed this well some decades ago:

> The philosopher's use of the term 'observation' has traditionally had a double aspect, and has played a double role. On the one hand there is the perceptual aspect: 'observation', as a multitude of philosophical analyses insist, is simply a special kind of perception, usually interpreted as consisting in the addition to the latter of an extra ingredient of focused attention.... On the other hand, there is the epistemic aspect of the philosopher's use of 'observation': the role that observation is supposed to play in leading to knowledge or well-grounded belief or in supporting beliefs already attained. For the empiricist tradition in epistemology proposed that all knowledge (or well-grounded belief) 'rests on experience', where 'experience' was interpreted as sense-perception. In that tradition, as indeed in most other philosophy, these two roles have been identified: the question of observational support for beliefs or knowledge was interpreted as the question of how *perception* could give rise to knowledge or support beliefs.
>
> In sophisticated areas of science, however, these two aspects have come to be separated, *and for good reason*. Science is, after all, concerned with the role of observation as evidence, whereas sense-perception is notoriously untrustworthy .... Hence, with the recognition that information can be received which is not directly accessible to the senses *science has come more and more to exclude sense-perception as much as possible from playing a role in the acquisition of observational evidence*; that is, it relies

> more and more on other appropriate, but dependable, receptors. It has broken, or at least severely attenuated, the connection between the perceptual and epistemic aspects of 'observation' and focused on the latter. And this is only reasonable in the light of the primary concerns of science: the testing of hypotheses and the acquisition of knowledge through observation of nature. [Emphasis in original]

(O-E) is a functional characterization of observation: observation is whatever it is that supplies the basic evidence of science. (O-P) aims to characterize the nature of observation—in terms of perception. In early science, the functional characterization of observation and the perceptual characterization, rather than being in tension, would have appeared to be complementary: perception did supply the basic evidence of science. A scientist would use his senses to discover what happens when light is passed through a prism or to find out what the surface of the moon is like. That information, the content of the perception, provides evidence that allows science to build a new theory of light or to refute the theory that the moon and planets were all perfectly smooth spheres. But in modern science the functional and perceptual characterizations do come apart. When a modern scientist sees symbols on her computer screen, the information contained in that perception does not provide her basic evidence—unless, perhaps, she is a computer scientist.

We saw that holding both (O-E) and (O-P) was inconsistent with even a minimal scientific realism that says that sometimes science leads to knowledge of the truth of a theory. There is no way of reconstructing scientific knowledge from perceptual knowledge alone. And clearly it is (O-P) that is at fault. For without (O-E) empiricism loses its point. If (O-E) is false, then there can be scientific evidence that is not observational and so is not perceptual. While that may be correct, it is not an empiricist picture of science.

And we have independent reason to reject (O-P) rather than (O-E). Scientists simply do not use 'observation' in a way that ties it to perception.[26,27] Think of a scientist using a Gauss meter to measure the magnetic field strength at a particular location: it is 150 mG (1.5 μT). The perception in such a case is of the meter's display; the content of the perception is that the display gives a reading of 150 mG. What is the observation? According to (O-P) it is the very same thing: a report of the content of the

[26] Fodor (1984) notes this, but disregards it as not relevant to the distinction he is interested in. It is merely an 'ordinary language' argument to let scientists' usage be a reason for ignoring his distinction, which, he says, 'it is also reasonable to call "the" observation/inference distinction, which is also of central significance to the epistemology of science'. Fodor's distinction takes observation to be perceptual. But he does not make out a case for its being 'of central significance to the epistemology of science'. He just assumes as much.

[27] It is worth adding that there is no historical or etymological link between 'observe' and perception. The Latin *observare* mean to be attentive to. So an observant Jew and an observant Catholic are people who are attentive to the demands of their religions. Someone who makes an acute observation is someone who makes a remark whose insight is due to their attentiveness to the subject matter (which may be abstract). It is this that leads to the notion of observation as used in science, since science requires careful attention to its subject matter in order to gather reliable data. For the reasons discussed in this chapter, that attention was historically often perceptual attention but it need not be.

perception. Yet a scientist would not normally regard such a report as an observation. If the instrument is held to be reliable, then the observation is held to be the fact that the field strength is 150 mG.[28] A survey of published research papers reveals many which report observations of magnetic fields. But none of these papers mentions perceptual interaction with an instrument. Scientists' usage of 'observation' directly conflicts with (O-P). Philosophers of science commonly regard magnetic fields as unobservable.[29] Those philosophers who insist that what is observable must be perceptible should recognize that this claim amounts to a proposal for a major innovation with regard to normal scientific use of 'observable'. Philosophers of science frequently talk of 'unobservable' entities. Scientists frequently observe them.

So, while in earlier science basic evidence was delivered by perception, modern science makes room for bifurcation between observation as perception and observation as basic evidence. And, as the example of the magnetic field shows, scientific usage sticks with the latter. For example, NASA reported 'Hubble Makes One Millionth Science Observation . . . the Earth-orbiting observatory logged its one millionth science observation during a search for water in an exoplanet's atmosphere 1,000 light-years away, . . . a spectroscopic measurement.'[30] Papers in experimental and other non-theoretical sciences, especially (but by no means exclusively) physics and astronomy, frequently contain 'observation' in their titles: 'Observation of the top quark', 'Experimental observation of the quantum Hall effect and Berry's phase in graphene', 'Observation of Bose-Einstein condensation in a dilute atomic vapor', 'Observation of gravitational waves from a binary black hole merger', 'Direct observation of the rotation of F1-ATPase', and so on.[31,32] In all these cases what is observed is some fact that cannot be known by perception, but which is important as evidence in the relevant field.

[28] If there is uncertainty about the reliability of the instrument then *neither* the statement about the field strength *nor* the report on the perception of the instrument are observations. The former might be a *datum* from which, along with other data, an observation is constructed.

[29] For example, Papineau (1978: 17–18) writes 'The modern physical sciences . . . seem to have discovered all kinds of perfectly real but unobservable entities, such as magnetic fields, radio waves, and subatomic particles.'

[30] http://www.nasa.gov/mission_pages/hubble/science/millionth.html, accessed 22 May 2013. We should also note that scientists more tend to use the term 'observation' to refer to an activity or episode of activity rather than to something with a particular propositional content. When NASA counted its millionth observation, this referred to the production of evidence from a set of related spectroscopic data.

[31] F1-ATPase is a sub-unit of the ATP synthase molecule, which is interesting for being a rotary molecular motor.

[32] Rowbottom (2019: 136–7) denies that there was any observation of gravitational waves or of a black hole merger, as asserted by Abbott *et al.* (2016) of the LIGO collaboration. Rowbottom must reject a lot of what scientists commonly say. His reason is that the observations in question require making inferences. That is an objection only if one has good reason to think that observation is inconsistent with making inferences, and that is an assumption only someone in the grip of an empiricist assimilation of observation to perception would make (even though perception itself is often tacitly inferential).

6.6.1 *The functional essence of observation*

If observation is not essentially perceptual, what then is the nature of observation? What we retain from the foregoing discussion is that observation must be able to play the evidential role: it supplies basic evidence that is the starting point for inferences leading to knowledge. With this in mind, my proposal is this:

> (O) Let scientists in field *F* be investigating subject matter *S*. An observation with the content that *p* is made when a representation of *S* is made that carries information that *p* that is causally dependent on the fact that *p* and which can fulfil the role of basic scientific evidence in *F*.[33]

Let us look at this in a little detail. First, an observation is a *representation* of some subject matter. Perception can be a source of suitable representation. And in early science suitable representations are almost always the mental representations of scientists resulting from perception. However, the development of instruments with mechanical recording devices, such as the barograph, eliminated the need for mental representation. The barograph, or recording barometer, is an aneroid barometer to which a stylus is attached. The stylus marks a roll of graph paper that rotates using a clockwork mechanism. The resulting graph plots the air pressure continuously over a period of time—the graph paper is a representation, therefore, of changing air pressure. In this case it is the graph paper that constitutes the observation. In the nineteenth century a human intermediary would be required if this observation is to be used to confirm or disconfirm a prediction. That person would read the plot and compare it with the prediction. But the role of perception in that act is somewhat incidental. It does not even have to be visual, since a Braille version of a barograph would do just as well. And in modern, automated science, we can see that the human intermediary can be left out altogether—the data and the prediction can be compared by computer.

Secondly, while the representation is the observation, it is the information it carries that is the scientific evidence. The mark on the barograph is the observation. This mark carries the information that atmospheric pressure at 06:30 was 1100 millibars. And that information is the evidence that may be used in some scientific investigation.

Someone could argue that the information plays the role of evidence once it enters the consciousness of some scientist, e.g. the scientist reading the barograph roll. My own view, in the light of Chapter 4, is that this information is evidence independently of the mental states of any scientist, so long as the instrument is suitably integrated into the scientific process. That allows for the information to be a case of social knowing, and, since knowledge is evidence, it is also 'social(ly possessed) evidence'. While that opinion is independent of this account of observation, I do ex-

[33] Shapere (1982) and Kosso (1988) both also discuss observability in terms of information processing in ways to which I am sympathetic. The information-theoretic approach does not on its own, however, provide a distinction between observation and other ways of producing knowledge/evidence in science, which can also be understood in terms of information transfer. Hence my condition concerning observation's role in providing inquiry-relative basic evidence.

amine how matters look when we consider the social epistemic subject (see Section 6.8 below).

Thirdly, the representation and information it carries should be causally dependent on the subject matter S and in particular on the fact that *p*.[34] Without this condition knowledge about S in the future could count as an observation, as would any knowledge gained about S by inductive inference on other things that are like S. As mentioned, perception can be a way of making an observation, but observation goes beyond perception in admitting representational states other than sense experiences. The causal relation between S and the representation is like that in perception. One could say that observation is perception writ large, and freed from dependence on the human. I shall go on to note that observation involves both this causal dependence and, very often, inference (which may or may not be automated). The presence of inference does not undermine the analogy. For perception itself involves a great deal of sub-personal and tacit inference, deploying background knowledge—as has been discussed by Helmholtz, Bruner, and the Gestalt psychologists, by Hanson, Kuhn, and others concerned with the theory-ladenness of perception, and more recently by cognitive neuroscientists, including the advocates of predictive processing (predictive coding). (Again, my account of social knowing helps here, though is not necessary. See Section 6.8 below.)

Fourthly, not just any representation can count as an observation. An erroneous or careless representation, because false or unreliable does not carry information and cannot supply evidence (since evidence is knowledge and so must be true and reliably formed). A scientist who mistakes what they see or a malfunctioning barograph do not make observations. That said, some false representations can be used in the scientific process. If we know that our barographs are not perfectly accurate, we might nonetheless use several of them to gain a good estimate of pressure. As I explain in the next section, the inaccurate representations do not supply evidence regarding the air pressure. But they do supply the raw data from which evidence can be generated. It is the result of the data processing—the information bearing (accurate, reliable) report regarding the pressure—that is the observation.

Fifthly, an observation is a representation *of the subject matter*. In the case of the imperfectly reliable barograph, one could argue that the following is accurate and reliable: this barograph has an ink-mark on the graph at coordinates '06:30h' and '1100 mbar'. But that report concerns the barograph and is not about the atmospheric pressure (the subject matter) and so is not an observation in meteorology (the field). (A representation of the barograph could be an observation in the scientific process of calibrating barographs.)

Sixthly, an observation supplies *basic* evidence *relative to a field of inquiry*. Part of the empiricist conception of observation is that it should be basic in that is should involve no inferences—rather it supplies the basic evidence from which inferences are made. We saw earlier that Mill rejected an objectivist conception of observation

[34] Some causal accounts of information and representation will imply that this condition is redundant. That may be right but I don't wish to assume it.

in favour of the phenomenalist conception precisely because he took the objective report to be the result of an inference from a phenomenal (subjective, internal) experience. (It is also mistakenly thought that inferences in science will tend to reduce certainty.) However, what are recorded as observations in science are often the products of quite considerable and complex inferences. For example, let us take a paper referred to above, 'Observation of gravitational waves from a binary black hole merger' (Abbott *et al.* 2016). This describes the process involved in, to use the words of the paper's abstract, 'the first direct detection of gravitational waves and the first observation of a binary black hole merger'. What are described as 'direct detection' and 'observation' are highly complicated matters, involving a considerable degree of data processing and depending on background information and theory—as I have attempted to capture in this pared-down description:

> Detection of gravitational waves is made by two LIGO detectors, which involve an interferometer of the kind used by Albert Michelson in the search for the electromagnetic aether. The changing lengths of interferometer's arms cause a change in the phase difference in the two, split light beams in the interferometer, and this produces an optical signal that is captured by a photodetector. The changing of the lengths of the arms is related to the strain on the arms, and so the basic data recorded by the LIGO detector concern strain. That the data record strain by detecting an optical signal is already theory-laden, and so the design of the instrument itself embodies various inferences. Since the strain is taken to be caused by gravitational waves, this strain data holds information about those waves. While the detector design eliminates much of the noise that might afflict detection of gravitational waves, it cannot avoid all noise. So an array of sensors records environmental disturbances. Any candidate signal for a significant gravitational wave is compared with the environmental disturbances to ensure that it could not be accounted for by the latter. Such signals are found by searching through the strain data in two ways. One generic search process looks for signals that are particularly improbable given the background noise, which is challenging since characterizing the background depends on various assumptions. The second search process looks for waveforms in the strain data that match the possible waveforms predicted by theory governing the coalescence of binary black hole system. The signal GW150914, detected on 14 September 2015, was identified as fitting an appropriate waveform template and being unlikely in itself. Using further theory, inferences were made regarding details of the binary black hole and its merger (such as the mass and spin of the initial and final holes), which could then be used to confirm the predictions of general relativity regarding such mergers.

So it is clearly a mistake to think that observing involves no inferences and that observation only ever provides the *input* to an inferential process. However, we should note that the inferences that go to producing the gravitational wave observations are inferences concerning the strain data. There are inferences built into the LIGO detectors that produce that data. And then there are the inferences from that data to find which signals in the data are those produced by gravitational waves. And finally there are inferences from the particular signals thus identified in the strain data to conclusions concerning the gravitational waves and cosmic events that caused them. So, although the observations are the consequences of inferences, those inferences do not start with evidence concerning the gravitational waves or the black holes. The latter come into the picture only at the end of the inferential chain. The paper does describe further inferences that can be made from the observations. These are inferences such as the confirmation of general relativity, testing various hypotheses concerning black holes, and constraining the value of Compton wavelength of the graviton. The latter, like the observations themselves, are propositions in the related fields of cosmology and fundamental physics. So we can divide the whole sequence of inferences into those whose premises do not concern cosmology and fundamental physics but instead concern something else (such as the strain data that concern the LIGO interferometers), and those inferences whose premises do concern entities that are the subject matter of cosmology and fundamental physics. It is the propositions that are both among the consequences of the former group of inferences and among the premises of the latter group that are the observations (see Fig. 6.1). So there is something right after all about the idea that observations are at the start of inferences—they are at the start of inferences *in the relevant field.* But they may be the consequences of inferences outside that field. So we can retain the idea that observations supply basic evidence—the basic evidence *relative to a field or topic of inquiry.*

One might question whether the term 'observation' is appropriate in such cases. Is it really correct to call a roll of paper, or even data on a computer disk an observation? Paul Teller (2001: 130), articulating van Fraassen's view, says, '. . . strictly speaking, observation of an empirical phenomenon only occurs when I look at the spectrograph, not when the spectroscope produces the spectrograph'. According to my view, that is exactly the wrong way around. Looking at the spectrograph or, more frequently these days, the computer screen, is not an observation. But the spectrograph *is* the observation, because it is the representation of the light emitted by a star that can be used as basic evidence in testing (for example) Hubble's law.

We saw that NASA was happy to say of the Hubble telescope that it was observing. The fact that no human is involved is not an obstacle to talking of 'observation'. Insofar as there is some residual awkwardness in using 'observation' in some such contexts, my proposal is (to that extent) a proposal for legislation—a proposal for how we ought to use the term. My conjecture is that the historical association with perception means that we are happy to use 'observation' for cases of non-human data gathering when the process is analogous to human perception. What the Hubble telescope does is like human visual perception.

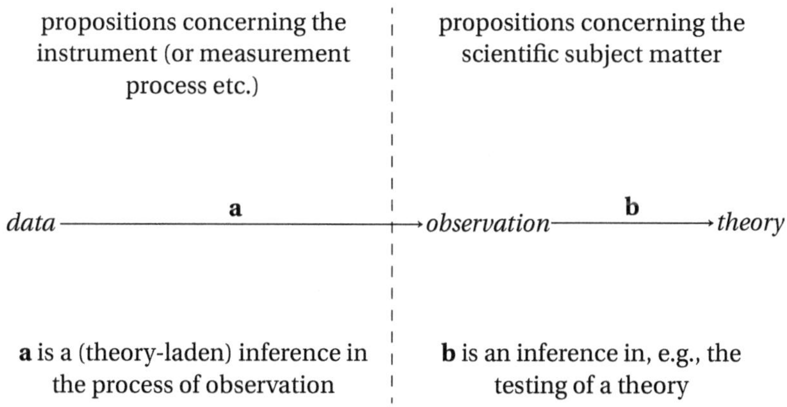

Fig. 6.1 Observations play the role of basic evidence relative to a field of inquiry. They are frequently the results of inference, **a**, from data. Data are reports of what is recorded by an instrument, or, more generally, what is recorded in the process of evidence production (such as responses to a survey in social science). In simple cases, such as data from a highly reliable instrument, the data will directly imply observations about the scientific subject matter. But often the data will require considerable processing and complex patterns of inference before a conclusion about the subject matter is reached. The conclusion thus reached is an observation, which can then be used, via further inference, **b** (also likely to be theory-laden), to test theories.

I do not suppose that the account of observation in (O) constitutes a water-tight analysis of the term 'observation' as used by scientists. For one thing, it would need improvement in order to rule out deviant causal chains. I do believe, though, that it captures much of what is important in the scientific concept of observation: that it does provide basic evidence relative to a field of inquiry, that it goes beyond perception but is analogous to perception, and that making an observation can involve a considerable quantity of inference.

Nonetheless, an account of observation is not itself that important. Ultimately what is significant is the fact that we engage in data gathering to build and to test our theories—data generate evidence. Automated science shows us that perception need play no part in this process. We *could* continue to link 'observation' to perception, in which case we must conclude that 'observation' is not necessary, nor any longer even important to science. Or we could link 'observation' to the idea of basic evidence. In which case 'observation' clearly is important to science. But then 'observation' need involve no element of perception.

6.6.2 *The distinction between data and observations*

My own view is that the second of the options given at the end of the preceding paragraph is the more useful way of thinking about observation, while also being closer to actual usage—hence my proposal (O) above. It is useful to have a concept that picks out the basic evidence that arises from experiments and the like. Does not the term 'data' fulfil this role? No, because basic data are not always observations in the sense proposed—basic evidence concerning the subject matter under scientific investigation. Basic data may fail to satisfy this description for two reasons: the basic data may not concern the subject matter under investigation; or the basic data may not be evidence.

Some basic data may concern not the entities under study but events concerning the measuring device. For example, a simple Geiger counter may only count ionizing events in the Geiger–Müller tube. Those counts will be the basic data, but do not directly provide the relevant basic evidence, which will concern the dose rate of the radiation emitted by the sample being studied. The former needs to be converted into the latter before we have evidence concerning the entity under study.

Other basic data may concern the entities being investigated, but fail to count as evidence. This is because the data collection may be imperfectly reliable in individual cases. A measuring device may be prone to error, so each individual data point fails to be evidence. Nonetheless, evidence may be generated from the data by taking multiple readings, so that the mean reading will be accurate within a small tolerance margin to a high degree of reliability.

Bogen and Woodward describe a case that nicely illustrates both points. In response to Ernest Nagel's (1961: 79) claim:

> ...many laws in the sciences formulate relations between things or features of things that are commonly said to be themselves observable, whether with the unaided senses or with the help of special instruments of observation ... the law that lead melts at 327°C [is a law of this kind].

Bogen and Woodward (1988: 308) remind us that:

> Despite what Nagel's remarks seem to suggest, one does *not* determine the melting point of lead by observing the result of a single thermometer reading. To determine the melting point one must make a series of measurements. Even when the equipment is in good working order, and sources of systematic error have been eliminated, the readings from repeated applications of the same thermometer to the same sample of lead will differ slightly from one another, providing a scatter of results. These constitute data. Given the absence of systematic error, a standard assumption is that the scatter of observed thermometer readings not only reflects the true melting point (the phenomenon in which we are interested), but also the operation of numerous other small causes of variation or 'error,' causes which cannot be controlled for and the details of which remain unknown. If one can make certain general assumptions about the character of these causes of variation (for example, that they operate independently,

> are roughly equal in magnitude, are as likely to be positive as negative, and have a cumulative effect which is additive), then it will follow that the distribution of measurement results will be roughly normal and that the mean of this distribution will be a good estimate of the true melting point. Standard scientific practice is to report this estimate along with the associated standard error, which is directly calculable from the variance of the distribution of measurement results. [Emphasis in original.]

What the scientist records in her bench notebook are the various thermometer readings. These statements should be regarded as just that—statements about what the reading on the thermometer is, not as statements about the temperature at which this sample of lead melted (let alone about the melting point of lead, the kind). We get to the latter from the former by statistical inference along with various assumptions about the experimental process. The statements in the notebook should not be regarded as evidence statements concerning lead or this lead sample, because, regarded as such, they are not all true let alone knowledge (since they are subject to error). They can be regarded as true statements regarding the thermometer on a particular occasion (if the experimenter was careful). However, regarded as such they do not bear upon any theory in which we are likely to be interested (e.g. theories about the melting points of metals) since such theories are not theories about thermometers. So the basic data in this case are the readings recorded in the notebook. Basic *scientific* evidence regarding lead's melting point can be inferred from the data—the inferred propositions are the basic evidence in (this field of) science since it is these that can test its hypotheses and theories.[35]

So while basic data can sometimes be basic scientific evidence, it need not always be. It is useful to have a concept for basic evidence (basic relative to the field of inquiry), and 'observation' is a natural term to use for this purpose.

It should be noted also that insofar as the thermometer is an instrument of observation, in relation to perception it is rather different from an ordinary optical telescope or a low-power optical microscope. For the latter can reasonably be thought to extend our perceptual faculties. But a thermometer does not do this. While humans normally do have a faculty of thermoception, thermometers do not extend this faculty, since we engage with thermometers with sight not with our thermoceptors. In which case, Nagel's claim that the temperature of an object is an observable feature of it implies that observation is not entirely perceptual, even in any (reasonable) extended conception of what is perceptual. Nonetheless, it is not uncommon to talk of observation in connection with temperature, as familiarity with the meteorological literature will reveal.[36]

Bogen and Woodward argue that insofar as observation is perceptual, such examples show that observation is not as important in science as it is standardly held to

[35] The basic data may still be evidence—since it can be known, but it is not basic scientific evidence, unless the hypothesis under consideration concerns the thermometer itself.

[36] As in, for example, this title of an academic paper 'Observations of temperature in the upper troposphere and lower stratosphere of tropical weather disturbances' (Davis *et al.* 2014).

be (e.g. as in the comment from Quine quoted above, 'It is common usage to say that the evidence for science is observation'). My view is different as regards the optimal use of the term 'observation'. They note, and say that they have no argument with, the scientists' willingness to talk of 'observing bosons'. I take that usage to guide my account of observation. It is underpinned by (O-E) which gives a significant role for observation in science, *viz.* providing representations of a subject matter that can perform the role of basic scientific evidence. Either way, Bogen, Woodward, and I are agreed that whatever does perform this role, it need not be perceptual, and in modern science it very rarely is. Does my preferred usage mean that Quine's claim is true after all, that observation (understood non-perceptually) is the evidence for science? Not necessarily, for reasons I examine in the next subsection.

6.6.3 *Evidence without observation?*

The claim (O-E) that observation has the function of providing basic evidence in science has usually been taken in a strong sense, implying that observation is *the* source of basic evidence in science—without observation there is no basic evidence. But (O-E) might—and should—be read in a weaker way, as asserting that observation has this function, while allowing that this function might also be fulfilled by means other that observation. We might find that there are cases that are not observations since there is no single object that is being observed; nonetheless, there may be basic evidence generated.

For example, in 1951 Richard Doll and Austin Bradford Hill wrote to all British doctors with a questionnaire about their smoking habits.[37] Two-thirds replied. Doll and Hill then followed this cohort and obtained information from the Registrars-General about subsequent deaths in the cohort, in particular recording deaths from lung cancer. This allowed Doll and Hill to establish a link between smoking and lung cancer (Doll and Hill 1954). In such a case it is difficult clearly to identify any observations. Would they be: doctors' self-observations of their smoking habits; or observations of their causes of death, in autopsy? These are not quite right for two reasons. First, the information gathered was summary information, so if a doctor reports his knowledge that he smoked 25 cigarettes a day for 20 years, that does not look like a self-observation but the summary of many self-observations over a long period; likewise, the report of a cause of death is not the report of basic evidence but the report of an inference from that evidence. The second reason is that in studies such as this the reports are generally treated as data (in the sense discussed above) rather

[37] Although I here question whether Doll and Hill's study employed observations as its evidence, it is true that their study is a paradigm of what epidemiologists call an *observational* study. Is there not a contradiction in the idea of an observational study that does not involve observations? No, because the meaning of the term 'observational' is different in epidemiological context from its meaning in the philosophical context. The epidemiological usage of 'observational' is intended to contrast with 'experimental'. In this sense, observations are passive whereas experiments involve the active manipulation of the objects of study—such as in a randomized controlled trial. As Bogen (2017) points out, logical empiricism transformed usage of 'observational' so that the distinction with experiment became elided and the contrast with theory became salient (though, as I describe above, I think that the beginning of the transformation can be traced back to earlier empiricists such as Mill).

than evidence. While Doll and Hill made efforts to ensure that their data was high quality (e.g. by following up the reports of causes of death), they were not interested in the questionnaires or reports as representations of individual doctors, nor could they assume that all those questionnaires and reports were accurate (in which case those reports would not be evidence). Rather, what they did care about were the statistics they gained from this data. The statistics, coming from a large cohort, could be relied upon to be true, understood as having a margin for error.

The best case for there being an observation in the Doll and Hill study is the proposal that what was observed was the *cohort* of doctors. That is not unreasonable, but might be thought to be stretching things if it is not clear that the cohort is a single object of study. And that suspicion will be that much greater for statistical evidence extracted from larger and more disparate, automatically generated 'big data' sets.

However, the important conclusion to be drawn from such considerations is not so much what the observation is or whether there is an observation. Rather, what is made clear is that the notion of observation is not especially significant. As far as the epistemology of science is concerned, it does not matter much whether it is correct to call a statistic regarding a population an 'observation' or not. What does matter is whether we can treat that statistic as evidence. That question is not settled by deciding the observation issue. It is settled by asking (questions such as) whether the data was generated from a sufficiently large and unbiased sample.

There are other cases where we have evidence that is not an observation nor derived from an observation. In making weather predictions meteorologists will run computer simulations using a model. They will run multiple simulations because their data is incomplete or imperfectly precise and because a good weather model is stochastic, allowing for small and chancy variations to lead to quite different weather outcomes. If a large proportion of many simulations run with a range of initial conditions predict a particular weather outcome (such as a storm), then one has strong evidence that such an outcome will arise. But this evidence, the agreement of multiple simulations, is not observational.[38] This is explained on my account of what observation is, because the evidence in this case concerns the outcomes of computer simulations and so does not concern the subject matter of the inquiry, the weather itself.

6.7 Empirical science without empiricism

I have just argued that perception is not essential to observation and that observation is not essential to science. Yet the claims I thereby reject, that observation is an essential foundation to science and that it itself is essentially perceptual, are core claims of empiricism and of empiricism applied to philosophy of science in particular. Indeed one might say that empiricism is the semi-official philosophy of the sciences. Someone sceptical of my anti-empiricism might point out that empiricism took on that semi-official role during the scientific revolution, a remarkable advance in science which empiricism played an important role in motivating and supporting.

[38] This example was suggested to me by Julien Dutant, who also noted that the outcomes of probabilistic procedures for checking the primality of a large number can provide evidence without any observation.

So how is it that science succeeded despite the falsity of empiricism? Why did it seem so obvious to the scientists and other thinkers of the seventeenth century that empiricism was the right philosophy for science? And how is it possible to maintain a respect for the achievements of science (which I do) while rejecting the empiricist basis on which many philosophers base that respect?

My answers in brief are the following. As Boyle (1661: 117) put it, 'There is great difference betwixt the being able to make experiments, and the being able to give a philosophical account of them'. What encouraged and supported the scientists in their work was not the full-on philosophical theory of empiricism, but something rather weaker (but important nonetheless). The most important element of that was the rejection of ancient authority. That in turn required making tests of the relevant claims, and that typically involved making observations. Given the state of science and technology, that itself was inevitably perceptual. Ultimately what makes science worthy of respect is that it is properly founded on evidence, not that it is founded on observation or perception. I shall enlarge on these answers.

6.7.1 *The scientific revolution without empiricism*

Let us look at some of the discoveries of Andreas Vesalius that corrected the opinions of Galen. Vesalius, as a result of dissections on human corpses, discovered that the human mandible has only one bone whereas Galen said it has two; he showed that the human sternum has three parts, as opposed to seven according to Galen. In these cases Galen's error was not a lack of observation but lay in the fact that Galen had limited access to human cadavers and carried out most of his observations on animals, in these cases on dogs and on apes respectively. The early Renaissance doctors and scholars whom Vesalius opposed did not reject the importance of observation. Rather they thought that the observations of Galen were sufficient for established knowledge and declined to test his observations.

Harvey famously showed that Galen was mistaken in thinking that blood was generated by the liver and flowed to the various parts of the body where it was consumed. Galen was, of course, entirely aware of the beating of the heart. Harvey's innovation was to measure the quantity of blood in the atria of the heart and to use this to calculate the quantity of blood passing through it, and so estimate the quantity of blood that would need to be produced by the liver. Since it was implausible that the liver could create such a quantity (almost 250kg per day), that theory was thereby refuted. What was important here was not so much observing what had not been observed before but was using quantitative reasoning from what was observed to refute Galen's hypothesis.

What really differentiated early modern scientists from their predecessors was not that the former respected observation and the latter did not. The differences lay elsewhere: in the moderns' disinclination to trust authority and in the new modes of reasoning that they used, quantitative ones especially. As we saw above, the motto of the Royal Society is 'Nullius in verba' which translates roughly as, 'take nobody's word for it'. Since that applied to both observation and theory, it implied new *acts* of observation and reasoning, not the replacement of reasoning by observing (as might be implied by the later *mistranslation* of the motto, 'there is nothing in words'). Recall that the alternative proposal for the motto was 'Omnium probate'—test everything. But this does not tell us to focus on observation or perception, and only for one who is already signed up to empiricism would it have such an implication.

The empiricism of the early modern period is typically contrasted with rationalism. Furthermore they are typically separated geographically as the 'British Empiricists' (Locke, Berkeley, Hume) and the 'Continental Rationalists' (Descartes, Spinoza, Leibniz), although this is a later and very misleading construction (Pierre Gassendi, for example, must be regarded as an influential and early empiricist (Norton 1981); Leibniz's relationalism is more attractive to empiricism than Newton's absolutism; etc.). Importantly, it is not the case that the rationalists or their philosophy were less supportive of or less well disposed toward science than the empiricists or their philosophy. The key fault-lines between the two camps—innate knowledge or not; innate concepts or not; whether intuition has epistemic value—do not play out in debates regarding science. It is not part of the scientific revolution that one rejects innate knowledge. It is part of the scientific revolution that one rejects (some of) the alleged empirical knowledge of the ancients. It is not part of the scientific revolution that one downplays reason. It is part of the scientific revolution that one uses reason in new (more critical, more quantitative) ways. Descartes did not deny the importance of perception in the production of knowledge; but as a rationalist he wanted a reasoned argument to tell us when one could rely on perception. Galileo did make important new observations. But his works are also full of new arguments as to why the ancient philosophers were mistaken. Some of those arguments use widely known observations—for example his use of our knowledge of the motion of a falling object in a ship to refute the claim that if the Earth rotates an object dropped from a tower should fall at a distance from its base. Other arguments from Galileo are largely apriori, for example his refutation of the idea that a heavier object must always fall faster than a lighter object.

6.7.2 *Science without perception*

So what remains of empiricism? Very little. Think of the scientist whose interaction with her apparatus is entirely via her computer. She does not strictly need to see the screen, since she could use braille or text-to-speech to get the information the computer provides. But presumably she will need *some* perceptual contact? All that seems left of empiricism is minimal empiricism: someone in a sensory deprivation tank would not be able to do any science.

But is even that true? Imagine a future where our natural long-term memory can be augmented by a computer chip integrated with the hippocampus. That chip can have its contents updated wirelessly. The addition of contents to long-term memory by this means would be unconscious (as it generally is). When our subject focuses her mind on some topic, the relevant facts appear in her working memory, just as if they had entered the long-term memory from some long-forgotten reading of a textbook or listening to a lecture. From the subjective perspective, there is no difference. She thinks to herself 'Lead—oh yes, melting point 327.5 °C'. It is conceivable that people might acquire knowledge in this way, and that they might come to know the outputs of experiments like this, instead of reading them from a computer screen. So long as the connection between computer and memory chip is reliable, this can be genuine knowledge. And that knowledge will be scientific evidence that she can use in her scientific reasoning. Yet this has happened without any perceptual experience.

So even minimal empiricism is only contingently true, not a necessary truth. Science does not necessarily require perception.

6.8 Social empiricism?

Earlier I urged not only that science should be understood as a social enterprise but also that we can see scientific and other communities as subjects with epistemic states. How does the foregoing discussion of empiricism look from this perspective? The earlier discussion argued that we need to appreciate two elements of the relationship between the social epistemic subject and the individual scientist. First, this relationship is characterized by the division of labour. Secondly, the social epistemic subject is not composed solely of individuals but depends also on the material means of information production, processing, and storage.

As this chapter shows, it is exactly these two factors that explain the failure of empiricism. However, for precisely that reason, it might be that the social epistemic subject is not itself prey to these objections to empiricism. Connections of testimony are connections between the parts of the social epistemic subject, rather like connections between an individual's cognitive sub-systems (e.g. between sight and memory). Likewise, instruments are internal to the social epistemic subject, rather as the lens of an eye is a part of the individual subject, even though it is not a component of any of the subject's neural information processing sub-systems. So, one might argue that the problems of testimony and instruments do not arise for social epistemic subjects. The social epistemic subject is an autonomous subject. Furthermore, we can regard the social epistemic subject's connection with the objects of science as akin to perception. Since we have taken Gauss meters, weather stations, and particle colliders as components of a social epistemic subject, that subject's connection with a magnetic field, with a weather system, or with a subatomic particle, is itself unmediated by any instrument and so can be seen as perceptual. This would allow for a reconnection between observation and perception: an observation is made when the social epistemic subject perceives some fact.

So, taking my claims about the social epistemic subject at face value would suggest that empiricism might be true after all, but at the level of the social rather than the individual subject.

While I think that this is true, it hardly amounts to a rehabilitation of empiricism. For the motivation for empiricism lies in (internalist) concerns about the certainty of the foundations of one's epistemic states. This is not a concern that arises for social epistemic states—empiricism is essentially an individualistic philosophy. Furthermore, there will be analogues of the problems of testimony for some social epistemic subjects. There are smaller and larger groups that can be epistemic subjects. Groups other than the maximally large one will communicate with one another, and so the problems of testimony will arise among groups just as they arise among individuals.

6.9 Conclusion

A very large proportion of the realism versus anti-realism debate has focused on the supposedly crucial question of whether we can have knowledge of unobservable entities and properties.[39] Almost all the realists involved in this debate have accepted the philosophical significance of the distinction, agreeing that if we cannot have knowledge of unobservables, then we know very little of what modern science appears to be teaching us. Why then, has there continued to be a realism debate, centred on the observable/unobservable distinction? The reason is that the realists in question are (optimistic) empiricists.

Another curiosity about the realism debate and its focus on observability is that there are so few anti-realists in the debate.[40] So why does the 'debate' occupy philosophers of science so much? There may be several reasons. An important reason is that if one is an empiricist, then establishing realism is difficult. Optimistic empiricists have a hard time convincing even themselves that their realism is tenable.

And that's because realism isn't tenable for an empiricist. As we have seen, empiricism faces insurmountable obstacles when it comes to accounting for crucial features of modern science: the reliance on testimony and the use of instruments and similar technology. If you want to be an empiricist, you should be a sceptical empiricist.

That said, sceptical empiricism is not a happy position either. It is unmotivated—the most obvious argument for it is question-begging. And worse, the constructive empiricist attempt to recast the aim of science so that science is both rational and empiricist fails. It renders science a largely pointless exercise, undeserving of the respect and interest that many of us accord it.

A very small number of philosophers (Maxwell, Shapere, Bogen, and Woodward) have pointed out that philosophers use 'observation' and its cognates in a way that is entirely at odds with the way that scientists use the term. Excluding the unobservable rules out very little, if, as scientists tell us, top quarks, magnetic fields, the core of the sun, and gravitational waves can be observed. For the large majority of philosophers of science, the scientists' usage may be an interesting linguistic fact (if they are aware of it). But, they would say, it does not tell us anything of epistemological significance, for which we need a more restricted notion of observation, one that is closely tied to perception.

However, if we abandon empiricism for the reasons given, then we can make perfect sense of the scientists' usage. When we want to find out about some object of study, we gather basic evidence about it—evidence not inferred from other evidence concerning that object (though that evidence might be inferred from other objects).

[39] And most of the remainder of the debate has concerned the challenge of the Pessimistic Induction, which we saw in Chapter 3 and address again in Chapter 9.

[40] That said, there has been a resurgence of anti-realist philosophy of science very recently, in particular Wray (2018) and Rowbottom (2019).

The representations that supply this basic evidence are observations. Or, more simply, observational propositions regarding X are propositions that are the basic evidence regarding X. From that (basic) evidence we make inferences that expand our knowledge of Xs. That expanded knowledge may make it possible for us to gather information about Ys for the first time. Hitherto they were unobservable, but now that we can know about them they have become observable, and the first facts we gain about Ys are our observations of Ys. These facts may involve inference (from knowledge of Xs), but they are basic evidence concerning Ys. What is observable is not fixed for all time (let alone fixed by the extent of our perceptual capacities), but is changeable depending on how the growth of knowledge make new facts about entities available.

7

ABDUCTIVE KNOWLEDGE

The task of this chapter is to provide an explanationist account of theoretical knowledge in science and of rational preference among scientific hypotheses. Knowledge and rational preference are two states that need separate treatment, though the former can be seen as a limiting case of the latter. Inference to the Best Explanation says that one hypothesis is to be preferred to another when it is a 'lovelier' potential explanation. But for knowledge we must employ Inference to the Only Explanation—competing explanations must be excluded by the evidence. I show that Inference to the Only Explanation is possible, as well as in fact found in science. And I give an account of the loveliness of explanations that allows loveliness to be a guide to the truth.

7.1 Introduction

By what means do we generate knowledge in science? And when knowledge in unachievable, how do we form rational preferences among hypotheses? These are the questions that will occupy this chapter and the next. Scientists use a plethora of techniques, methods, rules of thumb, heuristics, and so forth in formulating and accepting propositions of scientific interest. When it comes to the theoretical claims of science, one approach stands out: very roughly, we look for possible explanations of the evidence; to the extent that a proposed explanation is plausible, we are inclined to think that it is supported by that evidence. This mode of thought is *abduction* and the view that abduction is rational is *explanationism.*[1]

Palaeontological evidence shows that about 66 million years ago a wide range of animal and plant species became extinct in a (geologically) short period. It is estimated that 75% of such species died out at this time, including all the non-avian dinosaurs; the extinction was followed by an evolutionary development of and diversification from the surviving species, including the remaining mammals. Associated with this extinction event is a geological feature, a thin layer of sediment, now

[1] 'Abduction' is C. S. Peirce's term and it is a matter of some debate what exactly Peirce meant by it and how that relates to current usage. Mcauliffe (2015) shows that Peirce intended abduction to concern the selection of hypotheses for consideration, rather than an inference to their truth (see also Douven 2017*b*). It is thus confined to what below I call Stage 1. Even so, abduction in Peirce's narrower sense must still be directed towards the truth, if it is to be a rational component of science. The selection of hypotheses for consideration must select those that have a reasonable chance of being true in virtue of their explanatory features and ignore those that, because they are unlovely, are unlikely to be true. Here I propose to use 'abduction' more broadly, as a term to *describe* an approach to inference that takes the explanatory relations of a hypothesis as central; and I use 'explanationism' to refer to the *normative* view that abduction (in some form) is rational.

Knowing Science. Alexander Bird, Oxford University Press. © Alexander Bird 2022.
DOI: 10.1093/oso/9780199606658.003.0007

known as the K-Pg boundary.[2] Several hypotheses were, at different times, proposed about the cause of this mass extinction. One, initially controversial, theory, is now widely accepted. According to Luis and Walter Alvarez, the cause was a major asteroid impact. This would explain the extinction because the resulting dust cloud would have largely eliminated photosynthesis by blocking sunlight. The death of plants would lead to the death of herbivores and then other animals in the food chain. The hypothesis would also explain the iridium anomaly: the K-Pg layer has an unusually high concentration of the element iridium which is rare in the Earth's crust. It is however found in high concentrations in meteors and asteroids. The asteroid would also explain evidence for a shift in the angle of the Earth's axis of rotation and the existence of the Chicxulub Crater in the Yucatán Peninsula. According to explanationism, the fact that the Alvarez asteroid impact theory explains this evidence means that this evidence supports that hypothesis. While this theory has been accepted by most palaeontologists, others maintain that the correct explanation of the K-Pg extinction is a sequence of volcanic eruptions in the Deccan Traps. The supporters of this theory urge that it is more likely to be true because it is a better explanation of the fact that many kinds of life-form (such as types of foraminifera) were in decline, heading for extinction, before the date of the Chicxulub Crater. Explanationists regard the fact that one proposed explanation is a *better* potential explanation than another as a reason to have an epistemic preference for the former.

In this chapter I develop my own version of explanationism. I develop it in contrast to what I take to be the best alternative theory. The latter is Inference to the Best Explanation (IBE), as developed by Peter Lipton (2004). According to IBE we are entitled to infer that a possible explanation is true when (roughly speaking) it is a rather better explanation of the evidence than its rivals. My alternative, Inference to the Only Explanation (IOE), proposes that in order to gain knowledge a scientist needs to rule out the rival hypotheses. To rule out a rival hypothesis, a scientist's evidence must be show that hypothesis to be false. In implementing IOE I propose that scientists implicitly use a form of inference that I call 'Holmesian' in recognition of the form of inference attributed to Sherlock Holmes by Sir Arthur Conan Doyle. In promoting IOE I do not reject IBE. Rather I regard IBE as a rational basis for determining epistemic preferences among hypotheses—that is, for deciding which hypothesis is most plausible in the light of the evidence, even if we cannot rule the competitors out. We can see IOE as the limiting, optimal case of IBE, where evidence reduces the plausibility of competitors to zero. It is this case that leads to knowledge, as opposed to rational epistemic preference, which is what IBE without IOE delivers.

In the next section I will outline Lipton's account of IBE, to be followed by my account of IOE. Following that I return to the famous case of Ignaz Semmelweis and his reasoning regarding the prevalence of puerperal fever in two wards of the Vienna General Hospital in the 1840s. Lipton uses this case to support his IBE. I show that it fits my IOE rather better. I then look at and defend IOE in more detail. As I have

[2] 'K-Pg' stands for 'Cretaceous–Palaeogene'. This used to be known as the 'K-T (Cretaceous–Tertiary) boundary'. (Why is 'K' the abbreviation for 'Cretaceous'? Because 'C' is the geologists' abbreviation for 'Cambrian'.)

said, IOE is my explanationist account of abductive knowledge, while I accept IBE as an explanationist account of rational epistemic preference among hypotheses in those cases where the evidence fails to rule out all but one of them. So I then defend IBE's ability to achieve this in the final part of this chapter. (An epistemic preference among propositions naturally lends itself to a probabilistic treatment. So I consider the relationship between IBE and Bayesianism in the next chapter.)

7.2 Inference to the Best Explanation

Inference to the Best Explanation (IBE) is about choosing among explanations. It is a matter of choosing among potential explanations of some phenomenon the one that is the best by certain criteria. If there is a suitable best explanation, IBE says that we may infer that it is the *actual* explanation. If some hypothesis provides the actual explanation of a phenomenon, then that hypothesis is true.

How do we choose among potential explanations? According to Peter Lipton, IBE is a two-stage process, where both stages are filters of potential explanations (Lipton 2004: 56–64):

> Stage 1: The imaginative capacities of scientists generate all the plausible potential explanations and just leaves the remainder unconsidered. In effect the first stage is a filter that retains only plausible explanations while rejecting the implausible explanations from among the set of all possible explanations.[3]
>
> Stage 2: At the second stage, scientists investigate the live potential explanations that have passed through the first filter, and ultimately rank them according to their explanatory goodness. The top-ranking explanation is inferred to be *the actual* explanation.

Lipton explains that explanatory goodness, what he calls 'loveliness', must be distinguished from likeliness (i.e. how probable the hypothesis is), since the aim of IBE is to guide our estimates of likeliness on the basis of loveliness. I shall regard 'loveliness' as defined by its role in IBE: that is, 'loveliness' refers to the quality or set of qualities of explanatory hypotheses that scientists use to rank those hypotheses highly in an inference to the best explanation. In Lipton's view loveliness is a matter of potential understanding—a lovely explanation is one that would give us a high degree of understanding of the relevant phenomena were it to be true (and, I would add, known to be true).[4] Given my functional definition of loveliness, we need not agree with Lipton that this is what loveliness is.

[3] I find Lipton's filter analogy slightly misleading, insofar as it suggests that at some level all the potential explanations, plausible and implausible, are before the scientist's mind and she selects the plausible ones from these. But that is clearly not the case. Her imagination may touch on some implausible explanations and then ignore them, but most will not even receive that attention. A well-trained scientific imagination will be directed *towards* the plausible explanations, as I explain below.

[4] Since understanding is a species of knowledge, an explanation being true and being believed does not thereby constitute understanding, hence my addition of the need for the explanation to be known. See Section 3.5.

When we cannot use IOE we still have multiple potential explanations in play. IBE says that loveliness is the basis of our ranked preference among those explanatory hypotheses. For this to be rational, loveliness must be a guide to the truth—lovelier explanations should in fact be more likely to be true. Three objections raised by Lipton cast doubt on this correlation between loveliness and truth, the first of which targets Stage 1, while the other two concern Stage 2:

> *Underconsideration* Stage 1 delivers to us a set of hypotheses for our consideration. This is a subset of all possible explanatory hypotheses. The actual, true explanation may not be among those consciously or even unconsciously considered, but may remain among those not considered at all. Why is it likely that we even consider the actual explanation?
>
> *Hungerford's objection* Loveliness is an aesthetic property or akin to an aesthetic property in being dependent on the potentially variable response of the individual. In short, loveliness is a subjective quality. Whereas truth is an objective property. So loveliness and truth cannot be expected to correlate.
>
> *Voltaire's objection* Even if loveliness is objective, there will be many worlds where it does not in fact correlate with truth. So why think that truth and loveliness correlate in ours? In particular why should they correlate enough for the actual world frequently to be the most lovely (that is, the world where the most lovely potential explanation is the true explanation)?

(I will consider these problems in more detail in Section 7.6.1.)

Scientists do use Inference to the Best Explanation. But it is not how scientists gain theoretical knowledge—it is how scientists judge the relative plausibility of hypotheses in the light of the evidence, and so give themselves a rational basis for preferring one hypothesis to another. However, as long as there is more than one competing hypothesis with some plausibility, then no hypothesis can be known to be true. To know that a hypothesis is true, our evidence must rule out all but one hypothesis. This is Inference to the Only Explanation.

Inference to the Only Explanation (IOE) suffers from neither of the Stage 2 objections, Hungerford's objection and Voltaire's objection, because IOE does not employ loveliness as an indicator of likely truth. Rather it uses inconsistency with the evidence as an (infallible) sign of falsity. (The problem of underconsideration, which does not really concern loveliness, is still a prima facie problem for IOE.) I will next articulate IOE in more detail. Later I will return to IBE and show how it can be a guide to likeliness in those cases where IOE is not available—there is an answer to the three objections that allows a scientist to have a rational preference for one hypothesis over another, on abductive grounds, where both are still epistemic possibilities for her.

7.3 Inference to the Only Explanation

7.3.1 *Holmesian inference*

We are apt to classify the inferences ascribed to Sherlock Holmes (such as identifying a criminal on the basis of the mud on a man's boot, the analysis of a cigar ash, and so on) as inductive. Yet Conan Doyle described Holmes's method as *deductive*. This appears to be a solecism.[5] On the other hand, Sir Arthur goes on to provide details of the method which allow for a reconciliation of this terminology. On several occasions Holmes tells Watson, 'Eliminate the impossible, and whatever remains, however improbable, must be the truth.'[6] That clearly is deductive. If Holmes starts by knowing that one of ten hypotheses is true and by dint of further evidence gathered in the course of his investigations comes to know of nine of them that each is false, the deduction tells him that the tenth must be true.

As a procedure, what I shall call *Holmesian inference* has the following form. From initial evidence, e_i, Holmes gets to know that one of competing hypotheses $h_1,\ldots, h_n$ is true. These hypotheses are explanatory hypotheses; they explain some subset e_s of the evidence. Holmes then collects additional evidence, e_a, such that e_a (given e_i) rules out $h_1, \ldots, h_{n-1}$. Hence Holmes may deduce that h_n is true. Holmesian inference requires three premises:[7]

(i) The fact e_s has an explanation. (*Determinism*)
(ii) $h_1,\ldots, h_n$ are the only hypotheses that could explain e_s. (*Selection*)
(iii) $h_1,\ldots, h_{n-1}$ have been falsified by the evidence. (*Falsification*)

We need not see this solely in procedural terms. For we may say that at the end of the investigation Holmes's final evidence, $e_f = e_i+e_a$, deductively entails the one hypothesis, h_n. Pierre Duhem articulates what is Holmesian inference thus:

> Do you wish to obtain from a group of phenomena a theoretically certain and indisputable explanation? Enumerate all the hypotheses that can be made to account for this group of phenomena; then, by experimental contradiction eliminate all except one; the latter will no longer be a hypothesis, but will become a certainty. (Duhem 1954: 188)

Weber (2009: 21–2) understands Duhem to be intending this to be a deductive inference, one which Duhem rejects for two reasons. Duhem rejects Falsification, because of what we have come to know as the 'Duhem–Quine thesis'. And he rejects Selection because of the problem of underconsideration, which we have already encountered.

An ordinary IBE operates when the evidence leaves multiple hypotheses in play but makes one more lovely than the others. In a Holmesian inference the evidence renders all hypotheses but one knowably false. Being knowably false might be regarded as maximally unlovely, in which case we can see Holmesian inference as a

[5] This criticism is made, for example, by Peter Lipton (2001: 96) and by Walter Sinnott-Armstrong (2018: 185–6). ('Inductive' here is intended broadly, to include abductive reasoning.)

[6] For example in 'The Sign of Four' (Doyle 1953*b*: 94, 118) and 'The Adventure of the Bruce-Partington Plans' (Doyle 1953*a*: 1089). Kitcher (1993: 239) and Earman (1992) make favourable references to Holmes in support of eliminative induction.

[7] I take the names of the first two premises, Determinism, and Selection, from von Wright (1951: 131).

limiting case of IBE. Since Holmesian inference leaves only one explanatory hypothesis standing, I have called it Inference to the *Only* Explanation (IOE). Since Inference to the Only Explanation (*viz.* Holmesian inference) eliminates all but one of the hypotheses, we do not need to consider whether one hypothesis is lovelier than another. Loveliness drops out of the picture. In which case Hungerford's objection and Voltaire's objection do not arise.

Here is a simple but real example of IOE in action. In the 1830s Berzelius introduced the concept of isomerism: two distinct chemical compounds could have the same molar mass and chemical formulae. The phenomenon of isomerism supported the hypothesis of the existence of molecules structured by bonds between their atomic constituents. Further weight was lent to the hypothesis by Pasteur's discovery of enantiomers—compounds that are chemically identical but which have certain differences in physical properties. Tartaric acid, Pasteur found, existed in two crystalline forms that are the mirror images of one another, and which rotate refracted light in opposite directions. This led to the development of structural chemistry, the investigation of the molecular structure of chemical compounds. The Dutch chemist Jacobus van 't Hoff therefore wondered how the bonds of carbon are arranged in space, for example, in methane, CH_4. A natural hypothesis is that they are (co-)planar (Figure 7.1a). Van 't Hoff realized that this could not be the case, because the planar hypothesis is inconsistent with the lack of isomers for compounds such as difluoromethane, CH_2F_2. For a planar structure of bonds would allow for isomers in such cases: a molecule with the two fluorine atoms opposite one another will be different (Figure 7.1b) from one where they are adjacent (Figure 7.1c). Van 't Hoff was able to eliminate that possibility as inconsistent with the evidence that there are no such distinct isomers (Hendry 2018). Furthermore, the only arrangement of bonds that does not allow for isomers is a tetrahedral structure (Figure 7.2). So the evidence rules out all but one hypothesis in this case.

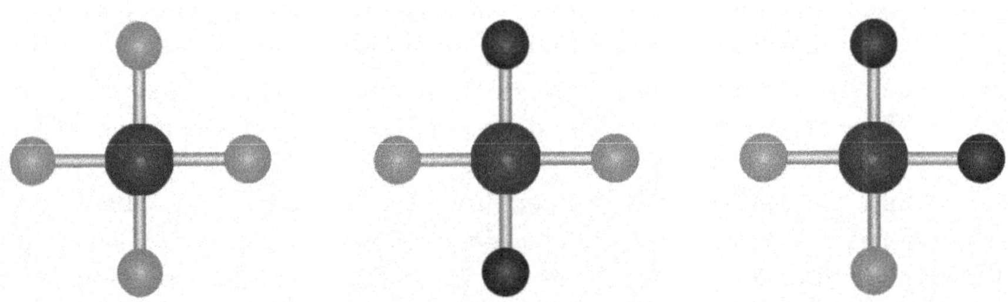

a: Planar structure for methane

b: One isomer for planar difluoromethane

c: A second isomer for planar difluoromethane

Fig. 7.1

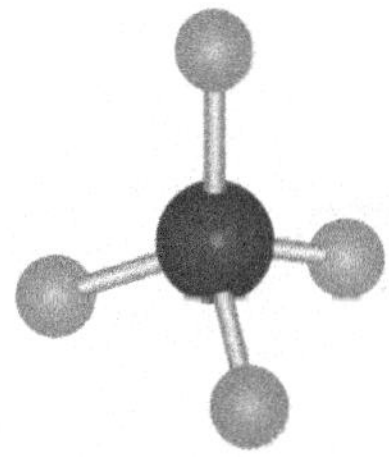

a: Tetrahedral structure for methane

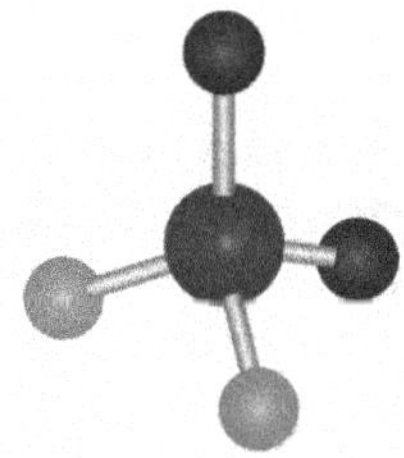

b: Only one structure is possible for tetrahedral difluoromethane

Fig. 7.2

The possibility of knowledge from Holmesian inference is undoubtedly controversial. As mentioned, it is rejected by Duhem among others. It requires not only the possibility of the truth of each of the three premises Determinism, Selection, and Falsification but also the possibility of (concurrent) knowledge of the three premises. I will comment briefly on each in turn.

7.3.2 *Determinism*

We need Determinism because ruling out all but one of the potential explanations is not quite sufficient for abductive knowledge by Holmesian inference. The evidence may not require an explanation at all. So Holmesian inference requires that the subject know that there is some explanation or other. The subject must have some evidence that rules out the null hypothesis, that there is no explanation.

The premise Determinism should not be too controversial. The assumption of Determinism is not the false assumption that universal determinism holds, but the assumption that the fact or phenomenon in question does have an explanation. In some cases the null hypothesis, that there is no explanation, may indeed be true. But in many cases the existence of some explanation is not in doubt. The decay of a single fissile nucleus may not have an explanation. But the detective knows that the bullet didn't just materialize out of nothing in the brain of the victim and that the entry wound didn't just come from nowhere. Similarly, that there is some correct explanation or other for the extinction of the dinosaurs is not up for question even if the nature of that explanation is. Van Fraassen (1980: 21) says that 'the realist will need his special extra premise that every universal regularity in nature needs an explanation, before the rule [of Inference to the Best Explanation] will make realists of us all'. But such a strong, universal premise is not required. It is sufficient for Holmesian inference that we know of some facts of interest that they have an explanation.

7.3.3 *Falsification*

On some occasions we are able to falsify hypotheses. If so we should sometimes be able to falsify all but one of the hypotheses in a finite set of mutually inconsistent hypotheses that includes the true hypothesis.

Falsification is not entirely uncontroversial. The Duhem–Quine thesis alleges that we always have the choice of avoiding a falsification by changing an auxiliary hypothesis. Knowledge from Holmesian inference requires that we know that we have falsified alternative hypotheses. Hence the Duhem–Quine thesis presents a challenge to Holmesian inference if the thesis is taken to be the claim that we cannot know that any hypothesis is falsified by the evidence. I address this objection below.

7.3.4 *Selection*

The premise Selection of the Holmesian inference is the most controversial. There are two reasons why. First, most philosophers of science are not willing to grant that the hypotheses that could explain some piece of evidence may be finite in number. They accept the thesis that theories are radically underdetermined by the evidence:

> (UD) For any given fact (phenomenon) *f* and any set of evidence *e*, there is always more than one plausible competing hypothesis consistent with the evidence that explains *f*.

I shall defend Selection against (UD) below. For present purposes, it suffices to say that defending (UD) requires adopting an implausibly sceptical stance that begs the question against Holmesian inference.

The second reason for rejecting Selection is the concern that the correct hypothesis might fail to be among the hypotheses we actually consider. So even if the number of possible explanatory hypotheses is finite, it is a fact about the imaginative capacities of scientists that they frequently fail to generate all those hypotheses, in which case Selection is false. This is the problem of underconsideration, already mentioned and which we will look at in more detail below (p. 190 and p. 237).

7.4 The Semmelweis case again

The story of Ignaz Semmelweis's discovery of the causal mechanism that led to puerperal fever in parturient women has been repeated many times by philosophers and historians. This at least has the advantage that the details are widely known. And since Peter Lipton has used the case as an illustration of his account of IBE, it is a useful case for my purposes, since I can show that IOE provides a better account than Lipton's IBE.

In 1844 Ignaz Semmelweis graduated from the Vienna Medical School and decided to specialize in obstetrics. He was appointed assistant to the professor of obstetrics, Johann Klein, first in 1846 and then again in 1847. Klein was responsible for one of the two labour wards at the Allgemeine Krankenhaus, the General Hospital in Vienna. Many poorer women came to the hospital to give birth and of these women a large proportion, up to one-sixth in some years, contracted puerperal (or childbed) fever, which was almost always fatal. It was widely known that the death rates were considerably higher in Klein's ward, Division I, than in the other ward, Division II, run by Professor Franz Xavier Bartsch. Semmelweis sought some feature of Division I that would explain its high rate of mortality. These are the principal hypotheses he considered initially:

(S1) Overcrowding in Division I.
(S2) Epidemic influences and climate.
(S3) Rough examinations by the medical students in Division I. (Medical students examined women in Division I, but not in Division II, which was the preserve of midwives.)
(S4) Psychological effect of the priest passing through the ward on his way to deliver extreme unction to dying women.
(S5) Women in Division I delivered on their backs.

In assessing these explanations, we must be careful in deciding what the explanandum is. The explanandum could be, among others:

(A) The existence of puerperal fever in Division I (and by extension the existence of puerperal fever elsewhere); or,
(B) The greater prevalence of puerperal fever in Division I.

The character of the inference is very different depending on which the explanandum is taken to be. In my view it is important to focus on explanandum (B), the difference in rates of puerperal fever and consequent mortality between the wards.[8] The explananda are connected: the principal explanations of the existence of puerperal fever in general might supply explanations of the difference between the two wards; and conversely a successful explanation of that difference might well provide insight into the cause of puerperal fever in general. But these are further inferences, and fraught ones at that, as I shall mention. In the light of this we should consider the explanations (S1)–(S5) as shorthand for explanations of the form 'X is the cause of the positive difference between the rates of Division I and Division II', e.g. (S1) should be understood as asserting that overcrowding in Division I is the cause of the greater mortality rate in Division I when compared to Division II.

Semmelweis (1861: 65–7, 69) noted that hypotheses (S1) and (S2) refer to features that were common to both Division I and Division II. Indeed, because of the desire of expectant mothers to be admitted to Division II rather than Division I, the former was even more crowded than the latter.[9]

[8] Our explananda concern the existence and rates of puerperal fever, but the data concern rates of mortality from puerperal fever. The mortality rates are good proxies for the morbidity rates since the disease was almost always fatal.

[9] In which case, one might ask, why would (S1) even have been raised? The principal answer is that while women were admitted to the two wards on alternating days Sunday through Friday, from Friday to Sunday afternoon, women were admitted to Division I. Furthermore, Division II was instituted in order to relieve overcrowding in Division I. So historically there had been a problem of overcrowding in Division I, until the difference in mortality became widely known. An additional reason is that overcrowding was a widespread problem in European hospitals, with several patients sharing a single bed being a common occurrence. In Vienna, however, one patient per bed was the rule. Nonetheless, the relationship between overcrowding and puerperal fever was a natural one for doctors to consider.

Lipton remarks, however, that the similarity between the wards is nonetheless consistent with one or other of those hypotheses being true. Since no one thought such factors to be sufficient for puerperal fever, those who maintained such hypotheses would think that they are only part of the explanation as to why any particular woman contracted the fever; a full explanation would refer to other factors as well, such as general state of health. Note, though, that this point holds only if the explanandum is (A) rather than (B). But as we have seen and will continue to see, Semmelweis's principal evidence concerns the differences between the two wards. Since Division II had puerperal fever, which could also affect women giving birth at home, Semmelweis was not in a position to directly infer the cause of puerperal fever *tout court*. Naturally, he was indeed interested in the cause of puerperal fever, as the title of his book (1861) on the subject demonstrates. But, as we shall see, the inference from an explanation of (B) to an explanation of (A) makes difficulties for Semmelweis.

According to Lipton's view of contrastive explanation, to explain the difference between the two wards, we must seek a feature in the history of Division I that is absent from the history of Division II. But hypotheses (S1) and (S2) do not identify such a difference (ignoring the lesser degree of crowding in Division I). Therefore they simply cannot be explanations of (B). Those hypotheses, construed as potential explanations of the difference between the two wards, are simply inconsistent with the evidence. The same goes for a number of other potential causal factors in a case of puerperal fever that are not mentioned in the list above: inadequate ventilation, excess blood in the circulation, stagnant circulation, disturbances caused by pregnant uterus, decrease in weight caused by emptying of the uterus, protracted labour, wounding of the inner surface of the uterus in delivery, imperfect contractions, faulty involutions of the uterus during maternity, the volume of the secreted milk, and death of the foetus (Carter and Carter 2005: 47).

Hypotheses (S3)–(S5) do mark a difference between the wards, at least at the beginning of Semmelweis's investigations. The hypothesis, (S3) that the difference in mortality is a result of rougher handling by the male medical students in Division I in comparison to the treatment of the women in Division II by female midwives, was not a plausible explanation.[10] For even if there were any such difference, the roughness of the handling by the students, as Semmelweis pointed out, was negligible compared to the trauma of childbirth itself. And the difference in roughness between the students and midwives would have been even smaller in comparison. Hence hypothesis (S3) seeks to explain a large difference between the two divisions, the fact that the mortality rate in Division I was three times that in Division II, by appeal to

[10] Lest one should imagine that the midwives were particularly gentle, consider the comment of Semmelweis's colleague Jakob Kolletschka, 'It is here no uncommon thing for midwives, especially in the commencement of their practice, to pull off legs and arms of children, and even to pull away the entire body and leave the head in the uterus. Such things happen, ganz und gar nicht selten, sondern kommen oft vor [they occur certainly not rarely, but often]' Balfour (1855) (quoted in Semmelweis (1861: 126, fn 5)). In mitigation of the midwives, one should note, as Carter and Carter (2005) do, that many of their patients were women from impoverished backgrounds who had suffered from rickets as children. Rickets can often lead to a malformed pelvis, resulting in difficulties in childbirth when adult.

what is at most a tiny marginal difference. Certainly, in some setups, incremental changes can have significant effects; but no doctor would suppose this to be such a case. In my view such a hypothesis is not merely implausible—Semmelweis could rule it out as inconsistent with what he knew about how trauma affects disease.

With respect to hypotheses (S4) and (S5) Semmelweis pursued the policy of eliminating the differences between the wards referred to in a given hypothesis. Thus the priest agreed to take a different route, avoiding Division I; and women in that ward delivered on their sides: again in both cases without any diminution in death. Semmelweis was thus able to generate evidence inconsistent with (S4) and (S5) and thereby eliminate them from his enquiries.

At this point, discussions of the Semmelweis case mention the fact that while on holiday in Venice in early 1847, Semmelweis's colleague Jakob Kolletschka died of a wound incurred during a post mortem examination. In his illness Kolletschka showed the same symptoms and on autopsy the same lesions as found in women who suffered and died from puerperal fever. This led Semmelweis to his final hypothesis:

(S6) Women in Division I were being infected with cadaveric matter transmitted by medical students from the autopsies that they had been carrying out beforehand.

I divide Semmelweis's hypothesis into two components:

(S6a) Women in Division I were infected during examination by medical students.

(S6b) The infectious agent was 'cadaveric matter' imported by the students after carrying out autopsies.

Kolletschka's death is often presented as a key piece of evidence, one that (S6) can explain in a lovely fashion whereas the other hypotheses cannot. Consequently (S6) is, in this respect at least, a better explanation than the others. I believe, however, that the importance of Kolletschka's death lies elsewhere. At this time, the leading explanation offered of puerperal fever, along with many other diseases, was the miasma theory, according to which diseases are often caused by bad airs that are themselves effects of geography and climate, and can be caused by stagnating water, rotting organic material, overcrowding, and the like. This is the theory covered by (S2). Note first that in terms of being able to explain other facts, supporters of the miasma theory of disease would argue that their theory explains a huge amount of data, such as the fact that some diseases, such as malaria, are common in low-lying marshy areas, why diseases such as cholera are more prevalent at sea-level than at altitude, why many diseases, such as typhoid and cholera are more prevalent in crowded, unsanitary cities than elsewhere, why improvements in sanitation lead to diminution in disease. With respect to the evidence concerning puerperal fever in particular, at the Vienna General Hospital and elsewhere, the miasma theory would explain why puerperal fever comes in epidemic waves and varies seasonally, being worse in winter than in summer. When considering explanations of (A), the occurrence of puerperal fever in general, the fact of Kolletschka's death counts for very little against this mass of evidence. And if we are considering (B), the difference in mortality between the two wards, Kolletschka's death is evidentially otiose. For the rival (S2), as an explanation of that difference, is already refuted by the evidence, as we saw above. And, more importantly, Semmelweis generated the crucial piece of evidence when he insisted on the students washing their hands in chlorinated water before examining the women, the mortality rate in Division I fell to equalling that in Division II. The headline figures are these: the mortality rates for the six years 1841–6 were: Division I—9.92%, Division II—3.88%, and for the twelve years 1847–58 were: I—3.57%, II—3.05% (figures from Gillies 2005). This crucial fact clinches the argument in favour of (S6a) independently of the evidence concerning Kolletschka.

The significance of Kolletschka's death is that it drew Semmelweis's attention to a difference between the midwives and the students that might otherwise have gone unnoticed—the fact that they attended autopsies and carried out dissections before performing examinations in the maternity wards. While he could not eliminate this difference, since he didn't control the students' timetable, he could isolate it causally, which amounts to the same thing, by insisting on hand-washing.

The evidence concerning Kolletschka enabled Semmelweis to do something else, to formulate a specific hypothesis concerning the infection, that it was due to 'cadaveric matter' being transferred from a dissected body to the uterus of an unfortunate mother via the hands of the students. Above I divided (S6) into a less specific claim (S6a), that some kind of infection from the medical students is responsible; and a more specific claim, that cadaveric matter from autopsies is responsible. This distinction is important because only (S6a) is verified by the evidence. Although (S6b) is rendered plausible, it is far from verified (and indeed it is strictly false). To be precise, the evidence verifies the claim that the explanation of (B) is some property of the hands of the medical students that is removed when they are washed. It strongly supports but does not verify the claim that this property is related to and a causal consequence of the presence of the students at the autopsies; and it lends some support, but much less, to the claim that the property in question is the presence of cadaveric matter. I should make clear that the notion of 'infection' I am using here is a very weak one, and does not imply any commitment to a modern germ theory. Rather it is intended to capture an idea that would have been familiar to Semmelweis's contemporaries, that of contagion, an idea that goes back to Frascatoro in the seventeenth century. The core of the idea is that diseases can be spread from individual to individual by the conveyance of some material medium between them. Frascatoro hypothesized the medium to be 'seminaria' (seeds), but tells us little about them, which is why I say that the core idea is that there is some material medium of transmission. Semmelweis's insistence on 'cadaveric matter' is a specific version of this theory.

I make these distinctions, even though Semmelweis did not, in order to make two points. First, as I shall go on to explain, this will allow me to demonstrate my principal thesis, that the evidence can lead us not simply to the best explanation of the evidence, but also, on occasion, to the only explanation of the evidence. (S6a) is a hypothesis of which this is true, but (S6b) is not.

Secondly, I suggest that one reason why Semmelweis failed to get his views accepted is that he did not distinguish between (S6a) and (S6b), and argued strongly in favour of (S6b) which was only partly supported by the evidence. Furthermore, Semmelweis did not clearly distinguish between explananda (A) and (B). Although Semmelweis's most effective evidence concerned the difference between the two wards, his ultimate aim was to explain the causes of puerperal fever *tout court*—all the cases in Division I, and in Division II, and elsewhere. This was because he insisted on a single cause for all cases. But his evidence did not support such a view. For example, it was unclear how cadaveric matter could explain the deaths in Division II and elsewhere. Semmelweis's explanation was that in such cases the women were self-infecting, due to internally decaying matter. Such an explanation seemed ad hoc. And while Semmelweis (1861: 81) noted that street births showed a lower rate of mortality than Division I, he could not explain why home births showed a significantly lower rate of mortality (circa 0.5%) than even Division II (over 3%)—if the cadaveric hypothesis implied that the deaths in Division II were unavoidable self-infections, then one would expect a comparable rate of self-infection among mothers giving birth at home.

Furthermore, the cadaveric matter hypothesis was not even novel. A commonly held alternative to the miasma theory of puerperal fever was the view that it is caused by internal putrescence, the rotting of the patient's own internal flesh and organs. For example, Dr John Clarke held that tight stays and petticoats and the weight of the baby in the uterus detained faeces in the intestine causing putrescence (Nuland 2003: 43). Getting people to believe a new theory may be difficult enough, but it is often even more difficult to get them to believe an old theory they regard as having been refuted. The principal piece of evidence against such a theory is the fact that puerperal fever was an epidemic disease which could afflict a population particularly severely for a number of years. Additionally it was seasonal, with winters being particularly bad. Semmelweis (1861: 122) explained the latter by reference to the greater diligence of the student doctors in winter months, and while that may have been an exacerbating factor, this seasonal variation was not limited to teaching hospitals (which is another reason to focus on explanandum (B) rather than (A)).

To conclude: if we, unlike Semmelweis, restrict our hypothesis to (S6a) and our explanandum to (B), then we see that the evidence forces us to accept that hypothesis by eliminating all potential alternatives.[11]

In Lipton's model of Inference to the Best Explanation, the loveliness of the hypotheses is central to their epistemic status: the rank order of their epistemic credibility should follow the rank order of their explanatory loveliness. One would have thought, therefore, that Lipton, in using the Semmelweis case to illustrate and support his model, would have identified and compared the loveliness of each of the hypotheses (S1)–(S6). But he does not do so. And that is because the epistemology of the aetiology of puerperal fever is not like this. Semmelweis did not rank the six

[11] Elsewhere I argue that IOE can be seen at work when scientists ruled out various hypotheses in order to reach the conclusion that AIDS is caused by a virus (Bird 2010*a*).

hypotheses according to their loveliness. It wasn't that infection via the doctors' and students' hands was a lovelier hypothetical cause of the difference in level of puerperal fever than the presence of a dolorous priest. That evidence didn't simply show the priest hypothesis to be unlovely, it showed it to be outright false. Likewise for all the other hypotheses considered by Semmelweis, with the exception of the infection hypothesis (S6a). Thus Semmelweis had no need to consider the loveliness of these hypotheses, and so it is no surprise that Lipton does not discuss their loveliness either.

7.5 Defending Inference to the Only Explanation

In this section I defend IOE's use of Holmesian Inference.[12] The objections are:

- The assumption of Falsification is false as is shown by the Duhem–Quine thesis.
- The assumption of Selection is false as is shown by the thesis of the underdetermination of theory. I discuss three versions of the thesis:
    * The *kind* or *quality* of the evidence is such that it cannot determine the correct explanatory hypothesis.
    * The *quantity* of the evidence is such that it is never enough to determine the correct explanatory hypothesis.
    * We should not ignore sceptical alternative hypotheses—the restriction to prima facie plausible hypotheses is unmotivated. These sceptical hypotheses cannot be falsified by the evidence.

7.5.1 *Defending Falsification—the Duhem–Quine thesis refuted*

We should start by reminding ourselves that scientists do think that hypotheses can be shown to be false: Harvey showed that blood could not continuously be produced afresh by the liver; Galileo's observations of the phases of Venus proved that the Sun and Venus did not both orbit the Earth; Pasteur's experiments with swan-necked flasks refuted the hypothesis that spontaneous generation is the cause of the spoiling of organic liquids such as broth or wine;[13] Rutherford's gold-foil experiments disproved the plum-pudding model of the atom; Wu's experiment ruled out the conservation of parity; and so forth. The pre-theoretic judgment of scientists is that in cases such as these it is possible to know that a hypothesis is false.

[12] David Papineau (1993) defends a version of IOE. His version does not employ Homesian inference because he denies that Selection and Determinism have to be consciously entertained as premises, and do not need to be known by the subject. It is sufficient that they are true in a particular case and are reflected in the subject's inferential dispositions. That is, the subject can argue directly from Falsification to the conclusion that the one unfalsified hypothesis is true. In my (2005*a*) I give my reasons for preferring the view in which the premises, if not explicitly stated, must be tacitly known.

[13] Pasteur is reported (Vallery-Radot 1926: 109) to have concluded a public lecture at the Sorbonne in 1864 by saying 'Never will the doctrine of spontaneous generation recover from the mortal blow of this simple experiment.'

Let us look at the challenge to such a judgment in more detail. Imagine a case where we are investigating a hypothesis *h*. From *h* plus an auxiliary hypothesis *a* we deduce a testable proposition *o*. We devise a suitable experiment and find that *o* is false. Normally we would regard *h* as refuted. But Duhem (1954) denies that we should do so:

> the only thing the experiment teaches us is that among the propositions used to predict the phenomena and to establish whether it would be produced, there is at least one error; but where this error lies is just what it does not tell us.

Duhem argues that it is always open to us to reject the auxiliary hypothesis and thereby retain the target hypothesis *h*.[14] The claim that there is no rationally required choice to be made is the Duhem–Quine thesis. Quine's (1951) contribution is to articulate this as confirmational holism: it is not single propositions that are confirmed (or falsified) but only sets of propositions; indeed the relevant set might be all of a person's beliefs.

Although the experiment does not itself tell us which proposition, *h* or *a*, to reject, that does not mean that there is no rational choice to be made. Presumably there is some uncertainty regarding *h* (which is why we are testing it). Whereas *a* may be very well supported by the evidence—sufficiently well supported that it is known to be true. In such a case the choice is between rejecting a proposition that is known, *a*, and a proposition that is not known, *h*. Rationality demands that we reject *h*, not *a*. Recall the dating of the hominin Zinj, described in Chapter 1. Louis Leakey hypothesized that Zinj was 600,000 years old. Mass spectrometry observations plus the theory of radioactive decay implied that Zinj was circa 1.75 million years old. Leakey could have held on to his theory by rejecting the theory of radioactive decay. But there is an epistemic asymmetry between Leakey's hypothesis and the auxiliary propositions. Leakey did not know that Zinj was 600,000 years old whereas the physicist behind radiometric dating, John Reynolds, did know that that ^{40}K has a half-life of 1.25×10^9 years. A rational individual should retain the proposition they do know and reject the one they don't. This is an instance of the principle of rationality: reject a proposition inconsistent with your evidence. Since *a* is known to be true, *a* is part of our evidence, as is the original, background knowledge *o*. Since *h* is inconsistent with $a \wedge o$ it is inconsistent with our evidence, and so we must reject *h*.

If one is to retain the Duhem–Quine thesis—and its challenge to Falsification and to Holmesian inference—one must reject the claim that the auxiliary hypothesis *a* can be known. Hence the thesis that we cannot know that a hypothesis has been falsified thus implies a general scepticism about the possibility of knowledge of (auxiliary) hypotheses. To the extent that the Duhem–Quine thesis is used to question the

[14] It may be that Duhem has a certain kind of theory in mind, specifically the kind exemplified by the physical theories being discussed in the late nineteenth century. Nonetheless, the thesis of holism now associated with his name is a general one. See Ivanova (2010, 2021).

possibility of knowledge from Holmesian inference, it begs the question by assuming a scepticism at least as strong as that which it seeks to establish.[15]

7.5.2 *Defending Selection—rejecting underdetermination*

> (UD) For any given fact (phenomenon) *f* and any set of evidence *e*, there is always more than one plausible competing hypothesis consistent with *e* that explains *f*.

Two clarifications regarding (UD):

- *Competing* What is intended is that the hypotheses in question are mutually inconsistent explanations of *f*. Clearly we would expect there to be multiple mutually consistent explanations of a phenomenon (e.g. by going further back into history).
- *Plausible* As before, we are only considering prima facie plausible hypotheses. I shall come back to this restriction later to discuss its consequences and justification.

It is clear that knowledge by Holmesian inference is inconsistent with (UD). What reason is there to believe (UD)?

There are two kinds of consideration that are typically cited in favour of (UD). The first states that there must be such a *quantity* of (distinct, mutually inconsistent) explanations, that however much evidence is gathered that rules out some of these, there will always remain more than one. The second consideration is that there is a *qualitative* difference between the evidence and the possible facts constituting the explanations. If the number of possible disjoint explanations were finite and not too great, (UD) might still be true if the only possible evidence were of such a kind that it could not rule out any of these histories. I shall look at the qualitative thesis before returning to the quantitative thesis. In both cases I shall argue that the underdetermination thesis is false.

7.5.3 *The qualitative thesis of underdetermination*

Evidence will qualitatively underdetermine theory if evidence propositions are all of one kind and theoretical propositions are all of another kind, such that the latter are epistemically inaccessible from the former. For example, if our evidence propositions consisted solely of propositions from pure mathematics and the theoretical propositions in question concern an issue in organic chemistry, then one would not expect to be able to get knowledge regarding the latter by any form of inference from the former. IOE and IBE are supposed to lead to conclusions about entities of a kind different from those mentioned in the evidence. For example, from the motions of observable pollen grains we are able to make inferences to the existence and nature of the hitherto unobserved atoms. On the one hand the explanationist regards this

[15] Susan Vineberg (1996) criticizes Kitcher's (1993) eliminativism on the ground that for him acceptable auxiliaries are determined by prior scientific practice. Here it is not merely prior practice that delivers the acceptable auxiliaries but rather that the relevant auxiliaries are known.

as a benefit of IOE/IBE. On the other hand it invites the sceptical empiricist argument we saw in Chapter 6, which asserts that while our evidence is observational our hypotheses are not.

(OBS) All evidence is observational;
(INF) From observational premises only observational conclusions may rationally be inferred;
therefore
(SCEP) Only observational propositions can be known.

If this argument were sound, it would prove that IOE cannot use evidence to pare down competing hypotheses to just a single hypothesis. The reply in Chapter 6 was to point out that given E=K, the premise (OBS) is equivalent to:

(OBS-K) All evidence is observational;

which just is the sceptical conclusion (SCEP). Hence this sceptical empiricist underdetermination of theory by evidence argument is question-begging.

This strategy for defending Selection generalizes. It provides a response to any argument of an analogous nature that appeals to underdetermination of theory by evidence on the grounds of a difference in kind between evidence propositions and theory propositions. A qualitative underdetermination argument may have the following form:

(EV) All evidence is of kind K;
(INF) From premises of kind K only conclusions of kind K may be rationally inferred;
therefore
(SCEP) Propositions of a kind other than K cannot be known.

Applying E=K to (EV) gives (SCEP) immediately. Any such argument will be question-begging.

7.5.4 *The quantitative thesis of underdetermination*

This consideration in support of (UD) is that the range of possible explanations is too large for any possible gathering of evidence to pare that range down to one. While theory actually is often underdetermined thanks to insufficient evidence, it needs substantial argument to show that it always must be. If, as in an empiricist qualitative thesis of underdetermination, one thinks that one's evidence is limited to the perceptual, then it is easy to imagine that very many hypotheses about the imperceptible are consistent with that evidence, however much of it we have. But we have seen that that argument is question-begging, and we have independently rejected that conception of evidence, in Chapters 5 and 6. The quantitative consideration is supported less by positive argument than by a sense, bred by familiarity with sceptical scenarios, that however much evidence one adduces in favour of an hypothesis, one could always imagine some competing hypothesis consistent with the same evidence.

Before proceeding further, it is worth remarking that there is at least one class of hypotheses for which (UD) is clearly wrong, those where *Mill's method of difference* is applicable. Let it be that we we are investigating possible causes of an outcome

E, under specific conditions, C. The positive hypothesis h_1 asserts that some factor F is the cause of E, while the negative (or *null*) hypothesis h_0 denies that F causes E. For example, h_1 might be the hypothesis that penicillin causes a reduction in streptococcal infection; h_0 says that penicillin has no effect on this bacterium. Mill's method of difference asserts that we can infer that h_1 is true if it is both the case that, under conditions C, F is followed by E and the absence of F is followed by the absence of E. The method of difference is a special case of Holmesian inference. First, it will often be clear that *Determinism* holds. While in some cases there is no explanation of a event (the spontaneous decay of a fissile nucleus), in many scientific cases we know that the phenomenon in question, E, must have some explanation. For example, the difference between the mortality rates in the two maternity divisions at the Vienna General Hospital was so great that it was possible for Semmelweis to know that this was not just a chance fluctuation. Similarly, when Howard Florey's team gave the dying Albert Alexander a dose of penicillin that rapidly and dramatically reversed his decline in a case of severe streptococcal and staphylococcal infection, they knew that this was not merely a spontaneous remission. Secondly, *Selection* applies, since one or other of h_1 and h_0 is true. And finally, *Falsification* holds, since the evidence is inconsistent with the null hypothesis, h_0. For if the factor F (e.g. penicillin) were irrelevant to the outcome E (e.g. reduced infection) then there would be no difference in the outcome measure between the cases where F is present and where it is absent. But the evidence shows that there is a difference. So h_0 is false. We can conclude that there is one class of cases where the assumption of *Selection* can be known to hold and so where Holmesian inference does apply.

So (UD) cannot be a universal truth. Nonetheless, the cases to which Mill's method of difference applies are a special class—those where we know the nature of the putative cause and our hypotheses concern whether or not it is the actual cause. So perhaps (UD) applies to those arguably rather more challenging cases where we are attempting to infer the nature of a hitherto unknown cause. It is true that in the past scientists had mistakenly thought that they had evidence that directed them to the correct theory, ruling out all the alternatives. Later those theories were found to be false and were subsequently replaced by new theories that were never even considered as potential rivals to the earlier theories. To take a simple but obvious example, Newton did not consider anything like Einstein's special or general theories of relativity when developing his own laws of motion and theory of gravitation. So might not scientists always be in Newton's position vis-à-vis relativity, focusing on one hypothesis or a small number of possible hypotheses while not even imagining a possibility that would turn out to be the truth (or at least considerably more successful)? This is what Peter Lipton (2004) called the problem of *underconsideration* and which Kyle Stanford (2006) calls the problem of *unconceived alternatives.* Stanford supports his argument that this problem is a threat to scientific realism by showing in detail how Darwin's pangenesis theory, Galton's stirp theory, and Weismann's germ-plasm theory all suffered from this problem. For example, the first of these was Darwin's attempt at providing a mechanism for inheritance. Darwin hypothesized that each part of the body generated information about its nature in the form of particles known as gemmules that concentrated in the gonads and so could be transmitted in reproduction. In developing the case for his theory, Darwin failed to conceive of what we take to be the correct theory of inheritance, classical genetics based on Mendel's and Bateson's work, supplemented by the Boveri–Sutton theory of chromosomes. (Of course, according to this view, what we take to be correct may not be, because *we* have failed to consider all the alternatives to today's orthodoxy.)

It is a weak inference to conclude from these cases that scientists *frequently* fail to consider the full range of plausible alternatives. And it is an even weaker inference to the conclusion that for any hypothesis (to a hitherto unknown explanandum) and any reasonably achievable set of evidence, there is always a plausible competing hypothesis.[16] Carl Hoefer (2020) compares the oxygen theory of combustion to the phlogiston theory. He argues that in the light of our current evidence we cannot seriously imagine the oxygen theory being rejected in favour of an alternative in a manner analogous to the way the phlogiston theory was rejected and replaced at the end of the eighteenth century. If we are to contemplate such a possibility, we must do so via the supposition of sceptical hypotheses, e.g. that we are all brains in vats being fed illusions of scientific experiments and so forth.[17]

I deny that the existence of sceptical hypotheses supports (UD) or undermines Holmesian inference in science—such hypotheses do not satisfy my requirement that the competing hypotheses are *plausible.* A scientist should not have to consider and refute Cartesian scepticism in order to establish the truth of a scientific hypothesis. I shall give a number of reasons why not:

- Any such claim is uninteresting in the philosophy of science. Sceptical hypotheses render whole areas of philosophy pointless (solipsism would make political philosophy vacuous, for example). So the questions we ask are posed under certain anti-sceptical assumptions. Even if epistemological philosophy of science makes room for serious consideration of some forms of anti-realism, it works with the assumption that our engagement in scientific activities (such as experimentation, dialogue, cooperation, etc.) is not entirely illusory. While I think that this point is correct, I think we can go further in our reasons for rejecting sceptical hypotheses.

- Sceptical hypotheses are not explanatory hypotheses. Note that Holmesian inference says that we can eliminate all hypotheses but one where each provides a potential explanation of some phenomenon f (hence 'Inference to the Only Explanation'). The premise *Determinism* states that f does have some explanation. (UD) would undermine Holmesian inference's assumption of *Selection* because it raises the possibility of unlimited hypotheses that explain f. Here are two distinct versions of a sceptical hypothesis:

16 I expand on this argument below, on p. 237.

17 The claim, which I endorse, that rejection of many scientific claims requires accepting some sceptical hypothesis is a central component of Hoefer's 'Tautological Scientific Realism' (see also Hoefer and Martí 2020).

  * Radical scepticism. This position maintains that we must consider but cannot eliminate the Cartesian hypothesis that everything we think to be part of science is an illusion brought about by an evil demon or mad scientist manipulating our envatted brains. In which case the phenomenon *f* (combustion, for example) is itself an illusion. The supposition that *f* does not exist but is illusory clearly does not to establish that there are multiple competing explanations of *f*.[18]
  * Constructive empiricism. The constructive empiricist is not a radical sceptic and so does not deny the existence of *f*. But the constructive empiricist may think that (UD) is correct because there is always a competing hypothesis that is not refuted by our evidence. If T is our preferred hypothesis, then the hypothesis T*—'T is false but the observational evidence is as if T were true'—is such an alternative. I have already argued that the claim that 'the observational evidence' exhausts the evidence is both false and question-begging.[19] My complaint here is that T* is not an explanatory hypothesis. To say that the evidence is as if the oxygen theory were true does not explain why combustion occurs. Such a hypothesis is therefore irrelevant to the current discussion of (UD).

- Confining ourselves, therefore, to cases where both the phenomenon to be explained and other evidence are as they appear to be, we can say that many (radical) sceptical hypotheses are ruled out. For if my evidence regarding combustion includes the experimental result that the mass of the residue on burning magnesium in air is greater than the mass of the original magnesium, then that evidence is inconsistent with the radical sceptical hypothesis that all apparent scientific activity is an illusion. Hence that hypothesis *is* ruled out by this evidence. In passing I note that I am not hereby depending on any disputed principle of closure, sine the incompatibility is a logical fact. I have not said that for Holmesian inference to work one has to deduce and thereby *know* that the competing hypotheses are false.
- Not all radically sceptical hypotheses can be eliminated by the move just made. For example, we might propose the following hypothesis for the outcome of Pasteur's swan-necked flask experiments: there is spontaneous generation in the flasks but the evil demon uses a special power to kill off the organisms thus generated before they can spoil the organic infusion. So Pasteur's evidence

[18] Of course, the sceptic may ask of some evidence proposition *e* how we can be sure that *e* is in our evidence—perhaps it only appears to be evidence? Can we distinguish between genuine and merely apparent evidence? While we may be able in many cases to make that distinction, the important point is that for present purposes it does not need to be answered. Being able to demonstrate that *e* is in S's evidence is not a condition of its in fact being in S's evidence—the KK principle is rejected. Our current question asks which hypotheses can be ruled out by a subject's evidence, not which hypotheses can be ruled out by what a subject knows to be her evidence.

[19] In particular, in this context, the supposition that it is always possible for T to be false while all the evidence holds is just what the advocate of Holmesian inference denies. As a reason to reject Holmesian inference, the supposition that there is an unrefuted theory of the form 'T is false but the evidence is as if T were true' is question-begging.

is just as we take it to be—there is no perceptual illusion caused by the evil demon—and that evidence is consistent with the sceptical hypothesis. But that does not mean that all my evidence is consistent with that hypothesis. It is difficult to tell without some specification of the putative 'special power'. Once specified an inconsistency might appear. For example, if it is specified that the organisms just disappear, then that would be inconsistent with the conservation of mass. Spontaneous death might be inconsistent with the conservation of energy or with biological principles that we know. These laws and principles, being known, are parts of my total evidence. Hence the alternative hypothesis is inconsistent with my evidence.

- My final reason for rejecting sceptical hypotheses as a reason to accept (UD) is perhaps the most important. Even if there are sceptical possibilities that are consistent with our evidence, knowledge by Holmesian inference may not require the exclusion of *those*. We may be entitled to ignore them because such possibilities are remote possibilities, in the sense that they are true only in remote possible worlds.[20] And it may be sufficient for eliminative knowledge of the kind being proposed here that all competing hypotheses true in close worlds are eliminated by the evidence. This might be endorsed by those who take knowledge to be tantamount to satisfying a requirement of safety (roughly: S knows that *p* iff S could not easily be mistaken in believing that *p*). Some contextualists will endorse a similar proposal, that in the scientific context it is proper to ignore (not consider) such hypotheses.

7.5.5 *Conclusion*

For there to be knowledge by Inference to the Only Explanation in a particular case, the premises *Falsification* and *Selection* of a Holmesian inference have to hold in that case. Two lines of argument familiar in the literature suggest that such premises never hold, in which case there is no knowledge by Inference to the Only Explanation. The first, the Duhem–Quine thesis, denies that we ever falsify a specific hypothesis: it is only a body of hypotheses, auxiliary hypotheses, background assumptions and the like, considered as a large conjunction, that is ever falsified. However, if the auxiliary and background propositions are known (and therefore evidence), then there is an asymmetry—it is the hypothesis under test that must rationally be rejected as it is the only proposition inconsistent with one's total evidence. The second line of argument maintains that our theories are always underdetermined by our evidence. But such arguments will either be question-begging (because they restrict the kinds of proposition they accept as evidence) or they will descend into scepticism.

[20] The condition that the possibilities in question are *remote* is important. One cannot ignore an unlikely hypothesis that is true in a close world. Police may suspect a known burglar who is suddenly rich of having pulled off a heist. But they should not ignore the possibility of a lottery win, since such a possibility (if the burglar bought a ticket) is a close possibility even though it is an improbable one. It seems that we will have to include epistemically possible but metaphysically impossible worlds in our domain of worlds. For some hypotheses need to be considered that are not metaphysically possible. For example chemists in the late eighteenth and early nineteenth centuries had to consider hypotheses about water that are metaphysically impossible, such as water being an element or being a compound with the chemical formula HO.

I conclude this section by remarking that the objections just considered seem most plausible when motivated by empiricism. The holism that Quine takes to be implied by the Duhem–Quine thesis is manifestly empiricist: 'Our statements about the external world face the tribunal of sense experience not individually but only as a corporate body' (Quine 1951: 41). If the tribunal against which hypotheses are judged consists only of sense-experience, it is clear that scientific hypotheses cannot be judged (and therefore falsified) singly. Whereas if, as we have done in the preceding chapter, we reject that empiricism and accept that the tribunal consists of all our current evidence (i.e. what we know), then there is no reason to accept that holism. The careful work of science allows us to know that certain auxiliary propositions are true, and so we can use these to test propositions about whose truth we are ignorant. Likewise the claim that our preference for a hypothesis is underdetermined by the evidence looks plausible when 'evidence' is construed as the empiricist does. Surely many different theoretical constructs can do a good job of predicting the pattern of sensory stimulation. Again, as we saw in the last chapter, that is not how science works and not how science conceives of evidence. Once we are released from the empiricist constraint on what our evidence is, it becomes much less plausible and certainly far from obvious that there are always multiple distinct theories that can consistently explain the evidence.

7.6 IBE and exemplars

So far, in this chapter I have proposed and defended the idea that we can attain knowledge through Inference to the Only Explanation. It is plausible that, on some occasions at least, the premises of a Holmesian inference are satisfied.

Often, however, scientists will find themselves unable to perform the Holmesian inference because *Falsification* is not known to be true—two or more potential explanations remain unrefuted by the evidence. Nonetheless, those scientists may well have a rational epistemic preference for one of the remaining live hypotheses over the others. In my view, IBE as described by Lipton is an account of this rational preference. In the remainder of this chapter I will examine the three challenges to IBE and show how we ought to respond to them.[21] I argue that the key to understanding how IBE can point towards the truth is to see how our standards of loveliness are responsive to the way the world is. In the next chapter I expand on this account by relating IBE to the use of probability in rational preference among hypotheses when we are not in a position to know which is true.

7.6.1 *Three objections to IBE*

Here are the three problems for IBE's use of loveliness as a guide to truth.

> **Underconisderation** The actual, true explanation may not be among the potential explanations we consider and rank. If so, the top-ranked explanation will not be true.

[21] Lipton provides his own answers to these challenges, but I do not find those answers entirely satisfactory; see Bird (2020*b*).

IBE cannot generate knowledge if we do not consider and assess all prima facie plausible explanations. Stage 1 selects for consideration only some of all the logically possible explanations, leaving many others unconsidered. What confidence can we have that the actual explanation is among the former? The Stage 2 ranking is of no use if the actual explanation hasn't made it through Stage 1 on account of the scientists' failure to think of it. Why should we think that we have considered all the relevant possibilities? As we have seen, Stanford's (2006) problem of unconceived alternatives argues that there is good reason, as demonstrated by the history of science, for thinking that we have indeed frequently failed to do so. We can articulate the problem as a problem about the truth-directedness of scientists' imaginations. The very idea of IBE concedes that the scientific imagination cannot be directed unfailingly on the truth. The scientific imagination produces multiple hypotheses for consideration, of which at most one is true. So, mostly, the scientific imagination produces false hypotheses. Given that the scientific imagination does not have a strong link with the truth, we ought not expect it to have the magical ability always or even very frequently to generate the true hypothesis alongside the false ones it generates.

> **Hungerford's objection** Loveliness is a subjective quality and so cannot be expected to correlate with truth, which is an objective property.

Lipton here borrows Margaret Hungerford's line in *Molly Bawn*, that 'beauty is in the eye of the beholder'. As Kragh (1990: 287) puts it, 'The main problem is that beauty is essentially subjective and hence cannot serve as a commonly defined tool for guiding or evaluating science.' It is unclear whether explanatory loveliness is itself an aesthetic quality. But even if it isn't, it is prima facie very plausible that it is akin to an aesthetic property at least as regards subjectivity, for five reasons. First, in making a judgment of explanatory loveliness, we seem to be reporting an *affect*—an inner response to the theory in question. The second aspect of loveliness that it shares with aesthetic properties is that we take a certain kind of *pleasure* in grasping a particularly lovely (i.e. revealing, unifying, apt, or economical) explanation. This pleasure is the same as or is very similar to the kind of pleasure one has on experiencing (with understanding) an artistically valuable sonnet, sonata, or still life. The third reason for thinking that loveliness is an aesthetic quality or is at least a subjective quality is that it is *ineffable*. It is difficult to say exactly what makes an explanation lovely. We are more able to recognize it than we are able to articulate what it is. Aesthetic qualities are ineffable in this way—many people will know that Mozart's oboe quartet is beautiful without being able to say in virtue of which properties it is beautiful. Fourthly, we find *variability* across different disciplines or across different times in the same discipline, regarding what counts as lovely. Even when we can agree that something such as simplicity is an aspect of loveliness, simplicity in physics looks to be a different thing from simplicity in physiology. Even within one field, astronomy, what counted as simplicity in one era (circular motion until the time of Kepler) was later no longer a determinant of loveliness. This variation in assessments of loveliness is what one would expect if loveliness is or is like an aesthetic quality, subject to changes in taste. And, finally, we tend to use *aesthetic terminology* to describe the good-making features of explanations and theories, whether Lipton's 'loveliness', or McAllister's 'beauty' (1999), or Glynn's 'elegance' (2010).

> **Voltaire's objection** If the loveliest explanation is to be true, then the actual world has to be lovelier than other possible worlds. But we should not expect the actual world to be the loveliest.

This objection says that the IBE enthusiast has an unjustified Panglossian faith that the actual world is the loveliest of all possible worlds. Even if loveliness is objective, there will be many worlds where it does not correlate with truth. So why think that loveliness and truth correlate in ours? Lipton regards Voltaire's objection as a version of Hume's problem of induction: there are worlds where apparent regularities are not true regularities, so why think that the apparent regularities are true regularities on our world? Voltaire's problem may be even worse. For induction to be a useful guide to the truth, it need only be that the actual world has a high enough degree of regularity. It does not have to be the most regular of worlds. Voltaire's problem is more demanding. For IBE to lead frequently to the truth, it has to be that the actual world is the loveliest of possible worlds, or at least among the most lovely worlds. And that just seems implausible.

7.6.2 *Exemplars and loveliness*

What demands do these three problems for IBE place upon a satisfactory account of loveliness, such that it can be a guide to the truth? Hungerford's objection tells us that loveliness must be objective: it must be a quality that could correlate with the truth. Voltaire's objection requires that we explain why the actual world is a lovely one, indeed a *highly* lovely world, a world where the loveliest explanations are often true. That suggests that our standards of loveliness cannot be apriori but are somehow answerable to the way the world is. Finally, answering Underconsideration requires that we show that the methods by which we select our hypotheses for consideration can direct us, in a good proportion of cases, to hypotheses that are likely to be true, thereby making it likely that the true explanation is among those we consider. Hypothesis selection cannot be a random walk in the logical space of possibilities, but must be directed towards the actual explanation.

When we say that loveliness must be objective, and so not subjective, we need to be clear what we mean by 'subjective'. In the relevant sense, a property Φ is subjective when the truthmaker for some *a*'s being Φ is a subject's opinion, attitude, or feeling (etc.) regarding *a* rather than some intrinsic property of *a* itself. When Mrs Hungerford wrote that 'beauty is in the eye of the beholder', she meant that whether Eleanor Massereene (in *Molly Bawn*) is beautiful is a matter of the attitude of the beholder, not an objective property of Eleanor. As Hume (1757) puts it, 'Beauty is no quality in things themselves: It exists merely in the mind which contemplates them; and each mind perceives a different beauty.' On the other hand, some objective, intrinsic properties of things might be detected by subjective experiences. So one way to understand secondary properties is as intrinsic dispositional properties of things to produce responses in observers. An instance of this view says that 'red' denotes an objective, intrinsic quality of objects, a disposition to cause certain ('normal') observers to have 'red' experiences.[22] So while responding to Hungerford's objection rules out loveliness being a subjective property in the former sense (like beauty, according to Hungerford and Hume), it does not need to rule out loveliness having the connection with subjectivity or response-dependence that redness has (when understood as a secondary quality).

In the remainder of this section, I give an account of loveliness that responds satisfactorily to the demands of the three objections. Drawing on a proposal by David Walker (2009, 2012), this account appeals to the idea that a scientific community's standards of loveliness (i.e. of explanatory goodness) are acquired in the process of training and learning which uses *exemplars*.[23] Those exemplars are themselves the products of science's engagement of the world, allowing us to explain how the standards of loveliness can be responsive to the world and so can correlate with truth.

Heuristics and gut feelings Goldstein and Gigerenzer (2002) discuss the recognition heuristic, a cognitive mechanism that is exemplified by people's answers to the following question: 'Which city has the larger population, Detroit or Milwaukee?' German students answered this question correctly more frequently than American students, despite knowing less about the cities than the Americans. In fact many of the German students had not heard of Milwaukee. And this is why they did better. Unconsciously they are using the recognition heuristic, which one might articulate as the unconscious rule: If you recognize the name of one city but not that of the other, then infer that the recognized city has the larger population. Because the heuristic is unconscious, its phenomenology is like that of intuition. When the heuristic is at work, the subject responds 'Detroit, surely' without knowing why she thinks that is the correct answer. Gigerenzer (2007: 16) uses the term 'gut feeling' to describe a judgment '1. that appears quickly in consciousness. 2. whose underlying reasons we are not fully aware of, and 3. is strong enough to act on'.

The relevance of gut feelings for my response to the problems of loveliness is this. Gut feelings have an element of subjectivity: they are judgments based on feelings or intuitions. They are nonetheless judgments that are aimed at determining an objective fact (e.g., which of two named cities is larger). They are able to connect the subjective feeling with the objective fact because the cognitive mechanism behind the judgment, the unconscious heuristic, is able to access what the subject knows (e.g., recognizing the name of one city, but not the other). So we know that it is possible for a cognitive mechanism to deliver judgments about objective facts with some degree of reliability, but which are also subjective in the weaker sense discussed above. A solution to the problems of loveliness should therefore not be impossible.

Cognitive psychologists from Bartlett (1932) onwards have described a number of mental mechanisms, such as schemata, scripts, frames, and analogies, that are in the same broad category as the heuristics that generate gut-feeling judgments. As with the recognition heuristic, the use of these mechanisms is partly or wholly

[22] In this territory lies the murky issue of response-dependent concepts and whether and in what sense they denote subjective or objective properties. See Rosen (1994), Wedgwood (1997), and Haukioja (2013).

[23] For a longer discussion of the following ideas, see Bird (2020*a*).

unconscious and draws on our knowledge and experience of the world to generate judgments and decisions. A further important feature of such mechanisms is that they are typically acquired through experience. While Gigerenzer is concerned to show that some heuristics are innate, being the product of evolution alone, others are developed in the course of our interactions with the world (the recognition heuristic could not be entirely innate, for example). So the judgments we form are not strictly intuitive, since they are learned responses—they are what I have called 'quasi-intuitive' (Bird 2005*b*; Bird 2007*a*: 21). They are nonetheless phenomenologically like intuition in that they lead at least partly unconsciously to a judgment of the form 'this feels right'.

Exemplars The mechanism on which I draw in answering the problems of loveliness is the *exemplar*. The term 'exemplar' is used by Kuhn (1970) to articulate one of the two senses in which he used the broader term 'paradigm'. An exemplar is an exemplary solution to a scientific problem that serves as a model for subsequent science in the relevant field. Young scientists acquire their understanding of exemplars in large part by practising solving problems with them—at first simple problems, then more complex ones. In so doing they acquire the ability to see a new problem as fundamentally similar to an exemplary puzzle and therefore requiring the same kind of solution (e.g. a vibrating string as similar to a pendulum and therefore as exhibiting simple or damped harmonic motion). This 'seeing' is essentially a matter of pattern recognition (where the pattern may be an abstract rather than a visual one). Pattern recognition, an ability typically acquired through repeated exposure and practice, is always in part and sometimes entirely an unconscious process as well as a cognitive one (think of recognizing a face, or correctly identifying the composer of an unfamiliar piece of music).

Kuhn (1977: 321–2) tells us that in acquiring understanding through training with exemplars, a scientist also acquires the shared values of her field. While there are, in the abstract, values that are shared across all of science (primarily accuracy, consistency, scope, simplicity, and fruitfulness, according to Kuhn), what features count as exemplifying these values, as well as how they are to be weighed against each other, varies from one field to another. My proposal is that this is how a scientist's sense of explanatory loveliness is acquired. (If it seems incongruous that I am using Kuhnian ideas to explain how it is possible for loveliness to be truth-conducive, then please see the accompanying footnote.)[24] Indeed, I suggest that the best way to understand loveliness is as a summary judgment of the value of a theory, to which the particular

[24] In *The Structure of Scientific Revolutions* Kuhn is very interested in the individual and social psychology of scientific change—there are more references to psychologists than to philosophers. Although he later expressed anti-realist views about truth, Kuhn does not discuss truth in the first edition of *Structure* nor does he say anything else there that directly implies anti-realism. The book was well received among practicing scientists. It is only after criticism from Lakatos and others that Kuhn came to be thought of as an anti-realist. And Lakatos himself had a quite specific, Popper-inspired, conception of what scientific rationality amounts to. In particular, Kuhn's interest in the psychology of scientific thought was especially excoriated by Lakatos. Times have since changed (even if the suspicion that *Structure* is anti-realist has not). What Lakatos (1970: 178) ridiculed as Kuhn's appeal to 'mob psychology' would now be regarded as naturalistic social epistemology (Bird 2005*b*).

values mentioned contribute. Thinking about loveliness in this way allows us to reply to the three objections.

Exemplars and Hungerford's objection Above I argued that Hungerford's objection, that loveliness is too subjective to correlate with truth, appears plausible because (i) we judge it on the basis of affect, an inner response to a theory; (ii) we take pleasure when we encounter it; (iii) it is ineffable; (iv) it shows variability across different fields and times; and (v) we use aesthetic terminology to describe it. While these are strongly suggestive of a subjectivity shared with aesthetic qualities, they are consistent also with the weaker kind of subjectivity that can correlate with objective properties. As we saw, the gut feeling of a judgment formed by an unconscious heuristic has the latter kind of weak subjectivity. Judging the loveliness of a potential explanation, informed by training with exemplars, is like a gut feeling in this respect—which explains feature (i). The feeling of rightness is also a positive one. The exemplar is held up as a paradigmatic example of how science ought to be; recognition that a hypothesis is like the exemplar provides intellectual pleasure—(ii). Because such judgments are based on a hypothesis 'feeling right', only partially informed by conscious reflection, it is difficult for a subject to articulate the basis of this judgment and to say what loveliness is—(iii). Because in different fields and at different periods, there are different paradigms (exemplars) in operation, what counts as lovely will vary over time and from one scientific field to another—(iv). These aspects, (i)–(iv) are also found in aesthetic qualities, and aesthetic qualities are the most prominent instances of properties displaying (i)–(iv), so it is not surprising that we use aesthetic terminology to describe them—(v).

Exemplars and Voltaire's objection Voltaire's objection was, put simply, that if our standards of explanatory loveliness are apriori, then it would be an implausible coincidence that our world, given all the ways a world could possibly be, meets those standards so closely that very frequently lovely explanations are true. The lesson to draw from that objection is that standards of loveliness are not apriori. And if standards of loveliness are gained from training with exemplars, then indeed they are not apriori. Of course, that our standards are gained from exemplars does not guarantee that they do correlate with the truth. For if our exemplars are themselves badly mistaken, the standards of loveliness gained from them will not be truth-conducive—much medieval science was in this position. However, if the exemplars of a field are true or highly truthlike, then the standards of loveliness they generate will be truth-conducive. And since it would not be an implausible coincidence that such exemplars are true, it does not require an implausible coincidence for true theories to have a high degree of loveliness.

Exemplars and Underconsideration Underconsideration raises the challenge of explaining why it is that our methods of theory selection at Stage 1 make it likely that the true theory is among those selected for consideration. Kuhn emphasizes the role that exemplars play in discovery, the process of generating hypotheses. Training with exemplars enables scientists to see the world in a particular way, which is to say that some features of the world become scientifically salient and others less so. It also means that certain explanatory schemata suggest themselves. In the Aristotelian paradigm motion is salient, and an Aristotelian scientist will naturally seek an explanation of motion in terms natural tendencies in the object itself. Whereas a Newtonian scientist will not find rectilinear motion itself salient, but will regard changes in motion or non-rectilinear motion as salient, and so she will look for the explanations of these in forces governed by general laws. So the explanations we reach for are not randomly chosen from among all possible explanations, but will be ones that are similar or analogous to explanations with which we are already familiar. When our exemplars are erroneous, as in the Aristotelian case, then we probably will not consider the correct explanation. But when our exemplars are correct, then they make it likely that the true explanation will be among those we in fact consider.

Exemplars exemplified—the discovery of Coulomb's law The discovery of Coulomb's inverse square law of electrostatic force provides an illustration of the effect of exemplars on scientific cognition and on judgments of loveliness in particular. In the context of Newtonian mechanics and the success of Newton's law of gravitation it was natural for a physicist confronted with an unstudied source of force to want to find its law also. So the exemplar of gravitation generates a research question to be investigated by those interested in electrical phenomena and the electric force. Furthermore, that exemplar directs us towards a solution: by analogy with Newton's law, an inverse square law naturally suggests itself; indeed Coulomb was not the first to propose an inverse square law. Franz Aepinus had proposed it without any experimental evidence and Charles Stanhope attempted a demonstration that was widely accepted despite being flawed. Gravity also supplied the means of testing this hypothesis. Coulomb employed a torsion balance in his experiment, much the same, though differing considerably in scale, as the torsion balance devised by John Michell and which Henry Cavendish used 'to weigh the Earth' (that is, to find a value for the gravitational constant *G*). Coulomb's experiment was not universally considered to have proven his case: other scientists had difficulty in replicating his results and some proposed different laws on the basis of results using the same apparatus; modern scholars differ on whether Coulomb really could have achieved the results he claimed he did (Heering 1992; Heering 1994; Falconer 2004; Martínez 2006). Nonetheless, Coulomb's proposed law was accepted by the majority, despite its empirical shortcomings. It is apparent that as far as Coulomb and many of his contemporaries were concerned, the very fact of the analogy with the law of gravitation played a powerful role in influencing their judgment in accepting the law. They took the inverse square law to be a lovely explanation. They regarded it as apt, displaying a pleasing symmetry with the law of gravitation—both are inverse square laws, their form differing only in the sign attached to the law ('–' for electrostatics, since like charges repel). This pleasing sense of aptness that disposed those scientists towards Coulomb's law is an inner positive affect of the kind I discussed above in connection with Hungerford's objection. Although in this case the analogy between this law and Newton's was consciously perceived, that analogy played the role of making Coulomb's explanation a lovely explanation—not the role of providing additional evidence for it. It is difficult to see how the fact that Newton's law is an inverse square law could be evidence for Coulomb's law, particularly in the context of IBE, since the form of Newton's law is not explained by Coulomb's law having that form (nor vice versa). Nor would it be true to the psychology of the scientists; the evidence they appealed to was what was known of electrical phenomena, in particular the various experimental results. Rather, the shared form of Coulomb's proposed law and Newton's well-attested older law made the former a lovelier explanation of the evidence.

7.7 Conclusion

Explanationism is the view that we can come to know a hypothesis to be true or to have a justified preference for one hypothesis by considering the ability of hypotheses to explain the relevant evidence. The standard explanationist view, Inference to the Best Explanation (IBE), holds that explanatory considerations allow us to rank hypotheses according to how well they explain the evidence and if one of them is a particularly good explanation and clearly better than the others, then we may be entitled to infer that it is true, and thereby come to know that it is true.

In articulating how we should respond to certain challenges to IBE, I argued that our standards of explanatory goodness ('loveliness') are fixed by our exposure to exemplars. This can be truth-conducive when our exemplars are true. Does that show how IBE can lead to knowledge? I do not think so, for two reasons. First, if evidence gives a hypothesis low plausibility, but does not rule it out, then the rational attitude should be to regard it as having low probability, not to reject it as false; whereas the standard IBE view implies that we must sometimes regard some such hypotheses as being false. In the next chapter I look in more detail at the relationship between plausibility and probability. Secondly, my exemplar-based account of loveliness suggests that while ranking according to loveliness can be a good guide to the truth, it is unlikely to be sufficiently reliable for knowledge.

Hence my preferred explanationist account says that to know that h is the true explanation of some relevant part of the evidence e_0, the total evidence e_t must rule out the competitors to h. That is, h is the only explanatory hypothesis consistent with the evidence.

8

PROBABILITY AND PLAUSIBILITY

Often in science our evidence falls short of what is required in order to accept a hypothesis outright. Instead we accord it some degree of probability. This chapter describes a theory of what such probabilities are. Standard, personalist Bayesianism is the best known theory of probabilistic reasoning in science. Some Bayesians reject explanationism, defended in the preceding chapter, on the ground that it inconsistent with Bayesianism. I start by developing my version of heuristic compatibilism. This line of defence maintains that explanationism is compatible with Bayesianism since it provides a heuristic for determining elements in Bayes's theorem. I outline three challenges for compatibilism. We see that answering these challenges motivates rejecting the personalism of standard Bayesianism. Instead I develop a 'super-objective' Bayesianism that takes probability evaluations in science to be assessments of plausibility in the light of the evidence. Super-objectivism allows explanationist considerations to inform our evaluations of the prior $P(e)$ and the ratio $P(e|h)/P(e)$, while also supporting my claim that explanationism needs Inference to the Only Explanation if it is to account for knowledge.

8.1 Introduction

In the last chapter we looked at knowledge of theories gained by thinking about their explanatory power. An eliminative version of abduction—Inference to the Only Explanation (which uses 'Holmesian inference')—can give us knowledge. Whereas, if the evidence does not eliminate all but a single relevant explanatory hypothesis, we cannot gain knowledge. Nevertheless, in the latter case, we can still hope for a rational preference among hypotheses, in accordance with Inference to the Best Explanation (IBE).

When we think about propositions that are less than certain, we often use probability in reasoning about them. So the question naturally arises, how does our assessment of a hypothesis in accordance with IBE relate to this use of probability?

This question is urgent. For I have suggested that IBE is not simply a correct description of an inference pattern that is frequently found in science. It is also normatively correct. That is, it is (or can be) a rational form of inference in cases where the evidence supports the hypothesis to some degree, but not to such a degree that we would be willing to assert outright that the hypothesis is true. At the same time Bayesians claim that Bayesianism is also a theory about how such evaluations should be made. Notable critics of IBE, such as van Fraassen (1989: 160–70) and Salmon

Knowing Science. Alexander Bird, Oxford University Press. © Alexander Bird 2022.
DOI: 10.1093/oso/9780199606658.003.0008

(2001), think that Bayesianism and IBE are not compatible. And because of this incompatibility, IBE cannot be rational, since Bayesianism demonstrably is. For example, van Fraassen argues that if explanationist considerations were relevant, then they would encourage a thinker to boost the credence in a hypothesis above that mandated by Bayes's theorem, when it is a better explanation of the evidence than competitor hypotheses while reducing the corresponding credences for those competitors. Such a thinker would be open to a Dutch book (a set of bets he takes to be fair but which will lose him money whatever the outcome). An account of reasoning cannot be normatively correct if it enjoins accepting a Dutch book. So IBE is not a normatively correct account of reasoning.[1]

The aim of this chapter is to show that Bayesianism and IBE are in fact compatible. I'll develop the idea that IBE provides a heuristic for implementing Bayesianism. My approach, however, requires a radical departure from standard Bayesianism.

8.2 Standard Bayesianism

Standard Bayesianism I take to involve the following commitments regarding reasoning in which some proposition, such as a scientific hypothesis, is evaluated in the light of evidence:

> *Personalism* A thinker's evaluations in such cases are *credences*, doxastic states of the subject that may be operationalized as dispositions to accept a set of bets on the truth of propositions.
>
> *Probabilism* The evaluations, if rational, can be represented as *probabilities*; they should obey the axioms of probability.
>
> *Conditionalization* The evaluations are subject to *updating* in the light of new evidence e, so that the new evaluation of a hypothesis $P_n(h)$, should be equal to the old evaluation of the hypothesis conditional on that evidence $P_o(h|e)$.

The standard Bayesian holds that credences are an analogue of belief *simpliciter*: traditional epistemology concerns what it is rational to believe outright while Bayesian epistemology concerns the rationality of partial belief. Various arguments (Dutch book arguments, accuracy-dominance arguments) are used to show that a personalist should be a probabilist, i.e. that rational credences are probabilities.[2] If probabilism is true, then the evaluation function, P, is also a probability distribution. And a consequence of the probability axioms, Bayes's theorem, holds of the evaluations of any propositions, e, h, and their conditional relations:

[1] This line of reasoning can be resisted. See Douven (2017*a*) for an articulation and defence of the boost or 'bonus points' view.

[2] See Joyce (1998) and Leitgeb and Pettigrew (2010*a*, 2010*b*). The argument that Bayesian conditionalization, in addition to Bayesian coherence (the probabilist requirement that a subject's credences satisfy the probability axioms), is required by rationality is thus a *diachronic* Dutch book argument. Whether a diachronic Dutch book implies irrationality (and epistemic irrationality in particular) is far from clear (Briggs 2009; Mahtani 2012).

(B) $$P(h|e) = \frac{P(e|h)P(h)}{P(e)}.$$

Conditionalization says that we should update those evaluations (probability distributions) in the light of new evidence; so if *e* is in fact new evidence, then:

(S) $$P_n(h) = P_o(h|e) = \frac{P_o(e|h)P_o(h)}{P_o(e)},$$

where P_o is the probability distribution before the evidence was acquired and P_n is the new probability distribution in the light of that evidence. In the context of conditionalization, $P_o(h)$ is the prior probability of the hypothesis, $P_o(e)$ is the prior probability of the evidence, also known as the expectedness of the evidence, and $P_o(e|h)$ is the likelihood; $P_n(h)$ is the posterior probability, the probability the hypothesis now has, the evidence *e* having been obtained. (S) is known as *strict* conditionalization. As we shall see, strict conditionalization is not the only updating rule that the Bayesian may employ.

Standard Bayesianism is normative, a view that asserts requirements about how reasoning ought to proceed, in addition to a descriptive kernel (personalism). A standard Bayesian is also a subjectivist if she holds that probabilism and conditionalization are the only rational constraints on credences. An objective Bayesian extends standard Bayesianism with further constraints, such as a principle of indifference.[3] Some Bayesians are non-standard in that they do not accept all the features of standard Bayesianism. For example, one might accept that there are counterexamples to Conditionalization.

8.3 Heuristic compatibilism

In my view there is in fact no incompatibility between Bayesianism and IBE. IBE is a *heuristic* we employ in lieu of directly applying (S). This heuristic view of compatibilism has been promoted in different ways by Peter Lipton (2004: 103–20), Ted Poston (2014: 149–81), and Jonah Schupbach (2016).[4] In this section I spell out how I think this idea should be understood.

A kind of problem may have an optimal method of solution, but that method may not be easy to implement properly. Implementing it properly might be costly in terms of effort. Or that attempt may be prone to error, and so lead to a mistaken answer. A heuristic for solving a problem is a method that in principle may be suboptimal but is more easily implemented. It may well be rational to employ the heuristic if the saving of effort outweighs the loss in reliability. Or the heuristic might in fact be *more* reliable than the optimal method as poorly implemented by us inadequate thinkers. For example, as we saw in the last chapter, the recognition heuristic is employed when subjects are asked to say which of two named cities has a greater population: when only one city is recognized, they judge that that city has the larger population.

[3] Other extensions to standard Bayesianism include commitments to Regularity (Kemeny 1955; Stalnaker 1970), Countable Additivity (Kolmogorov 1950; de Finetti 1974), the Principal Principle (Lewis 1980), and Maximum Entropy (Jaynes 1957).

[4] For other compatibilist views, see Weisberg (2009) and Henderson (2014).

This is easy to implement in the relevant cases; and it turns out also to be more reliable than the process of making inferences from relevant knowledge (subjects performed *less* well when they knew something about both cities)—this is the *less is more* effect. Lipton notes the considerable evidence that we find thinking with probabilities difficult and are prone to error. So it would be rational for us to employ a heuristic in place of Bayesian reasoning. I propose that the heuristic is this:

> (H) If hypothesis A is a lovelier potential explanation of the evidence than B, then judge A to be more probably true than B.

If (H) is a rational heuristic for Bayesianism then it should yield much the same results as (S) would. The heuristic compatibilist should aim to show that if (H) says that we should judge that A is more probable than B then (S) says that P(A) is greater than P(B). We could show that it does by arguing that elements of the explanationist heuristic correspond to components of the Bayesian updating in (S). According to Lipton (2004: 114), explanationist considerations do not neatly correspond either to the prior of the hypothesis $P(h)$ or to the likelihood $P(e|h)$, but they do inform both (as well as help determine the relevant evidence). My approach is different. First, consider the question: how good as an explanation is hypothesis *h*? Now some aspects of the answer will concern *h* alone, whereas others will concern *h*'s explanatory relationship to the evidence *e*. For example the simplicity of *h* will concern *h* alone. Newton's law of gravitation plus the laws of motion have a high degree of simplicity and strength, and that fact (other things being equal) makes it a lovelier theory than one with more laws or more complex ones. On the other hand, whether the explanation of *e* by *h* is lovely is a matter of the relationship between the two. For example, the hypothesis that an asteroid impact caused the K-T (K-Pg) extinction is a lovely explanation of the fact that the geological K-T layer coincides with an iridium anomaly, because the fact that asteroids have high iridium concentrations directly explains that coincidence. Following Einstein, I call these the *internal* and *external* explanatory virtues of a hypothesis.[5]

Armed with this distinction between the internal virtues of *h per se* and the external virtues of *h*'s explanation of *e*, we can articulate a set of correspondences between IBE and Bayesianism and argue that they are plausible.

First, the internal explanatory virtues of *h* correspond to $P(h)$, insofar as *h* is a novel hypothesis. Subjective Bayesianism specifies no constraints on the priors, and so there is no danger of an explanationist account of the internal explanatory virtues of *h* conflicting with the requirements of the subjective Bayesian. As far as the latter

[5] In his 'Autobiographical Notes' Einstein (1991, quoted by Salmon 1990: 181) contrasted:

> The first point of view is obvious: the theory must not contradict empirical facts ... this first point of view is concerned with the confirmation of the theoretical foundation by the available empirical facts. The second point of view is not concerned with the relation of the material of observation but with the premises of the theory itself, with what may briefly but vaguely be characterized as the 'naturalness' or 'logical simplicity' of the premises .... The second point of view may briefly be characterized as concerning itself with the 'inner perfection' of the theory, whereas the first point of view refers to the 'external confirmation'.

is concerned the priors can be anything—they could come from an ouija board, so they might as well come from explanationist considerations. However, the heuristic compatibilist aims at showing that the heuristic (H) (mostly) yields the conclusion that an ideal Bayesian would reach. And that would not be the case if the Bayesian were to choose unusual priors (which the subjective Bayesian will feel entitled to do). Nonetheless, as a matter of descriptive fact, our actual assessment of the probability of priors does correlate with our assessment of their explanatory value. In advance of the evidence we do think that simpler, more unifying explanations are more likely to be true—in the preceding chapter we saw that the form of Coulomb's proposed law of electrostatics gave it a high prior among his contemporaries independently of its ability to explain the data from his torsion balance experiments.

Indeed, not only do we in fact use explanationist considerations in fixing priors, Michael Huemer (2009) and Ted Poston (2014) argue that we need to if Bayesianism is to provide a good account of the rationality of our inductive practices. Consider the principle of indifference—one should accord equal probability to propositions regarding which one's evidence is symmetric (i.e. one's evidence bears the same relationship to each outcome; see Jon Williamson's (2010: 2) 'equivocation norm'). For example, nothing that one knows about a coin discriminates between heads and tails: it might be fair; or it might be biased, but if it is biased one knows nothing that says that it would be biased towards heads rather than tails or vice versa. If so, one should accord the hypothesis that the next toss will yield tails the same probability as the hypothesis that it will land heads. Now think about a sequence of ten tosses that one is about to make. Here are two approaches to thinking about the outcome by applying the principle of indifference. Approach 1: consider hypotheses each of which specifies how the sequence of tosses will turn out—there are 1,024 such hypotheses. Apply the principle of indifference to these, giving each a probability of 1/1,024. One can then calculate the probability of various proportions of heads in the ten tosses. Approach 2: consider first the hypotheses that assert the proportion of heads in the tosses—there are eleven such hypotheses (since there are ten tosses, the proportions are 0/10, 1/10, ... 10/10). Apply the principle of indifference to these eleven hypotheses of the second type, giving each a probability of 1/11. One can then use these probabilities to calculate probabilities for particular sequences. The two approaches lead to different probabilities for particular sequences: the first option says that a sequence of ten heads has a probability of 0.001 whereas the second option says that that sequence has a probability of 0.09. Likewise the two approaches lead to different probabilities for the proportions of heads/tails. More importantly, in cases like this one, Huemer shows that the choice leads to a difference between scepticism and inductivism: the first choice is sceptical, in that it does not allow the probability assigned to heads on the tenth toss to change, even if one has learned that the first nine tosses have all yielded heads, whereas the second choice does assign a high probability to heads in the light of such evidence. The second set of hypotheses are explanatory hypotheses whereas the first set are not. And that, argue Huemer (2009) and Poston (2014), shows that Bayesianism *needs* explanationism if Bayesianism is to be a satisfactory account of inductive reasoning. Descriptively, we do think that we should consider the hypotheses concerning proportions/probabilities first and allow the probabilities concerning particular sequences to follow as a consequence rather than vice versa. That strongly suggests that we are indeed moved by explanatory considerations in the assignment of priors to hypotheses. Put another way, we are as a matter of fact inductive thinkers, and since being a Bayesian inductivist requires (or at least motivates) explanationist thinking about priors, then we should expect that we are explanationists about priors.[6]

Secondly, the external explanatory virtues of h relative to e correspond to the *ratio* of the likelihood to the expectedness of the evidence, $P(e|h)/P(e)$.[7] External explanationist considerations do not correspond to the likelihood $P(e|h)$ alone. As Lipton appreciates, $P(e|h)$ can be very high even if h does *not* explain e well—if e is itself very probable: famously John Jones's taking a contraceptive pill does not provide a good explanation of his not getting pregnant, since it was certain anyway that he would not get pregnant (Salmon 1971); still $P(e|h)$ is very high. And, in reverse, $P(e|h)$ can be low even if h *does* explain e well. For example, Holmes suspects Professor Moriarty of having murdered Major Arbuthnot and of having disposed of his body, meticulously erasing evidence of the crime. Holmes then discovers a minute trace of the highly unusual Uzbek tobacco smoked by Arbuthnot in the cuff of Moriarty's coat. Now given the meticulousness that the hypothesis ascribes to Moriarty, finding that tobacco trace is itself highly unexpected, so $P(e|h)$ is low. But Holmes's hypothesis is still an excellent explanation. The latter fact is still consistent with the demands of Bayes's theorem, since $P(e)$ is even lower. I propose therefore that the external explanatory virtue of h in connection to e corresponds to the *ratio* $P(e|h)/P(e)$, i.e. the measure of how much more probable e becomes when conditioned on h. Holmes's hypothesis about Moriarty being the murderer is a very good explanation, and it corresponds to the fact that even if both $P(e)$ and $P(e|h)$ are low, $P(e|h) \gg P(e)$ and so $P(e|h)/P(e) \gg 1$. Whereas, in the case of John Jones's taking a contraceptive pill, the weakness of this as an explanation of his not being pregnant corresponds to the fact that while $P(e|h)$ is high, $P(e)$ is equally high, and so $P(e|h)/P(e) \approx 1$.

Bayes's theorem states that the ratio $P(e|h)/P(e)$ is equal to $P(h|e)/P(h)$, the ratio of the probability of the hypothesis conditional on the evidence to the prior probability of the hypothesis. So, given conditionalization, $P(e|h)/P(e)$ is equal to the proportion by which the probability of the hypothesis changes on learning the evidence. $P(h|e)/P(h)$, or some monotonically strictly increasing function thereof (such as its logarithm), is taken by some Bayesians to be a measure of the confirmation of h by e (Keynes 1921; Horwich 1982; Milne 1996). And so the proposal that (external) explanationist considerations correspond to $P(e|h)/P(e)$ and so correlate with this

[6] The assertion that Bayesian inductivism *requires* explanationism may be too strong in that there can be probability distributions that do allow for inductivism but which do not correlate with explanatory virtues. Nonetheless, such distributions would be unmotivated, and are not ones that we would be inclined to use.

[7] Leah Henderson (2014) complains that a typical compatibilist focus on explanatory considerations constraining the prior (e.g. Lipton) fails to consider how explanatory considerations influence the integration of new evidence. My proposal explicitly gives a central role to the latter.

measure of confirmation means that an explanationist account of confirmation corresponds to the Bayesian one.

The heuristic view holds that Bayesian reasoning and explanationist reasoning are distinct, although leading to the same or similar judgments reasonably frequently. We should therefore not expect every component in the Bayesian calculation to have an explanationist counterpart. So a subject forming a judgment on the basis of explanationist considerations forms a judgment that corresponds to a Bayesian value for $P(e|h)/P(e)$ without that being the result of preceding judgments corresponding to $P(e|h)$ and $P(e)$. That is an advantage of the view I propose, since it means that explanatory virtue can support a hypothesis even if we have no clear idea of what $P(e|h)$ and $P(e)$ should reasonably be: Holmes might have only an imprecise idea regarding the probability of the presence of Uzbek tobacco ash in Moriarty's cuff, both the unconditional probability and the probability conditional on Moriarty having carried out a carefully planned murder. Even so, he might be able to be rather more precise about the degree to which the Moriarty hypothesis *increases* the probability of the presence of the ash. So, in this case Holmes is able to estimate the ratio P(there is Uzbek tobacco ash in Moriarty's cuff, given that Moriarty is Arbuthnot's murderer)/P(there is Uzbek tobacco ash in Moriarty's cuff), even though he is not able to estimate the numerator or denominator individually. And his estimate of this ratio corresponds to and indeed derives from the fact that the Moriarty hypothesis is an excellent explanation of the tobacco fact—which is as my heuristic view proposes: if a hypothesis explains some evidence, then we think that the evidence proposition is more likely if the hypothesis is true than if we don't make that assumption.

The version of heuristic compatibilism I am developing here proposes that the internal loveliness (explanatory virtue) of an explanatory hypothesis corresponds to the prior $P(h)$ while the external loveliness of its relationship to the evidence corresponds to the ratio $P(e|h)/P(e)$. So, to complete this confirmation of the view, we need to show that our overall assessment of the probability of a hypothesis correlates with something that we can regard as the (arithmetical) 'product' of our judgments regarding the internal loveliness of a hypothesis and its external loveliness. Since our judgments of loveliness are not numerical, it is difficult to show this explicitly. On the other hand, the way we combine those judgments does seem to mirror multiplication rather than addition or averaging. For example, if a hypothesis is known to be false then even unlikely evidence it explains well will not make us think that the hypothesis is even moderately likely—that corresponds to the product of zero and a finite amount, whereas averaging would suggest that a known false hypothesis could become likely on favourable evidence.

To conclude: the challenge for compatibilism between explanationism and Bayesianism was that if the two give different results regarding the credence we should attach to a proposition or regarding the ordering of our preferences among propositions, then we would have a reason for thinking that explanationism is incoherent. If, on the other hand, the two always agree, then explanationism is otiose: Bayesianism gives a precise and well-motivated account of confirmation, whereas explanationism is imprecise and philosophically controversial. The heuristic approach advocated by

Lipton and Poston resolves this challenge, by showing that explanationism is a useful substitute for Bayesian reasoning. For various reasons it is unnatural and difficult for human thinkers to implement the probabilistic reasoning required by standard Bayesianism. So we use explanationist considerations as a heuristic—a manner of reasoning that will by and large get the same result as Bayesian reasoning but without its costs and pitfalls. I have added some detail to this picture, showing how aspects of explanationist thinking correspond to elements of Bayes's theorem. The internal explanatory virtues of a hypothesis correspond to $P(h)$; the external explanatory virtues correspond to $P(e|h)/P(e)$; and we combine these virtues in a manner that corresponds to their product. Explanationism does not give precise values for probabilities, but it does allow us to order our preferences among hypothesis. According to the heuristic approach, the ordering that the heuristic (H) recommends is for the most part the same ordering that explicit and accurate use of Bayes's theorem and conditionalization would have given us had we been in a position to use them.

8.4 Three challenges for compatibilism

In this section I raise three challenges to the compatibilism just articulated. These start from the assumption that explanationist reasoning can lead to knowledge. As we saw in the previous chapter, I think that IBE does *not* lead to knowledge when the evidence allows us only to rank hypotheses according to their loveliness and not to eliminate them. But in the limiting case of IBE—IOE—we can achieve knowledge, by an eliminative abduction using Holmesian inference.

One might reject the assumption that explanationist reasoning in some cases can lead to knowledge, but if explanationism is a correct descriptive account of some important inferential practices in science, then that rejection amounts to scepticism about those parts of science that employ those practices.

8.4.1 *Challenge 1: Knowledge and objectivity*

What does this compatibilist argument really show? It shows that if Bayesianism is a correct normative account of scientific reasoning, then explanationism need not be regarded as in conflict with Bayesianism. On the contrary, explanationism describes the heuristic mechanism whereby scientists typically implement Bayesianism. While this ought to satisfy the subjective Bayesian, it ought not satisfy the explanationist. For the latter thinks that there is more to epistemology than is found in subjective Bayesianism. The explanationist thinks that explanatory considerations are a guide to the truth: on some occasions explanationist reasoning can give us knowledge. So the explanationist cannot regard standard Bayesianism as a complete normative epistemology. For her, priors and likelihoods informed by explanationism are better than those not so informed. Explanationist Bayesianism must reject subjective Bayesianism, and a strong form of objective Bayesianism looks more suitable, with constraints on which priors are rationally acceptable.[8] Some priors—those guided by explanationist considerations—are better or more acceptable in that they will

[8] See Psillos (2004) and Huemer (2009) on explanationism as a rational constraint on priors.

lead to a posterior probability function that is more *accurate*—which is to say that hypotheses with high probabilities will more often be true than those with low probabilities.[9] So while the minimal compatibilism of Lipton and others can defuse the criticism from subjective Bayesians, the explanationist should hold that there is more to be said about the epistemological function of explanationist factors—specifically regarding their connection to truth and knowledge.

8.4.2 *Challenge 2: Knowledge, evidence, and credence*

Standard presentations of Bayesianism treat evidence, knowledge, and probability of 1 as equivalent.[10] If a subject has already acquired evidence *e*, then $P(e)$ is trivially equally to 1. If *b* is a proposition known by the subject, then $P(b)$ also 1, as reflected in the form of Bayes's theorem that takes background knowledge into consideration:

$$\text{(B}'\text{)} \qquad P(h|e\&b) = \frac{P(e|h\&b)P(h\&b)}{P(e\&b)}.$$

So Bayesianism treats evidence and background knowledge equivalently. (I take the significance of 'background' to be simply knowledge other than knowledge of the proposition *e*, which is knowledge in the foreground because when conditionalizing on new evidence *e* it is the probability of *e* that changes to 1 whereas the probability of *b* is already at 1.) In a sense, therefore, Bayesianism endorses the equation of evidence and knowledge, E=K (Williamson 1997 and Chapter 5 above). (Alternatively, if one accepts E=K for Williamson's or my independent reasons, then that implies that the propositions that make up *b* in (B′) really are the propositions *known* to the subject rather than, e.g., those believed by the subject.) Furthermore, lottery cases preclude regarding knowledge or evidence as having a credence of less than one (however well justified). For one cannot know, in advance of a fair draw, of a particular lottery ticket that it will lose. Nor is the proposition that the ticket will lose evidence for anything (Williamson 1997).

Yet the view that $P(e) = 1$ and $P(b) = 1$ has highly unpalatable consequences for personalism (the view that the probabilities in question are credences, canonically measured by betting quotients). For a credence of 1 in *p* means that one would be willing to take a bet that would yield a one penny gain if *p* is true and unending torture if *p* is false. But few if any propositions are such that reasonable people would take such a bet. Certainly the contingent propositions that we take to be knowledge or to be our evidence are not typically of that sort. That looks to lead to the sceptical conclusion that we do not know those propositions after all and that no such propositions can ever really be our evidence. Furthermore, that conclusion, apart

[9] Lipton himself needs to say something like this. Otherwise many of the prima facie problems that he raises for IBE (and addresses) would not be concerns at all. For example, Voltaire's problem asks why it is that we should think that the explanatorily loveliest explanation is the actual (true) explanation. Should we think that the actual world is the loveliest of all possible worlds? But that's irrelevant if *all* the epistemic work is done by Bayes's theorem while explanationist concerns are merely the source of the values of the priors and likelihood, without making any normative contribution. However, concern with Voltaire's problem is required if we want the resulting epistemic attitude to correlate with truth and for IBE possibly to lead to knowledge.

[10] See, for example, Glymour (1980).

from being sceptical, also means that Bayesian updating—strict conditionalization (S)—will never occur (because (S) requires $P(e) = 1$), thereby rendering it pointless.

8.4.2.1 *Jeffrey conditionalization* This concern might lead one away from standard ('strict') Bayesian conditionalization to Jeffrey conditionalization (Jeffrey 1983) (so long as one retains personalism—but see below). Jeffrey conditionalization does not conditionalize just on propositions that have probability equal to 1 (as in strict conditionalization), but also on propositions stating 'uncertain evidence'. So Jeffrey conditionalization may allow a way out of our problem. For it appears to allow the Bayesian machinery to continue working but without full-on evidence propositions (those with credence of 1). In this sub-section, I explain why this is unsatisfactory. There is nothing wrong with Jeffrey conditionalization *per se*. Indeed we need a rule much like Jeffrey conditionalization whenever we engage in Bayesian updating, even with strict conditionalization (S). That is because once we have updated some credences, C, with (S), there will be further propositions probabilistically dependent on C, and these will need to be updated too. And the rule for updating these will be much like Jeffrey conditionalization (a generalization of it, in fact). What is wrong with the Jeffrey picture is not Jeffrey conditionalization. Rather it is that it gives no satisfactory account of what processes rationally *initiate* updating.

If a proposition is uncertain, then so is its negation. Consequently, Jeffrey conditionalization conditionalizes both on the uncertain evidence proposition and on its negation, weighted according to one's subjective confidence. So:

(J) $$P_n(h) = \frac{P_o(e|h)P_o(h)}{P_o(e)}P_n(e) + \frac{P_o(\neg e|h)P_o(h)}{P_o(\neg e)}P_n(\neg e)$$

where $P_n(e)$ and $P_n(\neg e)$ are the weightings, *viz.* the new credences in e and its negation, resulting from some observation for example.

Strict conditionalization, (S), is the special case of Jeffrey conditionalization where $P_n(e) = 1$ (and so $P_n(\neg e) = 0$). So both strict conditionalization and Jeffrey conditionalization involve a process where a proposition, e, changes its credence in a way that is not itself the result of conditionalization, from $P_o(e)$ to $P_n(e)$, where $P_n(e) = 1$ for the strict conditionalization case or where $P_n(e) < 1$ for Jeffrey conditionalization.

Note that both Jeffrey and strict conditionalization can be seen as special cases of a more general Bayesian updating rule that is a consequence of finite additivity, the third of the Kolmogorov probability axioms:

(F) $$P(a) = P(a \wedge b_1) + \ldots + P(a \wedge b_i)$$

where $\{b_1 \ldots b_i\}$ is a partition, i.e. a set of mutually exclusive and jointly exhaustive propositions. Given the definition of conditional probability, (F) is equivalent to:

(F′) $$P(a) = P(a|b_1)P(b_1) + \ldots + P(a|b_i)P(b_i).$$

Let us say some evidence comes in that (rationally) causes the subject to change some of the $P(b_1) \ldots P(b_i)$, so they are now $P_n(b_1) \ldots P_n(b_i)$. Then the Bayesian will expect the subject to change $P(a)$ accordingly to $P_n(a)$:

(F″) $$P_n(a) = P_o(a|b_1)P_n(b_1) + \ldots + P_o(a|b_i)P_n(b_i)$$

where P_o is the old probability distribution P in (F′).

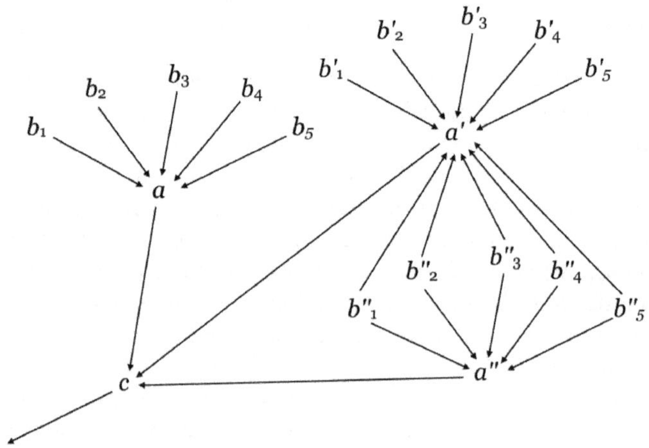

Fig. 8.1: A directed acyclic graph (DAG) showing that the credences given propositions such as a and c depend on those given to propositions such as b_1, b'_1, b''_1, etc. The latter are sources, since other credences depend on them, but their credences depend on no others. The Jeffrey conditionalizer takes 'uncertain evidence' to be these sources. Whether credences should then be modelled as a DAG, with sources whose credences are not dependent on other propositions, may be doubted.

The subject has updated her credence in a. But there will typically be other propositions regarding which S has a credence that will have to be updated as a result in the change in $P(a)$. The Bayesian typically models S's credence structure as a graph whose vertices are propositions to which S attaches a probability, between which arcs represent relations of probabilistic dependence, whose values are the conditional probabilities of the connected propositions (see Figure 8.1). An arc from p to q indicates that the subject's credence in q is dependent on their credence in p, so that when her credence in p changes, she should update her credence in q. This is a Bayesian network (or Bayes net). Such a network is a directed acyclic graph. Some vertices have arcs directed only away from them—these are *sources*. Changes in credences will propagate from changes at the sources throughout the network. So when a proposition somewhere in the middle of the network, such as a, changes its probability, $P(a)$, then propositions at the vertices of arcs emanating from a must have their probabilities updated. The proposition a will itself be a member of another partition, so these further, dependent propositions must be updated in accordance with the same rule (F$''$) (with these dependent propositions taking the place of a, and with a and the other members of its partition taking the place of b_1, b_2, etc).

So (F″) is the general updating rule we use to propagate changes in probabilities across the network of propositions. It is more general than (S) because in updating in the middle of the network we must update on propositions with probabilities less than 1, and it is more general than (J) because we should allow for partitions of several propositions, not just a proposition and its negation. Let us look further at (F″). Note that Bayes's theorem tells us that $P_o(a|b_1) = P_o(b_1|a)P_o(a)/P_o(b_1)$ etc. Hence (F″) is equivalent to:

$$\text{(G)} \qquad P_n(a) = \frac{P_o(b_1|a)P_o(a)}{P_o(b_1)} P_n(b_1) + \ldots + \frac{P_o(b_i|a)P_o(a)}{P_o(b_i)} P_n(b_i).$$

Since (F″) is an updating rule so is (G). An updating rule identical to Jeffrey conditionalization, (J), is the special case of (G) where we update using the partition $\{e, \neg e\}$.

So what is distinctive about Jeffrey conditionalization is not the updating rule, which is in use whenever there is a change in the probability distribution. For example, as a result of strict conditionalization, some probabilities will change and (G) and its kin will be used iteratively to update further propositions. What is at issue, therefore, is not the updating rule but instead which changes in the probability distribution should rationally *initiate* updating.

The strict conditionalizer has a clear if incomplete answer to that question. Updating is initiated when a proposition achieves the status of evidence. This answer is incomplete in two respects. First, Bayesianism tells us almost nothing about when a proposition is evidence—the only constraint is that evidence propositions are those propositions that have a probability equal to 1. Secondly, for updating to be a rational requirement, the change of a proposition's status to evidence must itself be rational. If a subject who starts with a rational probability distribution changes the probability he attributes to a proposition p from less than 1 to 1 without any rational cause, then it is not rationally required that the subject update the rest of his probability distribution in line with strict conditionalization. Another option (and the one that does seem rationally required) is to reverse the change. Put another way, if the subject keeps the rest of his probability distribution fixed (which he ought, given that it was rational to begin with and the subject has undergone no rational change), then he is rationally obliged to change $P(p)$ again, in line with Bayes's theorem, which will set $P(p)$ back to where it started.

Consequently the Bayesian has to outsource some important epistemological questions: Which propositions are evidence propositions? and Why is it rational to update when a proposition becomes an evidence proposition?[11] A response can be given that answers both questions. As argued in Chapter 5, evidence propositions are precisely those that are known; and as already discussed, known propositions are indeed accorded probability of 1. And clearly if one's *knowledge* changes then it is rational to change one's probability distribution accordingly. As we shall see, this answer requires dropping personalism, but we will come to appreciate that this is a price well worth paying.

[11] See Williamson (2000: 217) on the need for supplementing Bayesianism with an account of evidence.

If we turn now to Jeffrey conditionalization, we see that analogous questions arise, but answering them becomes that much more difficult. If 'evidence' propositions whose changes in probability are ones that rationally initiate updating, then which propositions are they? The answer that evidence propositions are the known propositions is not available. The new probabilities of these propositions are less than 1, whereas known propositions have probability equal to 1. Jeffrey conditionalization allows one to retain as epistemically possible (probability >0) propositions inconsistent with the evidence, but that makes no sense if the evidence propositions are known: propositions inconsistent with the subject's knowledge are not epistemically possible. That is really just to say that there is no natural place for an account of evidence as knowledge in Jeffrey conditionalization. However, that means that the explanation given above about why it is rational to update on changes in evidence ('because it is rational to update on changes in what one knows') is not available.

So Jeffrey conditionalization needs an account of evidence that allows evidence to be less than knowledge, yet explains why it is rational to update in response to changes in evidence. So far no such account is forthcoming. Jeffrey conditionalizers might aim to get by with something weaker: an account of evidence such that updating on evidence propositions is rationally permissible, even if not rationally required. Which propositions are those then? I suggest that something like the following thought is part of the Jeffrey conditionalizer's outlook. It would be rationally *impermissible* for a change in the probability of p to *initiate* updating, if that probability is fixed by its relationships to the probabilities of other propositions, via (F$''$) and (G) and the like. For a change in $P(p)$ in such a case would have to come about *as a result of* changes in those other probabilities. So the propositions that are permitted to initiate changes in a probability distribution are those whose probabilities are not dependent on the probabilities of other propositions. Einstein's GTR is a proposition whose probability depends on the probabilities of other propositions, concerning the outcomes of various experiments and observations as well as other theories, etc.; so it would not be permissible to update in response to a change in the probability accorded to GTR if that change comes about directly as opposed to as a result of changes in the probabilities accorded to those experiments, observations, and other theories. In terms of the Bayes net model of credences, some propositions, those in the middle of the network, have incoming arcs. These should have their probabilities changed only by changes in the probabilities of the propositions from which they emanate. Some propositions—the sources—have no arcs leading to them; arcs only lead way from them. So, on this picture, we may and should update on changes in the source propositions, and on only them.

So it is not surprising that the propositions that Jeffrey and others use as examples for e in (J) are propositions arising from sensory perception: famously the subject who looks at a piece of cloth by candlelight and who is uncertain as to its colour and so gives various probabilities to the propositions that the cloth is green, that it is blue, and that it is violet (Jeffrey 1983: 165). (Jeffrey conditionalization invites and indeed is motivated by the empiricism considered and rejected in Chapter 6.)

Nonetheless, even those propositions do not meet the requirement that their probabilities are not dependent on the probabilities of other propositions.[12] Those probabilities, Williamson (2000: 217) points out, will depend on what I believe about the reliability of my eyesight, the nature of the lighting conditions, the character of the cloth, etc. It is difficult to see that *any* proposition will meet this independence condition. One might think that reports of sensory appearances do meet the condition ('It seems to me that this cloth is probably green'). However, if these internal reports are fallible, then they will be dependent on my assessment of my own reliability in delivering such reports. On the other hand, if internal reports are infallible then we may ascribe probabilities of 1 to those reports, in which case we do not need Jeffrey conditionalization.

One might alternatively think that updating should be initiated by involuntary or spontaneous changes in credences. In some circumstances my credences just do change, even if partly determined by other credences. On examining at the cloth, I just have a credence of 0.8 that the cloth is green—that is how things strike me. But it does not follow that one should therefore update ones other credences in line with (J) or (G). We know that we often have irrational involuntary beliefs and credences—beliefs and credences formed by wishful thinking or self-deception and so forth. I wake up one morning with a new theory, of whose truth I am irrationally convinced just because it is *my* theory. Should I update on such a belief? Scientists point to evidence that apparently refutes my theory. Should I give a high credence to the proposition that the scientists involved were incompetent or fraudulent, in order to maintain probabilistic coherence among my credences? No—optimally I should change my initial irrationally high credence in the theory. But if it is stubborn, then it would be rational to bracket that credence and prevent it from infecting my other credences. To maintain probabilistic coherence at all costs leads to the species of irrationality characteristic of the conspiracy theorist.

Jeffrey conditionalization looked liked a plausible option to take account of the fact that if personalism is true, then hardly ever, if at all, will a proposition have a probability of 1—certainly not the propositions of interest to science. Bayesian updating, whether strict or Jeffrey, needs an account of the circumstances in which a change in a probability distribution may rationally initiate updating of other propositions, since coherence can always be achieved by reversing the initial change. Such an account will come from outside the Bayesian framework. Strict conditionalization can be supplemented by a natural account of when that is: when a proposition becomes known, that proposition may (and should) initiate updating.[13] No such account is available to the Jeffrey conditionalizer.

8.4.3 *Challenge 3: Knowledge and explanationism*

One important difference between explanationism, whether my IOE or 'standard' IBE, and Bayesianism is that the former tells us which propositions to infer from oth-

[12] In which case the credence structure should not be modelled as a DAG.

[13] For completeness, it should be added that updating should also be initiated when a proposition loses its status as knowledge.

ers whereas Bayesianism does not. That is so say, explanationism, but not Bayesianism, includes a rule of detachment. Standard IBE, as employed for example by Lipton (2004), might be summarized in a rule of detachment such as:

> (DET-IBE) When H_n is a much lovelier potential explanation of the evidence than all its competitors (and it is in itself a good explanation, there is a good quantity of evidence, and alternative explanations have been sought assiduously), then infer that H_n is the actual explanation of the evidence and so is true.

Bayesianism might be supplemented by a rule of detachment based on the Lockean Thesis (Foley 1992), that rational belief in a proposition is (or is legitimate when one has) a rational credence above a threshold (e.g. 0.95). So the Bayesian rule of detachment would state:

> (DET-BAYES) Infer those propositions that when conditioned on the evidence reach the threshold probability.

But such rules are highly problematic unless the threshold is equal to 1.[14]

In this section I raise a related problem for compatibilism. Standard IBE believes that using the rule (DET-IBE) can lead to knowledge of the inferred proposition. This leads to a problem in the Bayesian context if this proposition is true:

> In a typical explanationist inference, the evidence is consistent with more than one of the competing hypotheses, and no hypothesis has a prior equal to 1.

According to standard IBE, a typical explanationist inference—one using (DET-IBE)—will produce a preferred hypothesis whose probability is less than 1. We saw above that (B′) entails that known propositions have a probability equal to 1. These together entail that a typical explanationist inference will not lead to knowledge. So explanationism seems to condemn us to scepticism. (Note that this problem is distinct from the challenge raised in Section 8.4.2. The latter arises from personalism. This problem is independent of personalism and arises simply from probabilism in the form of (B′), even if we take the probabilities to be something other than credences, as this chapter will propose.)

One might attempt to resolve this with the following. Bayes's theorem gives one the value of $P(h|e)$. Conditionalization says that on receipt of evidence e one should alter the probability attached to h to be equal to this value. However, if the epistemic circumstances are favourable then the subject gets to know, and the probability of the proposition rises to 1. That is, under such circumstances one can depart from conditionalization. This, however, is just a special case of the boost view of explanationism. While there are counterexamples to Bayesian conditionalization, this departure is undermotivated.

[14] For example, such a rule might permit the inference of each of two propositions individually but not their conjunction. In other cases, such as the proposition asserting that a given lottery ticket will not win, we might think that the proposition, though reaching the threshold, is not one that should be inferred.

Jeffrey Dunn (2014) raises the same problem in connection with evidence. Can a proposition that is learned by inductive inference itself be evidence? Let p be the proposition that a new Pope has been elected, and s be the proposition that there is white smoke coming from the Sistine Chapel. My credences are $P(p) = 0.5$, $P(s) = 0.5$, and $P(p|s) = 0.9$. I then see white smoke over the Sistine chapel, so s is now in my evidence set. So, with conditionalization, my credences are now $P(p) = 0.9$, $P(s) = 1$, and $P(p|s) = 0.9$. Can I now use p as evidence (e.g. with regard to some proposition concerning the immediate future of the Church)? According to Bayesianism, I cannot. For if p were now evidence, then it would have to have a probability of 1, whereas its probability is 0.9.

(Dunn concludes that p is not in my evidence, and so this makes trouble for E=K, since I can get to *know* that a new Pope has been elected by seeing white smoke. I don't think that this is really the problem. For as remarked above, the Bayesian is committed to E=K. If I really do know p, then p now becomes part of my background knowledge, and so, thanks to (B′), $P(p) = 1$. So the probability assigned by conditionalization, $P(p) = 0.9$ is inconsistent both with the proposition that p is now in my evidence and with the proposition that I now know p. E=K is preserved.)

8.4.4 *Three challenges summarized*

To summarize, the three challenges to compatibilism between Bayesianism and explanationism are:

1. Mere compatibilism with subjective Bayesianism answers Bayesian objections to IBE but fails to articulate how IBE can lead to the truth and so be a source of knowledge.
2. Bayesianism requires that evidence propositions and other known propositions have probability equal to 1. If that probability is a credence then, since no-one reasonably has credence 1 in most non-logical propositions, very few propositions will ever be known/evidence. The challenge is a consequence of Bayesianism commitment to personalism
3. If IBE selects among hypotheses that are all consistent with the evidence, Bayesian compatibilism implies that the inferred hypothesis has a probability of less than 1. But a known proposition cannot have a probability of less than 1. The challenge is a consequence of Bayesianism's commitment to probabilism plus the claim that IBE may select among hypotheses that are all consistent with the evidence, *viz.* the rule (DET-IBE).

I have proposed that the explanationist should prefer some strong kind of objective Bayesianism. So in the next section I consider whether objective Bayesianism provides all that we want from the compatibilist rapprochement between explanationism and Bayesianism. I suggest that it does not. Objective Bayesianism will not suffice to answer 2, since objective Bayesians are personalists—the probabilities in question are credences. Below I consider a move beyond objective Bayesianism that rejects personalism and embraces a different conception of probability. This conception associates the probability of a hypothesis with its *plausibility*. I then return to challenge 3, which I answer by retaining probabilism but rejecting the rule (DET-IBE). As we saw in the preceding chapter, the explanationist route to knowledge, IOE, denies that the explanationist either may or needs to select among hypotheses that are all consistent with the evidence.

8.5 Probability and plausibility

Explanationism requires objectivity since it holds explanatory considerations to be a guide to the truth. So we must rule out subjective Bayesianism. Should explanationism therefore endorse objective Bayesianism, accepting further constraints on conditionalization?[15] Typically, objective Bayesians will deploy a version of the principle of indifference. Objective Bayesians will often hold that one's prior probabilities should be calibrated according to evidence of frequencies: if all one knows about some possible outcome *o* is that in the relevant circumstances such outcomes occur with frequency *f*, then one should accord *o* the probability *f*. Nonetheless, these kinds of constraint (indifference, calibration) do not suffice to capture what is needed to provide a form of objective Bayesianism that marries well with explanationism. For often scientists are faced with competing hypotheses that are not neatly symmetric, for example: the hypothesis that the K-T extinction was caused by an asteroid collision versus the hypothesis that it was brought about by a period of rapid evolution. Furthermore, it is difficult to bring frequentist considerations to bear on such hypotheses—this is most especially obvious if the competing hypotheses concern the laws of nature. The other principles to which objective Bayesians typically appeal are, like indifference and calibration, principles that have an apriori appeal—Jaynes's (1968) principle of maximum entropy for example. The problem we faced was that compatibilist explanationism needs explanatory considerations to inform or parallel the use of Bayes's theorem in such a way that there is a correlation between higher posteriors and the truth. Further apriori principles seem too thin to be able to do that—if they did, that would amount to an apriori justification of induction. So objective Bayesianism, as it is typically articulated, turns out to be an unsatisfactory answer to challenge 1 after all.

We need therefore to move beyond objective Bayesianism's supplementation of standard Bayesianism with additional apriori principles. Instead, I propose a 'super-objective' Bayesianism where good priors and likelihoods are determined by explanatory considerations that cannot be captured as apriori principles but instead have a *de facto* reliability. Super-objective Bayesianism would be a significant departure even from objective Bayesianism. If the relevant constraints on credences are apriori and assuming that credences, being doxastic attitudes, are introspectively knowable to the subject, then the subject is in a position to know apriori that her credences obey all the relevant constraints—the subject can know that she is being rational.[16]

[15] For a thoroughgoing articulation and defence of objective Bayesianism, see Jon Williamson (2010).

[16] I deny that a subject is always able to know her own credences, but the thought that credences are transparent to introspection and that Bayesianism allows us to police our own rationality is a common one (see Savage 1954: 20; Egan and Elga 2005).

So super-objectivism would undermine one motivation for personalism, that it permits the subject to assess the rationality of her evaluations. At the same time, super-objectivism does not itself solve challenge 2, since that challenge arises from the commitment to personalism. A natural step therefore is to renounce personalism in addition to adopting super-objectivism. Having renounced personalism, we need an alternative account of the evaluations of hypotheses. I agree with Salmon (1990: 182) in holding that 'Plausibility considerations are pervasive in the sciences; they play a significant—indeed, *indispensable*—role.' Accordingly, I propose that our epistemic evaluations of hypotheses are evaluations of plausibility in the light of the evidence—such evaluations I shall call 'plausibilities'. Salmon's objective Bayesianism is an important attempt to link Bayesian confirmation to Kuhn's five criteria of theory choice via plausibility considerations, to which the proposals in this chapter have some similarity. Salmon, however, regards plausibility arguments as supplying priors within the context of personalism, whereas I follow Williamson (1998, 2000) in regarding plausibilities as an alternative to credences. (As we have seen, I also regard plausibility considerations as determining the ratio of the likelihood to the expectedness.)

We retain probabilism regarding plausibilities: the plausibility of a hypothesis in the light of the evidence can indeed be modelled by the probability calculus.[17] Insofar as we can quantify plausibilities, we can represent them as ranging from a minimum, 0, the plausibility of a hypothesis inconsistent with the evidence, to a maximum, 1, the plausibility of a hypothesis entailed by the evidence. Thus represented the plausibility of a disjunction of mutually exclusive hypotheses is the sum of the plausibilities of the hypotheses individually; in particular, the plausibility of a disjunction of a hypothesis and its negation is maximal. Thus finite additivity is satisfied.[18] If my evidence allows me to know, for example, that the Earth orbits the Sun, then that hypothesis has maximum plausibility and could not gain additional plausibility from additional evidence. Likewise, if a theory entails that water is frozen at 80°C, then my evidence shows that this theory has zero plausibility. Since plausibilities for empirical propositions can be given values of 0 and 1, this approach avoids challenge 2.

In Section 8.3 I identified two aspects to the explanatory loveliness of a hypothesis: an internal component, which is independent of the evidence, and an external component, which concerns the ability of the hypothesis to explain the evidence in question. The internal component corresponded to the prior probability of the hypothesis, that is to say, the plausibility of the hypothesis independent of its eviden-

[17] This approach therefore builds on Williamson's (1998, 2000) notion of *evidential probability*.

[18] A further question asks how plausibilities relate to credences. Clearly there is no neat relationship, in that some propositions will have a plausibility of 1 but a credence of less than 1. On the other hand, if a subjective Bayesian does treat a subject's evidence/knowledge as having credence equal to 1 and has priors coinciding with those of the plausibility theorist, then conditionalization will lead to the same posterior probabilities.

tial support.[19,20] In science we do find assessments of plausibility of this sort. In the preceding chapter, I described how in 1785 Charles-Augustin de Coulomb proposed an inverse square law to explain the forces between charged objects. Even before the torsion balance experiments that he then carried out, this hypothesis carried a high degree of (evidence-independent) plausibility because of its analogy to the inverse square law of gravitation. This is evidence-independent plausibility in that it is not the product of any process of conditionalization (e.g. on the law of gravitation), not even in an idealized way. Why should we not treat the structural similarity between Coulomb's hypothesis and Newton's law as a piece of evidence (since that similarity is a known fact) in favour of the former? Doing so for a Bayesian would require setting the expectedness of the evidence, a prior for the fact of there being a similarity between Coulomb's hypothesis and Newton's law. It is very unclear where that could come from. Furthermore, this influence on the plausibility of Coulomb's proposed law can come about even if we hold that the law of gravitation is not itself known (and so not evidence). More generally, experience can influence the assessment of plausibility in ways that cannot simply be reduced to an evidential relationship (rather as one may correctly recognize the facial similarity of two members of the same family without being able to articulate the basis of that similarity).[21] So a hypothesis can gain plausibility by its connection to elements of the scientific context in a way that is independent of conditionalization on evidence. Analogy with well-confirmed hypotheses is an example of how this occurs. In the context of the heuristic approach, we should see such factors as influencing the priors for hypotheses. And so we have one answer to the question: where does evidence-independent plausibility come from? (There may be other sources of evidence-independent plausibility.[22]) The remaining question concerns the objectivity of this plausibility.

8.6 Judging objective plausibility

I have just argued that we make judgments of the plausibility of a hypothesis that are independent of conditionalization on the evidence—in so doing we respond to internal explanatory virtues, and these may be influenced, for example, by analogy with other theories. Similar remarks may be made about external explanatory virtues,

[19] The 'internal/external' terminology here does *not* align with a distinction between factors that are acceptable to an internalist epistemology and those acceptable only to an externalist epistemology.

[20] This internal explanatory virtue is what Williamson (2000: 211) calls the 'intrinsic plausibility [of the hypothesis] prior to investigation'. It does not seem correct, though, that plausibility can be a truly *intrinsic* property of a hypothesis. If it were, there would be a plausibility that is independent of the psychological make-up of even an ideal human scientist, including their innate psychology. Such a notion looks as if it would be close to the apriori objectivism we considered but rejected above.

[21] Suppes (1966: 203) says something along these lines when he describes priors as summarizing the heterogeneous ways in which experience, hunches, and the like can bear on a hypothesis beyond the 'narrow framework of the experiment itself'.

[22] For example, another source might be the epistemological conservatism advocated by Poston (2014). That a proposition is believed by a subject suffices to give it some degree of plausibility. Poston provides arguments for conservatism from an internalist perspective, while I can see externalist, naturalistic reasons in its favour.

which allow us to assess the factor by which the degree of plausibility changes when the evidence is obtained (i.e. the ratio $P(e|h)/P(e)$). This too may be informed by exemplars of successful explanatory relations between theories and evidence.

In the preceding chapter I argued that we can draw on a central lesson of the work of Thomas Kuhn (1970), which is that scientists' assessment of any hypothesis is crucially influenced by their prior experience of other successful theories.[23] According to Kuhn, such theories, which he called 'exemplars' or 'paradigms', are models that inform the development and assessment of newer hypotheses. Because Newton's theory of gravitation was so successful, it became a model (in multiple ways) of what a good theory might look like, and so, as we saw, Coulomb's proposed law acquired plausibility because of its similarity to that model.

I linked this point to explanationism by noting that scientists acquire their sense of explanatory goodness (loveliness) from their training with and use of exemplars. Let us assume some modest degree of scientific realism—the theories in question have a high degree of truth-content. Let us also assume that theories similar to true or highly truthlike theories are more likely to be truthlike than less similar theories. If these assumptions are correct (and they are themselves at least plausible), then the practice of allowing one's preferences among theories to be influenced by their loveliness will be a truth-conducive practice. Exemplars change over time, and so what counts as lovely (the explanatory virtues acknowledged by a community) will change over time also—we can see this as a process of the evolution of a scientific community's values in the direction of greater reliability.[24] We should not expect these values and their applications to be the same in all scientific fields. Different fields have different exemplars and so values (or, more frequently, what counts as instantiating a given value) may differ between fields.[25]

So now let us imagine a scientist who considers some candidate hypothesis in advance (or independently) of the evidence. Her assessment of its plausibility is influenced in part by the degree to which it possesses the virtues acknowledged by the community. In the light of the foregoing, and if the two assumptions are correct, we may expect there to be some correlation between the plausibility assessments and truth. Consequently, probabilities interpreted as (quantified) plausibilities will be objective in the sense relevant to challenge 1: probabilities will relate to truth, in that hypotheses assessed as more plausible will be more likely to be true.

Note that it does not need to be claimed that this correlation is perfect. From the Bayesian perspective the assessment of prior plausibility might be only rough and ready yet good enough. For the phenomenon of the washing-out of priors will mean that imperfections in the prior probability will be increasingly insignificant as

[23] I argued that we can accept this result of Kuhn's work without taking it to have any anti-realist consequences, as this section proposes. See Bird (2000). Salmon (1990: 196–9) does likewise, though concentrates on Kuhn's five shared criteria of theory choice, which Salmon takes to play a role in determining the plausibilities of hypotheses and thereby the priors of those hypotheses.

[24] By 'values' here I mean the features of a theory that the community regard as lovely.

[25] As Salmon (1990: 186) remarks, 'simplicity ... [is] a highly relevant characteristic, but one whose applicability varies from one scientific context to another'.

ever more evidence comes in. The washing-out of priors argument does not help the subjective Bayesian since that position accepts no epistemologically relevant constraints on priors, meaning that priors are acceptable that are so extreme that they cannot be washed out by a reasonable quantity of evidence. If plausibility is a rational constraint on priors, then extreme priors will be excluded (unless warranted in particular cases by extreme plausibility or implausibility).

8.7 Plausibility, knowledge, and eliminativism

Since plausibilities are not credences, a subject who assigns plausibility of 1 to a proposition is not obliged to accept wild bets on the truth of that proposition. So the second challenge is not a problem for this approach. A probability of 1 means that the proposition is maximally plausible *given the evidence and background knowledge.* Rationally assessing a proposition as maximally plausible does not seem unachievable in the way that a rational credence of 1 looks unachievable for most propositions.

However, $P(h) = 1$ does mean that any proposition inconsistent with h (e.g. a competitor hypothesis) must have zero probability, that is to say maximum implausibility in the light of the evidence. This looks to be in tension with IBE, since the latter is typically understood as making inferences on the basis of one explanation being a *better* explanation than the competitor hypotheses. The latter implies that one can make a reasonable IBE when the best explanation is just rather *more* plausible than its competitors; the latter thus retain some degree of plausibility. If that is right, then the inferred proposition will not be knowledge, since knowing h requires $P(h) = 1$ and $P(h') = 0$ for all competitor hypotheses h'.[26] This was challenge 3.

So, in order to avoid challenge 3, knowledge of a proposition inferred on the basis of explanationist considerations requires that the competing hypotheses must be *ruled out* by the evidence and background knowledge. This is Inference to the Only Explanation (IOE), the eliminative abduction that uses Holmesian inference, and which I have described in detail in Chapter 7. While standard IBE-focused explanationism appears to doubt that explanationist considerations can ever achieve this, I argued that elimination of all but one of a set of prima facie plausible competing hypotheses is achievable.

The compatibilism developed in this chapter therefore requires us to reject standard IBE and its rule of detachment (DET-IBE). In their place we require IOE, and its rule of detachment:

> (DET-IOE) Let $h_1, \ldots, h_n$ be all the hypotheses that are (mutually exclusive) potential explanations of some phenomenon described by evidence e_s, a subset of the total initial evidence e_i, and which have some degree of plausibility. When further evidence, e_a, is collected such that $h_1, \ldots, h_{n-1}$ are ruled out by $e_i \wedge e_a$, then infer that h_n is the correct explanation of e_s.

[26] Since a subject's known propositions and evidence propositions are the same, and evidence propositions have a plausibility of 1, so also do known propositions. $P(h) = 1$ is consistent with fallibilism, when the probabilities are plausibilities. If the probabilities are *credences* then fallibilists will want probabilities to be less than 1 even for knowledge.

8.8 Conclusion

The heuristic approach to compatibilism between Bayesianism and IBE can see off the challenge from the subjective Bayesian. Explanationist considerations do not supplement Bayesian considerations by giving a boost to the probability of a hypothesis when conditionalized on the evidence. Rather we can see explanationist factors as corresponding to the various elements of Bayes's theorem, in particular the prior $P(h)$ and the ratio of the likelihood to the expectation of the evidence, $P(e|h)/P(e)$. Still, that cannot be all that explanationist considerations achieve, for explanationism reasonably aspires to explain how we have knowledge, whereas subjective Bayesianism does not. This may motivate a move from subjective to objective Bayesianism. But that position faces the problem that no scientific proposition can reasonably be accorded credence of 1, in which case no proposition can ever be known or be evidence. So we need to move beyond the personalism of both subjective and objective Bayesianism. What I have dubbed 'super-objectivism' says that we should understand the probabilities with which we are concerned not as credences but rather as something else. That something else must allow it to be possible to have evidence and knowledge and to accord the relevant propositions probability 1. I have proposed that scientific judgments of the plausibility of hypotheses in the light of the evidence are just this (and these may be identified with Timothy Williamson's evidential probabilities). We have seen that a hypothesis is held to be plausible in virtue of its explanatory loveliness. A hypothesis is held to be a lovely explanation in part because of its similarity to successful, highly truthlike theories. Hence plausibility can be objective (in the relevant sense) and may be correlated with the truth. In the limiting case, IOE, all initially plausible competing hypotheses are eliminated by the evidence and the target hypothesis achieves the maximum plausibility of 1.

9

METASCIENTIFIC KNOWLEDGE?

Does science give us knowledge of the world? To claim to know an answer to this question, whether positive or negative, is to claim metascientific knowledge. The two most famous metascientific arguments are the No Miracles Argument and the Pessimistic Meta-Induction. I argue that neither argument can give us metascientific knowledge of the global sort, concerning all *of science, that each aims to achieve. One line of argument I pursue points out that science is too heterogeneous to permit a successful global metascientific argument. So, normally, one cannot use a metascientific argument to bypass engagement with or to gainsay the conclusions of the first-order scientific arguments. At most, local metascientific arguments might adjust the confidence we have in the methods—and so results—of some particular field.*

9.1 Introduction

One of the two key themes of this book is that *knowledge* is central to what science is about. Science aims at knowledge, scientific progress is a matter of accumulating knowledge, evidence is knowledge, and so on. The second key theme is the rejection of empiricism. In rejecting empiricism I reject the various kinds of scepticism or anti-realism based on empiricism, such as instrumentalism and constructive empiricism. These might both indicate that I am an enthusiastic advocate of scientific realism. But that is not the case. In fact I reject both realism and anti-realism. I reject the anti-realist view that knowledge of the truth of scientific theories is not possible. And I reject the realist view that there is a general argument whose conclusion is that the leading theories of the mature sciences are highly likely to be true.

Consider Jamie who is wondering whether to believe theory T. He thinks to himself, 'Jane tells me that theory T is believed by all scientists in her field. She has explained in detail what T says, and has given me all the evidence, namely E, that those scientists regard as supporting T, as well as their reasoning from E to T's truth. Still, I wonder whether I should believe T. And the same goes for other theories presented by other scientists, which, they say, are also well supported by their evidence. Should I believe them too?' Whispering in Jamie's left ear is Philosopher Delta. Delta says, 'Don't believe T! Scientists have believed other theories in the past that have been successful in accounting for the relevant evidence; but those theories have subsequently been found to be false. So you should expect T to be false as well.' Whispering in Jamie's right ear is Philosopher Gamma. Gamma says, 'Of course you should believe T and other theories that are today believed by scientists. Look how successful those theories have been in explaining their evidence, which includes confirmed

Knowing Science. Alexander Bird, Oxford University Press. © Alexander Bird 2022.
DOI: 10.1093/oso/9780199606658.003.0009

novel predictions from those theories. It would be a miracle if science could be so successful but false.'

Delta presented Jamie with the Pessimistic (Meta-) Induction (PMI) while Gamma sought to persuade Jamie with the No Miracles Argument (NMA). These are two well-known global metascientific arguments. They are *metascientific*, because they are about science and its theories; they start from data concerning the track-record of scientific theories and draw conclusions concerning the probable truth or falsity of current and future theories. They are *global*, in that both concern the nature of science overall (rather than about some specific field of science)—they both attempt to answer what van Fraassen (1980: 219) calls 'the global question of the success of science'.

I argue that we cannot have knowledge of the global metascientific kind that both arguments purport to deliver. My central reason is that there is no methodological unity to science that would allow for science to be either likely to be true across the board or likely to be false across the board. Science is methodologically disunified—the methods used to test the Higgs hypothesis are quite different from those used in biochemistry which are different again from those used in social psychology (one could go on . . .). Some of these methods are more reliable than others. One might have reason to believe the best theories of one field while remaining sceptical about the leading theories of another field.

9.2 The No Miracles Argument Mark 1

9.2.1 *NMA as an inference to the best explanation*

The No Miracles Argument is so called thanks to a succinct and famous version of the argument due to Hilary Putnam (1975*a*: 73):

> The positive argument for realism is that it is the only philosophy that doesn't make the success of science a miracle. That terms in mature scientific theories typically refer (this formulation is due to Richard Boyd), that the theories accepted in a mature science are typically approximately true, that the same term can refer to the same thing even when it occurs in different theories—these statements are viewed by the scientific realist not as necessary truths but as part of the only scientific explanation of the success of science, and hence as part of any adequate scientific description of science and its relations to its objects.

Note that Putnam is concerned to establish claims both about reference and about truth. The point of the claim about reference is to distinguish realism from versions of anti-realism that deny that our scientific terms ever refer to the sorts of thing to which we normally think of them as referring (sub-atomic entities, extra-solar planets, events which occurred many millions or billions of years ago, etc.), but hold instead that if they refer at all, they refer to our sense-data or to the outcomes of operations with measuring instruments. Here I shall assume that this semantic component is accepted on all sides—*if* scientific theories are true, *then* their terms refer to entities, properties, kinds, etc. that may (depending on the theory and background knowledge) not be available to perception or measurement. The latter is an imprecise statement of the referential element of scientific realism, but since we are setting it aside, further precision is not needed. What is at issue is the truth element. Do we have any reason to believe the antecedent of the conditional claim just made? Should we believe any scientific theory to be true? Putnam intends his argument to give a positive answer to this question too.

It is natural to interpret Putnam's argument as an inference to the best explanation, along the following lines:

> (NMA-1) Science shows a high degree of empirical success. That success is best explained by the hypothesis that the theories of science are at least approximately true.

(We will assess another, more sophisticated version of the NMA in a later section; hence 'NMA-1' for the argument currently under consideration.) Arguably Putnam's argument is stronger than (NMA-1). He is suggesting that any other explanation of the success of science, an explanation consistent with the falsity of scientific theories, would in effect be to appeal to a miracle, and so would have the implausibility attaching to the adhocness of any such explanation. That implies that the truth explanation is not just better than the alternatives, but also that the alternatives are very poor explanations indeed. So if Putnam's argument is a good one, then (NMA-1) ought to be a good argument too.

The explanandum in (NMA-1) is the fact that scientific theories are empirically successful. Let us consider (NMA-1) regarded as an argument concerning some specific theory, T_A. 'T_A' denotes the theory that has content abbreviated by 'A'; in what follows we may replace 'A' by the claims of the theory. So T_A might be the Alvarez asteroid theory of the K-T extinction (now also known as the K-Pg extinction), in which case where 'A' occurs we should understand this as shorthand for '66 million years ago an asteroid about 10km in diameter struck the Earth. This impact produced a dust cloud that caused the death of plants and of the animals dependent upon them, and so caused the extinction of many dinosaur species. It also deposited a high concentration of iridium . . . '. Note that the propositions 'Theory T_A is true' and 'It is true that A' are (logically) equivalent, and so what is explained by the fact that T_A is true is explained by the fact that it is true that A. So (NMA-1) applied to T_A is:

> $(\text{NMA-1})_A$ Theory T_A has a high degree of empirical success. That success is best explained by the hypothesis that T_A is at least approximately true.

which is equivalent to:

> $(\text{NMA-1})_A$ Theory T_A has a high degree of empirical success. That success is best explained by the hypothesis that it is at least approximately true that A.

(The scientific realist takes their view to be confirmed rather than refuted by cases of scientific theories that are strictly false but are approximately true. For example, a theory that gives the correct mathematical form for some law but specifies a value for a constant in that law that is false but close to the truth should be regarded as a scientific success. Hence the reference to 'at least approximately true' in the above. For convenience, in the remainder of this section I drop this qualification since the difference between exact truth and approximate truth is not germane to the more general discussion of the explanatory power (or otherwise) of truth itself.)

9.2.2 *Does the truth of a theory explain its success?*

I now consider the assumption of (NMA-1) that truth explains success. I argue that (NMA-1) cannot support an argument to a metascientific conclusion, because the claim that 'truth explains success' amounts to no more than the scientific (first-order) claim that we should believe such-and-such a theory on the basis of the evidence that supports it. We will consider the explanandum and explanans of $(\text{NMA-1})_A$ in turn.

The *explanandum* of $(\text{NMA-1})_A$ is the empirical success of A. What is that empirical success? There is some dispute about this—for example not all agree that evidence that played a part in directing the construction of the theory or is otherwise accommodated by the theory counts among its successes.[1] Nothing in what follows depends on accepting or rejecting this.[2] All will agree that the empirical successes of a theory include the fact that the theory makes certain predictions and that these predictions are found to be true. Let the conjunction of the predictions of A be E_A, where, for simplicity, E_A is deduced from A alone. To say that A is successful is to assert both (i) $A \vdash E_A$ and (ii) E_A. Of this conjunctive assertion it is (ii) that stands in need of explanation. The fact in assertion (i), that E_A follows from A, needs no explanation: that's a fact of logic. So the explanandum is really just E_A.

The *explanans* is that A is true. (NMA-1) thus takes truth to be an explanatory property of theories. I argue that even if the truth of a theory is explanatory, that truth does not add any explanatory content not already present in first-order facts referred to by the theory. Michael Levin (1984) also doubts the No Miracles Argument on similar grounds.[3] He says that the reason that airplanes stay up is not (a): 'the pressure on the underside of a moving airfoil is greater than the pressure on its overside' is true; rather it is (b): the pressure on the underside of a moving airfoil *is* greater than the pressure on its overside. My argument is entirely sympathetic to Levin's claim that the explanatory power is to be found in (b), but unlike Levin I don't deny that (a) can be explanatory also. We ought instead say that insofar as (a) is explanatory that is only because (b) is explanatory, so there is no distinct or additional explanatory power in

[1] See Harker (2008) for discussion.

[2] Although some of the examples in this chapter include accommodated evidence, those who acknowledge only novel predictions can choose examples of novel predictive success in their place.

[3] Ghins (2002) draws on Levin in his critique of the NMA, but allows that there is some weak explanatory power in truth, if the NMA is not construed naturalistically as as scientific argument but as a philosophical argument. Alai (2012) makes the case that the NMA is partly scientific. Either way, the NMA does not add reasons to believe a theory is true that is not present in the first-order scientific argument. Alai reconstructs the NMA as starting from the first-order arguments and then proceeding by generalization and semantic ascent. As a summary of first-order arguments, the second-order argument does not strengthen the first-order arguments. It might be thought that the process of generalization—'conglomeration' as Lipton (2004: 198) calls it—could add such strength. I address this below in Section 9.4.

(a). To be confident in this conclusion and since the nature of truth is disputed, we should examine the various theories of truth (see Glanzberg 2016) to show that each delivers the verdict that (a) does no more explaining than (b). I now do this (but if you do not need convincing, then you may prefer to go straight to the next section, 9.2.3).

Deflationary theories of truth The redundancy theory of truth says that 'A' and 'A is true' mean the same (Ramsey 1927). If the redundancy theory is true, then the explanans is equivalent to A *simpliciter*. Other deflationary theories of truth (e.g. Horwich 1990) do not hold an exact equivalence in meaning, but, because they deny that truth is a substantive property of propositions, they do agree that there is no more to the fact that A is true than there is in the fact that A. Consequently the truth of A explains precisely what A explains. In our example of the K-T extinction, whatever is explained by its being true that there was an impact by an iridium-rich meteor is explained by the impact of an iridium-rich meteor.

The identity theory of truth Though not intended as a deflationary account of truth, the identity theory of truth (Moore 1899; Moore 1902; Russell 1903) nonetheless leads us to the same conclusion. For on that view the truth of the proposition that A just is the fact that A. So whatever is explained by A's being true is explained by A. The truth of the proposition that there was a meteor impact just is the fact that there was a meteor impact. Therefore those things explained by its being true that there was a meteor impact are explained by there having been a meteor impact.

The correspondence theory of truth The correspondence theory takes truth to be a substantive relation between a structured proposition and a fact (Russell 1919). There is therefore no simple reduction of the truth of A to A. When it is true that A, 'A' denotes the fact that A while 'the truth of A' denotes the fact that a correspondence relation holds between the proposition that A and the fact that A. So the content of the truth of A includes but goes beyond the content of A. So, prima facie, the truth of A may be able to explain the things that A cannot explain. However, on further inspection, we can see that the additional content of 'A is true' over 'A' adds no explanatory power, and hence whatever is explained by the truth of A is explained by A. The additional content, over and above the fact of A, is the holding of the correspondence relation between the structured proposition that A and that fact. The holding of this relation adds nothing to the explanatory power of A itself.

Let us illustrate this claim with our example of the theory of the K-T extinction. As we discussed above, 'the success of the meteor impact theory' can be replaced without loss by the relevant evidence that is deducible from the theory; in this case that is the iridium anomaly in the K-T layer (for sake of argument we will take that to be the sum total of the relevant evidence). So we have:

> The iridium anomaly is best explained by the truth of the meteor impact theory.

According to the correspondence theory, this amounts to:

> The iridium anomaly is best explained by the impact of an iridium-rich meteor plus the correspondence between that fact and the proposition that there was an impact by an iridium-rich meteor.

When we consider this claim, we can see that the component following 'plus' adds no explanatory power. And why should we expect it to? Why should a relation between between a fact and a proposition explain things such as the iridium anomaly?

Anti-realist theories of truth We have not yet considered the theories of truth that are associated with anti-realism: the coherence theory and the pragmatist theory. We may be entitled to ignore these, since it might be thought that they would violate the referential element of scientific realism that is being assumed in this chapter. It would be odd to use an anti-realist account of truth in the defence of scientific realism. But some versions of anti-realism or quasi-realism may conceive of themselves as supporting scientific realism. So it will be instructive to see that these anti-realist theories of truth fail also to give explanatory power to the truth of A not possessed by A.

Consider first the pragmatist theory of truth (Peirce 1902) in the version that says:

> The proposition that A is true iff that proposition is among the propositions believed at the end of an enquiry conducted by the ideal enquirer.

In this case it is difficult to see how the fact that the proposition that A would be believed by an ideal enquirer explains anything. Consider this theory as applied to our earlier example:

> The iridium anomaly is best explained by the fact that at the end of enquiry, the ideal enquirer would believe that there was an impact by an iridium-rich meteor.

But that is clearly not correct, and if anything gets the explanation the wrong way around: it is because of the iridium anomaly that the ideal enquirer believes that there was an impact by an iridium-rich meteor.

Now consider the coherence theory of truth. How this plays out depends on what 'coherence' is taken to be. But it must already be apparent that coherence with other propositions is not going to give a proposition the relevant explanatory power. The most plausible account of coherence is from Bradley (1914) who takes coherence to be a matter of standing in explanatory relations. Thus we get:

> The iridium anomaly is best explained by the fact that the proposition that there was an impact by an iridium-rich meteor stands in explanatory relations to other relevant propositions.

That looks false; furthermore, which are those other relevant propositions? It might be that the only relevant proposition to stand in an explanatory relation to the proposition that there was a meteor impact is the proposition that there is an iridium anomaly. So we get:

> The iridium anomaly is best explained by the fact that: an impact by an iridium rich meteor explains the iridium anomaly.

Again that looks false. Perhaps one can construe it so that it looks true, but only at the cost of making it tautologous.

So the anti-realist theories of truth fail to give truth the required explanatory power. Assuming such accounts, we see that the truth of A has, if anything, less explanatory power than A. That may seem counterintuitive. But the reason is clear. If we ask 'what is explained by the truth of A [where truth is considered in an anti-realist way]?' then our answers are inevitably constrained to be the thin, explanatory weak sort of facts that the anti-realist is willing to contemplate. But we are not so constrained in answering 'what is explained by A?'

9.2.3 *No meta-argument from success*

So no theory of truth provides the fact of a theory's being true with more explanatory power than the facts referred to by the theory. Therefore, in the explanans, we can replace:

explained by the hypothesis that T_A is true

by:

explained by the facts ... asserted by T_A.

or equivalently:

explained by the fact that A.

We started with:

$(NMA\text{-}1)_A$ Theory A has a high degree of empirical success. That success is best explained by the hypothesis that A is true.

Putting together our conclusions concerning the explanans and explanandum of the above, we see that $(NMA\text{-}1)_A$ is true if and only if the following is true:

The evidence E_A is best explained by the fact that A.

Thus the metascientific claim in $(NMA\text{-}1)_A$ is reducible to the scientific claim made by scientists in support of the theory that A. The No Miracles Argument, applied to a particular theory, provides us with no additional reason to think that the theory is true beyond the reasons provided by the scientists in favour of believing the facts asserted by the theory.[4] Let us return to our subject Jamie, who wants to know whether to believe Jane, the scientist. Gamma sought to persuade Jamie by deploying the No Miracles Argument. If Jamie does not accept Jane's reasoning, then the argument of this section tells us that the No Miracles Argument cannot provide Jamie with any additional reason. Jamie should find Gamma's argument no more persuasive than Jane's. In effect, our argument shows, Gamma's No Miracles Argument is really just a shorthand for Jane's scientific argument.

I conclude therefore that the No Miracles Argument, as applied to individual theories provides neither a way of supplementing the scientists' arguments for their theories nor a way of by-passing them. The No Miracles Argument is presented as

[4] Lipton (2000: 197) also notes that the NMA adds nothing to the first-order evidence for a scientific theory.

a metascientific argument, but in effect reduces just to the relevant first-order scientific argument. Putnam presented his argument as an argument for scientific realism at large, not about individual theories. Nonetheless, scaling up to science as a whole cannot avoid the objections raised at the level of individual theories. So Putnam's global scientific realism amounts to a shorthand affirmation of all the first-order scientific arguments—as an argument, this NMA does not have any distinctively philosophical component. Putnam (1975*b*: 178) himself acknowledges this: 'If realism is an *explanation* of this fact [that science is successful], realism must itself be an overarching scientific hypothesis'. Since the first-order theories are scientific, quantifying over them all generates a claim that is also scientific in the sense that it could only be verified by doing the first-order science (a lot of science). (In the more sophisticated version of the NMA I consider in Section 9.4, scaling up to all of science does make a difference.)

Furthermore, the blanket conclusion of (NMA-1) is not very plausible. Should we really endorse all or even most first-order scientific arguments? Surely they are not all of the same (high) quality? Indeed, as we shall see, we have reason to think that the scientific arguments in some fields are not as good as in others, for example as shown by the replication crisis in clinical biomedicine and in social psychology.

9.2.4 *What is the target of the NMA?*

It might be complained that I have ignored the challenges to which the NMA is a response.[5] These challenges come from outside science. Within the context of science, some theory T might be the best explanation of its evidence. But when we consider the realism versus anti-realism debate, then the range of potential explanations is wider. For example, we should consider van Fraassen's proposal that the success of science is explained by the Darwinian processes by which it selects its preferred theories (van Fraassen 1980; Wray 2007). Since science vigorously tests theories, the favoured theories of a mature science will be those that have survived such testing whereas other theories lie on the dustheap of history, having been refuted by these tests. Inevitably, the current theories of a mature science are successful. To reduce 'the best explanation of success' to the first-order science ignores such alternative, metascientific explanations. Hence the NMA is needed to counter them.

To this complaint I have two related replies. First, this kind of reply implies a species of scepticism or relativism. For it implies that when the first-order science argues that the best explanation of E is given by T, and so T should be regarded as the correct, actual explanation, that is a mistake. For there remain explanations of T's success that the scientist has not considered and which are genuine potential explanations that need to be rejected before T can be regarded as true. And that implies that the scientist's first-order reasoning is not reliable, at least not until supplemented by the NMA to rule out these extra-scientific arguments. Hence the scientist does not in fact know that T is correct. An alternative to this scepticism is a relativism that states that the scientist knows only 'in the first-order context', but does not know

[5] Though Putnam (1975*a*) himself says nothing on this point. Indeed, he gives the NMA very briefly in the context of introducing his version of mathematical realism.

when the context is broadened to include the realism versus anti-realism debate. If we reject this scepticism and relativism, then it cannot be that such extra-scientific 'explanations' of the success of science are genuine competitor explanations to the first-order scientific explanations.

My second response therefore argues that the alleged metascientific explanations of the anti-realists do not provide alternative, 'better' explanations than the first-order scientific explanations. We should reject the idea that their 'explanations' really do explain the success of science. For example, consider van Fraassen's selectionist explanation of the alleged success of science. It is not at all clear that it does explain why any scientific theory is successful. It is consistent with no theory being successful at all—the selectionist story does not tell us why there is any theory good enough to pass the tests of science. It is consistent with our best theories being 'successful' only transiently, with a rapid turnover of theories that are devised in an ad hoc manner in the light of the refutation of their predecessors. The 'explanation' does not explain what the scientists regard as success (even if we confine success to confirmed novel observational predictions). What gives physicists confidence in the existence of the Higgs boson is not just that the theory was tested and not refuted, but that its degree of success is so significant. The ATLAS experiment didn't just find evidence of the Higgs particle, as predicted. Rather, this evidence was evidence at the 5σ level. That is to say, the evidence was an experimental outcome such that if the Higgs hypothesis were false, then that outcome would occur on one occasion in 3.5 million repetitions of the experiment. Nothing in van Fraassen's selectionist story explains the remarkable extent of the theory's success. This example illustrates the importance of focusing on the first-order science. When we focus on the first-order science then we can see, for some theories at least, that what the scientists regard as the successes of those theories cannot be explained by the selection explanation. That is a conclusion that holds for *those* theories—it isn't a blanket conclusion for all of science. (By contrast, as we shall see, in some fields current theories are often successful only in van Fraassen's minimalist sense. And these theories provide only a poor (scientific) explanation of the evidence and should *not*, in my view, be regarded as probably true.)

9.3 The Pessimistic Induction

The PMI has been resisted on several grounds.[6] One set of responses focuses on the inductive base. The evidence of failure, the eventual refutation of theories that were once successful, may be insufficient to support any strong conclusions concerning current or future failure. The cases of failure should be contrasted with cases of continuing success. In Chapter 3 I listed a number of scientific claims that it is not scientifically plausible to imagine will ever be refuted (blood circulates pumped by the heart, water is a compound of hydrogen and oxygen, light is electromagnetic radiation, smoking causes cancer, etc.). I noted that of all the scientific Nobel prizes

[6] The term 'Pessimistic Induction' has been associated with Laudan's (1981*a*) well-known argument. The latter is not a Pessimistic Induction in the standard sense—an induction on scientific failure.

only one has been awarded for work we now regard as erroneous. The cases of failure cited by Laudan (1981*a*) come from points in history when the relevant sciences were in their early infancy. As Fahrbach (2009) argues, this may not be apparent when looking at a timeline of the histories of those sciences. But the measure of success cannot be the length of time that a theory survives testing; rather it is the quantity (and quality) of the testing undergone (see above p. 44).

Furthermore, insofar as older successful theories have been subject to a process of refutation and replacement, an inductive argument to future failure of successful theories is inappropriate; the falsity of earlier theories is the very reason for developing the new ones—with a view to avoiding that falsity (Lipton 2000: 197ff.). It would be folly to argue that because no man has run 100m in under 9.5 seconds no man ever will. On the contrary, improvements in times spur on other competitors, encourage improvements in training techniques and so forth, that make a sub 9.5 second 100m quite a high probability in the near future.[7] Later scientific theories are not invented independently of the successes and failures of their predecessors. New theories avoid the pitfalls of their falsified predecessors and seek to incorporate their successes. Even if the successor theory is false also, we cannot apply a simple enumerative induction. It might be at least as rational to think that the successive improvements suggest that the true theory is not far off. Of course, successive improvements would be consistent with an asymptotic approach to the truth. The point of this part of the argument is that we cannot make any good inference from the premise 'the sequence of theories $T_1, \ldots, T_n$ are all false' to the conclusion 'later theories in this sequence will also be false' without additional information. To make an inference of this kind requires knowledge also of the content of those theories and of the details of their relationship to the evidence. Which implies that there is no global Pessimistic Induction. The latter is the conclusion I will argue for in detail in the remainder of this section.

9.3.1 *Science as a fallible oracle*

The PMI appears to be a straightforward induction, arguing from past failure to future failure. The form may not be 'all observed Fs are Gs *therefore* all Fs are Gs'. Depending on how the PMI understands its inductive basis, the inference may have the form of Reichenbach's straight rule of induction 'a high proportion of observed Fs are Gs *therefore* a high proportion of all Fs are Gs'. The proponent of the PMI can maintain that it doesn't matter much quite what the form of the PMI is, in order to reach a sceptical conclusion. Imagine that we treat science as analogous to an oracle, telling us which theories to believe (assume we have no independent reasons to believe or to reject the theories in question). Even if the oracle gets things wrong in only a fairly modest proportion of cases, that undermines our claims to knowledge even in those cases where the oracle is correct. An oracle that gets things wrong in one-third of cases is not sufficiently reliable that it is true to say that those who

[7] When I first wrote the preceding sentence, in early 2007, the record stood at 9.77 seconds set by Asafa Powell on 14 June 2005. The record fell to 9.58 seconds shortly thereafter, set by Usain Bolt on 16 August 2009, thus reducing the gap to 9.5 seconds by more than two-thirds.

believe the oracle's pronouncements *know* those pronouncements to be true, even in the two-thirds of cases when the oracle gets things right.

The PMI treats science like an oracle in that it concerns only the track record of science—the PMI ignores the scientific reasoning behind the theories of science, just as we are in ignorance of the oracle's reasons for making its predictions. According to the analogy, the following are both strong inductive arguments:

> (O) A sizeable proportion of the predictions of the oracle have been refuted in the long run, therefore we should not trust the oracle's predictions in the future.
>
> (S) A sizeable proportion of the successful theories of science have been refuted in the long run, therefore we should not trust the successful scientific theories of the future.

The oracle analogy is, however, a poor one for two (related) reasons.

First, it assumes that science is a single unified entity. We know from Goodman's new riddle of induction that not any old predicates may take the place of 'F' and 'G' in an inductive inference of the form 'all observed Fs are Gs *therefore* all Fs are Gs'. In particular, if 'F' is a predicate that does not pick out a single property or kind but is instead disjunctive in nature, then the induction should be rejected.

Intuitively there is such a unity in (O), at least until further information is forthcoming, provided by the predicate 'prediction of the oracle'—it is a single oracle, after all. Should we assume that there is also an induction-supporting unity in (S) regarding all those things classified as 'successful scientific theory'? A positive answer is far from obvious.[8] The terms that best translate 'science' in different languages are not synonymous: famously the German 'Wissenschaft' includes the humanities whereas the English 'science' does not. Variance among these concepts of science may suggest that there is no distinctive feature of science that would form the appropriate basis for the PMI's induction. Perhaps the concept of science is a family resemblance concept (Simon 1969): in German the relevant resemblances are taken further, in English they are more restricted. If that hypothesis were correct, then 'scientific theory', even in English, will not pick out a set of entities with shared properties that would form the basis of a satisfactory induction. Failure to find an agreed demarcation principle may be explained by the lack of any single such principle, as the family resemblance proposal would predict.

Secondly, the oracle's failures deprive us of knowledge, even in cases where the oracle is correct, precisely because we know nothing about the oracle's process of making predictions. We have no more information relevant to any given prediction than that it is a prediction of the oracle. Hence induction on the oracle's track record

[8] It is true that our predicate is 'successful scientific theory' and things falling under this predicate do have something in common—they are empirically successful. But we should not think of that as providing the required unity. It would be odd if the successful theories had something unifying them that made them susceptible to the *Pessimistic* Induction that was lacking for scientific theories in general. The PMI doesn't want to claim that—after all it thinks that things are even worse when we think about unsuccessful theories. The reference to specifically successful theories is intended to strengthen the conclusion by making it specific: we can expect *even* successful current theories to be falsified.

is the best we can do. But that is not the position we find ourselves in regarding science. On the contrary, we know in detail how science comes to favour its theories. In particular we are able to see the arguments and evidence that are the scientists' reasons for favouring those theories. Once the oracle starts giving reasons for its assertions, the rational response is not to ignore those reasons and continue with an induction on the oracle's track record; the rational response is to assess those reasons.

Let us say, then, that we do consider the oracle's reasoning, while continuing to note the oracle's track record. And we find that we accepted the oracle's predictions and reasons on many occasions when it later turns out that the oracle is in fact in error. We were therefore at fault in accepting the oracle's reasons. Thus the PMI is saying in effect that scientists' reasons are inadequate, that there is a systematic failure in scientific rationality. Although the reasoning of scientists *seems* cogent to us at the first-order level, when we engage in higher-order reasoning (the PMI) we find we have reason to doubt the truth-conduciveness of that first-order reasoning.

I conclude from this subsection that the PMI needs to argue that:

- There is a genuine unity to science that allows an induction on the predicate 'successful scientific theory'. For the induction to hold good, that unity should be non-accidentally related to the induced property, the future falsification of the theory.
- The induction on past falsification can give us reason to doubt the cogency of our own scientific inferences. One might expect the nature of the unity of science to give some insight into the failure of scientific reasoning to deliver truth.

(I note, in passing only, that there is some irony in the PMI using an inference with the form and character of a scientific inference to cast doubt on the cogency of scientific inferences. I do not think, however, that this means that the PMI is self-refuting, although it does put the supporter of the PMI under some obligation to explain why the second-order reasoning succeeds where the first-order reasoning fails.)

9.3.2 *Underconsideration and scientific method*

In this subsection I consider three related proposals that meet the two requirements on the PMI laid out at the end of the preceding subsection. First, I examine the proposal that science is unified by its use of the scientific method and it is because the scientific method is flawed that successful theories are frequently refuted. Secondly, I look at the thought that it is because scientific inferences are ampliative that they are falsified. Finally, I consider the more specific accusation that problems arise because scientific inference requires the comparison of competing theories: scientists are frequently unable to even think up the correct theory in a given field.

A natural proposal for what unifies science is that scientific theories, successful ones at least, are developed, evaluated and so adopted as a result of the application of *the scientific method*. If that were so, then the explanation of the ultimate failure of initially successful theories is the fact that the scientific method, used to select all preferred theories, has a flaw that means that those theories will not withstand empirical scrutiny for ever. Theories are then rather like cars manufactured on an assembly line with a poor design that builds a fault into each car produced. The cars may work satisfactorily for a while, but in due course the fault will cause the cars to break down. For example, one might say that the fault is that the scientific method is unable to distinguish between true theories and those false theories that are adequate to current evidence.

If this were so, then one might expect that one should be able to identify the scientific method. However, what is striking about the scientific method is that despite its obvious significance for science and for philosophy of science (should it in fact exist), neither scientists nor philosophers have been able to say what the scientific method is. James Blachowicz (2009) shows that there is no agreement among authors of scientific textbooks about the scientific method or its principal ingredients. Furthermore Blachowicz also shows that scientific *fields* differ significantly from each other on the elements of the scientific method that they regard as important. For example there is some tendency in biology textbooks to emphasize the continuity of science and everyday thinking whereas psychology textbooks reject the assimilation of science and everyday activities. Philosophers of science have fared no better in articulating what the scientific method is.

One response to the criticism raised here would be to point out that even if there are multiple methods of science, they will all have in common the fact that the hypotheses they generate or support (such that they become the leading theories in their fields) all go beyond the evidence. There is an ampliative inference from the evidence to theory. Ampliative inferences can go astray, and the evidence of the PMI shows that they frequently do, so the argument goes.

My own view is that not all inferences from evidence to theory are straightforwardly ampliative—Inference to the Only Explanation is not. But let us put that on one side, accepting that we do often make ampliative inferences in science and that our best theories are often (logically) underdetermined by the evidence. In that case, the PMI should run thus:

> Ampliative inferences, because their conclusions are underdetermined by the evidence, are fallible. Falsification of well-supported theories shows that such inferences frequently do fail. We should therefore be sceptical concerning the truth of current and future theories based on ampliative inferences.

The previous proposal erred in looking for a feature, the scientific method, that is common to all or most science and is also specific to science. But there is no such feature. However, even if there are many distinct methods of science, all or most of them embody or imply inferences that are ampliative. So the current proposal avoids the previous problem by declining to be constrained by the specificity of science—the ampliativeness of scientific inferences is shared by many non-scientific inferences.

This, however, would prove too much. After all, we use ampliative inferences, such as simple enumerative inductions, on a daily basis. And while they do sometimes fail, we do not regard that as a reason to be sceptical about every conclusion we draw from ampliative reasoning. Of course, one might be persuaded by general inductive scepticism, but quite apart from undermining the inductiveness of the PMI, that would render the PMI otiose and uninteresting.

We are able to distinguish between different kinds of ampliative reasoning, distinguishing the less fallible from the more fallible: inferences concerning the near future of the middle-sized dry goods in my immediate surroundings are typically secure whereas those concerning complex events in the more distant future (such as the outcome of an election next year) are highly fallible.

So the PMI needs to find a property on which to base its induction that does not claim to be coextensive with science and yet which encompasses the theoretical claims of science while not extending so far beyond science that claims to everyday knowledge are included also. For example, the argument just considered, the one appealing to the ampliativeness of scientific inferences, would need to be supplemented with the additional claim that the ampliative inferences used in science are of the more fallible kind (again as evidenced by a record of falsification).

An argument that seeks to do this is that which Lipton (1993; 2004: 148–63) calls the 'argument from underconsideration' and which Stanford (2006) calls the 'problem of unconceived alternatives'. As we have seen, in many cases the inferential relationship between evidence and theory is one of Inference to the Best Explanation (IBE) or Inference to the Only Explanation (IOE). In the former competing theories are ranked according to how well they explain the evidence. In the latter evidence rules out all but one of the competing hypotheses. In either case, the inference is only any good if the true hypothesis is among those actually considered and evaluated. As van Fraassen (1989: 143) says, our selection may well be only the best of a bad lot. Van Fraassen and Lipton raise this as an apriori objection whereas Stanford treats it more as an empirical objection, deploying case studies that show that scientists have failed to consider true alternatives.

The argument has been resisted on various grounds. When we discussed the argument previously (p. 190), I maintained that Stanford's case studies were a weak basis for a generalization across all of science. In some instances, such as the oxygen theory of combustion (Hoefer 2020), it is very difficult to see how we could have failed to conceive of the correct explanation, even if in other instances we might admit that there could be other explanations, even if we cannot say exactly what they are. Here I wish to expand on that response by pointing to the *variety* of the possible scientific problems to which IBE/IOE can be applied. In my view this variety means that the problem of underconsideration is a genuine concern in some cases but not in others. At one extreme, we may take cosmology and fundamental physics, where the goal is to come up with theories of universal applicability: such laws aim to cover all things at all times in all places. Our evidence for such theories, while impressive in itself, is thin compared to the domain the theories cover. Much of it comes from a small corner of the universe at a particular time. Astronomy allows evidence of the deeper past and further reaches of the universe. But still the domain of applicability of the theories covers regions and times as well as scales regarding which we have little or no evidence. Hence in such cases it is plausible that our theory choice is underdetermined by the evidence and in particular that there may be theoretical possibilities that we have not conceived of and perhaps cannot (in the present state of knowledge) conceive of. Therefore we may not be especially surprised if in due course we find evidence that does refute a currently favoured hypothesis. On the other hand, not all theories in other branches of science are quite so bold in their ampliative ambitions. A palaeontologist's theory about the evolutionary relationships between certain extinct species will not go so wildly beyond the data. Indeed, concerning some important questions of this kind, it is easy to see what all the competing theories are: *a* is the ancestor species of *b*, *b* is the ancestor species of *a*, and *a* and *b* have a common ancestor. We may be able to find evidence that decisively chooses between these three hypotheses. In another kind of case, we may be able to conduct a controlled trial in order to establish or refute a causal hypothesis. If we are able to refute the null hypothesis then the causal hypothesis is established. The point here is not that underconsideration is never a problem. Rather it is that underconsideration is a problem to different degrees in different parts of science. It is also plausibly not a significant problem at all for some particular scientific inferences. Furthermore, such differences in susceptibility to underconsideration are to some degree detectable by inspection of the first-order scientific inferences.

That said, it is may well be that scientists in a particular field might be overconfident about their ability to conceive of a full range of plausible hypotheses. This overconfidence might subsequently be revealed by evidence that they failed to conceive of the full range of possibilities. However, that itself need not be a basis for a strong PMI. It is important to note that theoretical physicists are typically cautious about claims to truth for the more universal of their hypotheses. If their late nineteenth-century counterparts were overconfident about their ability to think of the true universal theories of gravity, motion, and electromagnetism, that very history is one reason why today's physicists make more restricted claims about the absolute truth of their best theories (e.g. claiming that the theories are good approximations to the truth regarding certain domains of application). Consequently the PMI might be ineffective against the more cautious claims of today's physicists. Moreover, the physicists' caution does not carry over to palaeontology and epidemiology and their very different theories, nor is there any good reason why it should.

9.3.3 *Conclusion—no global Pessimistic Induction*

For the PMI to be a credible inductive argument for a serious scepticism about science, it needs first to argue that science has the appropriate kind of unity that will allow all of science be tarred with the same brush of fallibility as the cases cited in the inductive sample. Moreover, whatever provides that unity must be something that we may plausibly think has a non-accidental link with a propensity to future falsification. At the same time, that should not be something so general (such as 'being an ampliative inference') that the PMI would in effect become an argument for a scepticism that reaches well beyond science.

The proposal that comes closest to meeting these conditions states that theories in science must be selected from among competitors and that this process is subject to the problem of underconsideration. But even this fails to show that the problem is a global one for all theories in all fields of science. The PMI applied to a particular kind of theory in a specific field might have a local force. That force will not translate to theories of different kinds or in other domains. And it may undermine the applicability of the PMI to the extent that scientists themselves are more circumspect regarding the epistemic claims they make about their leading theories.

9.4 The No Miracles Argument Mark 2

9.4.1 *The explanationist defence of scientific realism*

Richard Boyd (1981: 617–18; 1985: 59–60; 1989: 8; 1990*a*: 360; 1990*b*: 181) puts forward an argument for scientific realism that is developed by Stathis Psillos (1999), who generously calls it the 'Putnam–Boyd explanationist defence of realism'. The argument is, however, different from Putnam's and a separate treatment is called for.

Putnam's argument focuses on theories and their predictions—the best explanation of the predictive success of theories is the truth of those very theories. But Boyd, on the other hand, is concerned not with the theories that make the predictions, but with the theories that are bound up in the methodology of science. Psillos (1999: 78) puts the argument thus:

> Scientists use accepted background theories in order to form their expectations, to choose the relevant methods for theory-testing, to devise experimental set-ups, to calibrate instruments, to assess the experimental evidence, to choose among competing theories, to assess newly suggested hypotheses, etc. All aspects of scientific methodology are deeply theory-informed and theory-laden.

Psillos (1999: 79) goes on to explain: our explanandum is 'a general feature of scientific methodology—its reliability for yielding correct predictions.' The best explanation of this is that 'the [background] theories implicated in this methodology are relevantly approximately true.' Psillos takes a further step. He notes that these background theories have themselves typically been arrived at by the use of Inference to the Best Explanation. And so we may infer that IBE is generally reliable at leading to the truth.

What I would prefer to call the 'Boyd–Psillos argument' differs from Putnam's in that it concentrates on 'scientific methodology' and the background theories that are part of that methodology. Background theories we may contrast with 'foreground' theories. The foreground theories are the theories that are being tested by the scientific methodology. That methodology is laden with the background theories. The focus on background-theory-laden scientific methodology, rather than the foreground theories, might seem puzzling. For scientific methodology does not itself make any predictions. What makes predictions is the foreground theory (along with auxiliary assumptions). So I take it that what Boyd and Psillos mean by saying that scientific methodology is successful in leading to correct predictions, is that the methodology endorses and recommends certain foreground theories to scientists, and these foreground theories make correct predictions. For example, foreground theories in palaeontology are tested using radiometric dating. That method employs theories of radioactive decay as background assumptions. The palaeontological theories that are selected as a result of this testing make predictions (or retrodictions) and these are successful. Now those successful predictions will be cited by the palaeontologists as further evidence for their theories. But Boyd and Psillos regard that success as evidence for the background theories of radioactivity.

Why do Boyd and Psillos focus on the background theories? There are two reasons. Boyd is explicit about the first. A key target is the defence of realism against instrumentalism, and in particular van Fraassen's constructive empiricism. A second reason is that the argument for the truth of the background theory is stronger than the argument for the truth of the foreground theory.

NMA and instrumentalism A key purpose of the Boyd–Psillos explanationist defence of NMA is to see off a challenge from van Fraassen's version of instrumentalism, according to which the aim of science is the acceptance of theories that are empirically adequate (van Fraassen 1980). Van Fraassen thinks that we can expect science, in good cases, to satisfy this aim. To assert simply that science can and does also meet an aim of truth would be to engage in a stand-off with the instrumentalist. The Boyd–Psillos argument points to the theory-laden methods that selected the theories that the instrumentalist thinks are empirically adequate and so are successful by instrumentalist standards. The best explanation of that success is that the selection methods are reliable, as regards selecting empirically adequate theories. And this reliability means that the background theories involved in those methods must by and large be true.

If that is correct, Boyd and Psillos cannot be accused of begging the question against the instrumentalist. For they emphasize the instrumental success of theory-laden methodology. But that instrumental success is best explained by the *truth* of the background theories with which that methodology is laden.

Background theories are better established The background theories implicated in the methods of science are typically better confirmed than the foreground theories. Both foreground and background theories can take confirmational credit for the predictive success of the foreground theories. However, the foreground theory has just these predictions to its credit whereas the background theory has to its credit not only its own predictions (when regarded as a foreground theory) but the successes of all the foreground theories that it, as a background theory implicated by our methodology, helped select. For example, the background theory of radioactivity is employed in many instances of the use of radiometric dating in selecting palaeontological theories (not to mention many other uses besides). So the subsequent success of all the theories that radiometric dating was used to select provides confirmation of the theory of radioactivity underlying radiometric dating. And that is a much greater quantity of confirmation than any one of the individual foreground palaeontological theories in question.

It might appear that the move to background theories provides an argument that is similar to the first-order arguments for foreground theories except stronger—more evidence, more reliable inference to the truth. We should, however, note that Boyd and Psillos want to draw a broader conclusion, one concerning the reliability of scientific methodology in general. And that conclusion is not just a generalization over many cases such as this. Psillos is explicit on this point—the argument aims at establishing the general reliability of IBE (as mentioned above). (This mirrors the PMI's attempt to draw a general conclusion about the unreliability of scientific methodology from the fact that it produce false theories in a variety of different cases.)

9.4.2 *An aside on foreground and background theories*

There is an important normative dimension to the distinction between foreground and background theories. The Duhem–Quine thesis in effect says there is no sharp distinction between these categories: all theories involved in testing, foreground and background, are themselves being tested. An experimental observation or other evidence that contradicts the predicted outcome casts doubt on all those theories. That doubt need not be equally distributed, for the reason given two paragraphs back—some propositions are better connected in the web of belief and are so better confirmed overall. Even so, the background theories involved in the testing receive some of the doubt too. The refutation of the Duhem–Quine thesis in Section 7.5.1 pointed to a crucial asymmetry between the target hypothesis under test and an auxiliary hypothesis assumed in the test. In a typical case, the truth (or falsity) of the target hypothesis (the foreground theory) will not be known but the truth of the auxiliary hypothesis (the background theory) is known. Hence, in such cases, falsifying evidence falsifies the target hypothesis specifically, not merely the conjunction of that hypothesis and the auxiliary hypothesis. So if one is aiming to test a hypothesis and to be able to assert that it is false if there is a falsifying outcome, the background theories employed in the testing must be known. More generally, the following principle holds:

> If a test is to supply evidence that will falsify or confirm a target hypothesis, then the background theories (auxiliary hypotheses) used in the test should be *known*.

The rationale for this principle is that to be evidence a proposition must be known. But if the theories upon which our acceptance of that proposition is based are not known, then it too will not be known.

Scientists will therefore want their background theories or auxiliary hypotheses to be known. So we can say that the auxiliary hypotheses assumed in the methods of science include the hypotheses that scientists believe that they know. These are, therefore, among the hypotheses that the scientific community is entitled to assert, if what is assertible is what is known (Williamson 1996*b*). This leads to a sociological point regarding the significance of background theories for scientific realism. While scientists often say that they regard their theories as tentative, they clearly have in mind only their foreground theories, the theories that are at the cutting edge of science. They do not treat as tentative the theories that are used as background theories in the theory-laden methods used to test the cutting-edge foreground theories. The background theories are treated not as tentative but as established. This is important for scientific realism, for it can be unclear which are the theories that a realist should be realist about. Certainly one should not be a realist about everything that is published, even in a reputable journal. This would be far too inclusive, since even the best journals will publish research that falls short of the highest standards of confirmation if the idea is intrinsically interesting.[9] There is no clear criterion by which we can tell what it is that 'science is telling us'. However, a sufficient condition for telling what the scientific community really believes (but not, I think, a necessary condition) is that the community has a shared willingness to allow the theory to play the role of background theory in its agreed methods of theory assessment.

9.4.3 *Evaluation of the NMA Mark 2*

My principal criticism of the Boyd–Psillos argument concerns their treatment of scientific methodology and of IBE as epistemic unities that can be validated by a single argument. Quite how strong my criticism is depends on whether one conceives of IBE in the standard way, as *always* using explanatory goodness to select among hypotheses, or as including Inference to the Only Explanation as a (not infrequent) limiting case in which ranking by explanatory goodness is replaced by outright falsification of all but one hypothesis. The Boyd–Psillos argument is on stronger ground if the latter account is correct. Even so, my criticism retains some force.

Before coming onto that, I note that it is not obvious that the instrumentalist should be moved by the argument, for two reasons.[10] First, an instrumentalist thinks that the observational evidence for any foreground theory can never warrant belief in its truth but can warrant belief in its empirical adequacy. If one thinks that, then one can also think that the reliability of scientific methodology in producing empirically adequate theories warrants belief in the empirical adequacy of the background theories but not in their truth. It is not clear why the move to scientific methodology and background theories demands truth in a way that consideration of the foreground theories and their evidence does not.

Secondly, the inference from the fact that many of *this* set of true theories (the background theories) were produced by IBE to the conclusion that IBE is reliable is a poor one, at least without further discussion. For the truth of most IBE-produced background theories is consistent with IBE being fairly unreliable. It could be that IBE as typically employed encourages scientists to accept a high proportion of false theories. Such false theories do get weeded out in due course, before they have been around long enough to become established as background theories that may be safely used in testing other theories. So it might be that IBE is in fact fairly unreliable (but better than guesswork), but that the combination of IBE plus the weeding out effect

[9] This can be taken too far, since it was the justification given by the editor (Horton 1998) of the *Lancet* for publishing the infamous paper by Wakefield *et al.* on a supposed connection between the MMR vaccine and autism.

[10] Another concern is that this argument is circular, since it uses IBE to establish the reliability of IBE. I reject this concern for the epistemologically externalist reasons that Psillos and others give.

of time is reliable. That proposal is not just a conceptual possibility. It is not absurd to suggest that a 'first pass' IBE directs us to cutting edge theories many of which are false and then the majority of the latter are refuted by subsequent research. That is why scientists think that even well-confirmed cutting-edge theories are only tentative. And for that same reason, scientists are only willing to allow theories to be used as background theories in testing once they have stood the test of time. If that is correct, then there isn't much to be inferred about the reliability of IBE in general from the success of science or about 'scientific methodology' more generally. One might try to draw the conclusion that IBE+the-test-of-time is a reliable methodology. That will be addressed by my principal concern with Psillos's proposal, which now follows.

So the NMA Mark 2 need not be persuasive to the instrumentalist or even to someone who is neutral on the matter. But that does not make the NMA a poor argument (even if, as I suspect, the authors would like it to have broad persuasiveness). For it cannot be a condition on a correct response to a sceptic that the response must persuade the sceptic. I will now move on to a challenge to the NMA to which I think Boyd and Psillos do need a reply. I will first present the argument in terms of IBE as standardly conceived—which is how Boyd and Psillos understand it. And then as regards IOE.

The NMA and Inference to the Best Explanation An inference to the general reliability of IBE from cases where IBE has been used and truth has been the result, depends on there being sufficient uniformity in those cases that such an inference is warranted. As we saw with the PMI, the inference won't be warranted if what goes by the name 'IBE' differs significantly from one field of science to another. Which it does, because what counts as a 'good' explanation varies across the sciences. For example, simplicity is widely regarded as an explanatory virtue. But what physicists regard as making a theory simple (simple equations, symmetry) is different from simplicity in a theory in biology (unification of phenomena, causal mechanisms with fewer steps and components). So the standard model of particle physics and the germ theory of disease are not sufficiently alike that one can say that they were both better than their rivals *in the same way*.

For this reason I do not think that there is a single inference procedure or method for forming preferences among hypotheses that one can call IBE.[11] *A fortiori* I don't think that there is a single uniform 'scientific methodology', as Boyd and Psillos conceive of it, employing IBE at its heart. In Section 7.6 I explained how the standards of explanatory goodness employed within a scientific field are derived from the exemplars, the established, successful theories that are models of good science in that field. And since different fields have different exemplars, the standards of explanatory goodness will differ between fields. By a difference of 'standards' I mean a difference in what sort of thing counts as good, what goodness looks like—which is why simplicity in physics looks different from simplicity in biology. We should also expect,

[11] Douven (2017*a*: 11) says that Inference to the Best Explanation is a slogan that should be fleshed out differently in different circumstances.

therefore, standards to differ between fields in another sense. Which is to say that the effectiveness of the standards of goodness in directing us towards the true explanatory theory can be expected to vary also. There is no reason to suppose that simplicity-in-physics is as good at directing us towards the correct physical theory as simplicity-in-biology is at directing us towards the correct biological theory. So we should not think that there is a single, uniform inference procedure, IBE, about which we can make inference regarding its general reliability across all of science.

The NMA and Inference to the Only Explanation The argument I have just presented concerns IBE as standardly conceived: as providing standards according to which we can rank competing hypotheses, such that the higher-ranked hypotheses are more likely to to be true and the topmost can be inferred, if sufficiently better than the rest, to be *the* correct explanatory theory. While Boyd and Psillos take the standard view of IBE, in Chapters 7 and 8 I have articulated a view that differs in important respects. While 'standard IBE' does play a role in scientific thinking, in generating rational credences less than 1, it cannot deliver knowledge. And as I explained above, it is important that our background theories (the subject of the Boyd–Psillos NMA Mark 2) are known. So they are not obtained by standard IBE. Instead they are obtained by the Holmesian inference, Inference to the Only Explanation. If that is correct, then the argument of the preceding paragraphs loses some of its force. For, when we use IOE, we do not rank hypotheses according to standards of goodness that vary across the sciences. Rather, the evidence rules out all but one hypothesis. So the reliability of IOE is logical rather than contingent and the objection raised does not apply.

Even so, I do not think that we can simply resuscitate the Boyd–Psillos argument by substituting IOE for IBE, for several reasons. Note that when we here talk about the reliability of IBE or IOE, we are talking about the reliability of our deploying IBE or IOE. It is apriori that IOE leads to true conclusions from true premises. Nonetheless, while the *validity* of IOE is not in question, the *soundness* of any particular instance of IOE can be questioned. In some cases the premises (Selection, Determinism, Falsification) of an IOE are clearly true, in other cases they may not be. We might deploy IOE without knowing that the premises are true and so fail to get to know the conclusions.

First, recall the structure of the Boyd–Psillos argument. It intended to show that the use of IBE in selecting foreground theories is reliable because IBE is shown to be reliable in selecting background theories. While I have said that IOE must be at work in the selecting background theories, it is far from always the case that when we are considering cutting-edge foreground theories we are in a position to use IOE—presumably we are often in the position of not having excluded all but one hypothesis and so can have only a preference generated by IBE. So the reliability of IOE (in any case guaranteed by logic) is no guide to the reliability of the use IBE to form a preference among foreground hypotheses, which in many cases will be the best that scientists can do (at that point in the development of the science).

Secondly, although IOE obviates the problems about ranking (Hungerford's objection and Voltaire's objection) it does not *per se* solve the problem of underconsideration. Above (p. 190 and p. 237) I discussed the problem of underconsideration for

IBE, in the context of its being a general reason to reject IBE's reliability. There I noted that the strength of this concern varies from one field of science to another. So one could not draw a general conclusion about IBE's *unreliability*. For the same reason one cannot draw a general conclusion from the Boyd–Psillos argument about the *positive* reliability of either IBE or IOE. Scientists' ability reliably to draw up competing hypotheses that include the truth may be high in some areas (geology, palaeontology) and lower in others (some subfields of cosmology).

Thirdly, the preceding paragraph in effect pointed out that it easier to establish the premise Selection of IOE (see Section 7.3.1) in some fields of science than others. The same is true of the other premises. For example, Determinism is the premise that the phenomenon in question has an explanation. In some cases this is clear—that the mass extinction at the K-Pg boundary has some explanation (possibly a combination of causes) is not in doubt. In other cases, it can be hard to determine whether a phenomenon, such as an interesting correlation, has an explanation or is just a stochastic artifact. Falsification is the premise that the hypotheses other than the one inferred are falsified by the evidence. Again this can be easier to establish in some cases than in others.

Fourthly, although IBE and IOE are important elements of scientific reasoning, they are not the sum total of scientific methodology. Science is replete with myriad methods, procedures, rules, and heuristics that play a part in influencing scientists' beliefs and preferences among hypotheses and other scientific propositions. Some of these are common to more than one field, whereas others are more specific in their application. Radiometric dating is a method for determining the age of rocks and other matter that is useful in geology, palaeontology, and archaeology. Null hypothesis significance testing is an inference procedure of rather wider application. As in the case of radiometric dating, the methods scientists use are very frequently themselves the products of science. As science grows, so will the number of such methods. As discussed, scientists want it to be the case that the science from which such methods derive is sufficiently well-established that it is knowledge. But on some occasions the methods might not derive from reliable science. Again, there will not be any kind of unity to the methods that allows us to infer from the reliability of some to the reliability of others

9.4.4 *Conclusion*

Boyd and Psillos *do* talk of theory-laden methods (in the plural) while nonetheless focusing on a single inference procedure, IBE. This implies that IBE is not the only method of science—there are other means of generating scientific conclusions. Being theory-laden these other methods may be of varying reliability. And they will not be of general applicability but will be specific to certain applications or subfields. For this reason alone, we cannot think of all of science's conclusions as falling within the domain of a single No Miracles Argument.

The picture of 'scientific methodology' that I have been drawing describes many methods, rules, and techniques that are typically heavily dependent on prior scientific results, both theoretical results as well as singular empirical propositions. For example, the design and so also the scientific use of a resistance thermometer (or RTD) depend both on the theory of the temperature dependence of electrical resistance and also on propositions concerning the value of constants, such as the value of the temperature coefficient of resistance of platinum and nickel. While in some cases the relevant background scientific facts are well established, in other cases that may not be so. In some cases, indeed, we might come to discover that a background theory or proposition which we thought was well established is in fact in error. Where there is doubt there may be methodological dispute. These disputes will typically be local, important in establishing what is a reliable method in some fields but irrelevant to others.

Such methodological disputes are more frequent in immature sciences. The rejection of Galileo's telescope as a tool for astronomy by Cesare Cremonini and Guilio Libri is a well-known (if often over-stated) example.[12] The phrenological theory of Franz Gall was used to generate 'data' on the psychological and behavioural dispositions of many individuals and types of people. In due course the theory upon which those false observations were made fell into disrepute among scientific psychologists (though not among others). Remaining with psychology, Wilhelm Wundt's method of experimental self-observation (introspection) assumed that a subject's own mental states are transparent to a suitably trained and careful observer.[13] Psychologists no longer accept this assumption, and introspection has consequently been relegated as an experimental technique. Because such methods are themselves the products of science, as science grows there are more methods and so more opportunities for methodological debate. So methodological disputes may be found in mature sciences also. Molecular clocks are a key technology for establishing phylogeny and dating evolutionary changes. But the assumptions underlying their reliability have been questioned, so that the standard reading of the evolutionary development of the bilateria and its relationship to the fossil record might need radical revision (Budd and Mann 2020).[14] Candidate gene studies aim to establish a causal link between prespecified allelic variants and phenotypes (e.g. that a certain allele increases the risk of depression). This methodology turns out to be rather less reliable than was initially supposed (Hirschhorn *et al.* 2002; Sullivan 2007).[15] The debate between Bayesians and frequentist statisticians is a debate of broader interest in science, though many of the methodological debates and disputes in statistics concern specific techniques and inference procedures. Bayesians tend to reject altogether the dominant method-

[12] There is evidence that Cremonini did look through a telescope, claiming that it gave him a headache. But it is clear that his reluctance was also theoretically motivated. Debate over related instruments continued for some time—Margaret Cavendish rightly cautioned against the use of weather glasses (my thanks to Sophie Allen for this case). Cavendish also questioned the value of the imperfectly reliable microscopes of her day. For a detailed assessment of Cavendish's views on instruments and experimentalism see Wilkins (2014).

[13] My thanks to Celia Heyes for this example.

[14] My thanks to Peter Godfrey-Smith for this example. Andrej Spiridonov has suggested another example from palaeobiology where a much-used method for avoiding sampling bias in the assessment of biodiversity in the fossil record has come under question (Sakamoto *et al.* 2017).

[15] My thanks to Marcus Munafò for this example.

ology used to assess the evidence of controlled trials, *viz.* null hypothesis significance testing (NHST). But even some of those who accept the methodology believe that it needs to be amended in order to be reliable, by changing the significance level (α) for such studies from 5% to something smaller (Benjamin *et al.* 2018). Below I mention the replication crisis. My analysis (Bird 2021) of the causes of the crisis suggests that NHST with an α of 5% can be reliable in some fields but unreliable in others.

What we learn from such methodological disputes and reversals in science is that they are almost always very local. They turn on issues specific to some subfield of science. Their resolution affects that subfield and may have some, limited consequences for other subfields. Even the broadest methodological disputes (e.g. between Bayesians and frequentists) are not universal. Consequently the success or failure of such methods tells us little that has a bearing on the global claims of scientific realism and anti-realism.

9.5 Local metascientific knowledge

My argument is consistent with there being metascientific knowledge that is local in the sense that it applies to particular fields or sub-fields of science. For these can be unified by shared features that may be the basis of a meta-induction.

For example, some fields within biomedicine and psychology are said to be suffering from a crisis of replication (Baker 2016). What seemed to be well-established experimental results cannot be confirmed by subsequent studies of similar or improved quality. Either the reported effects cannot be found or they are found but with much smaller magnitudes than were first reported. This suggests that hypotheses are accepted in these fields more readily than they should be.

The track record of hypotheses in these fields is a reason to suspect the reliability of the methods that they use, but not a reason to doubt the methods employed in other fields. While various explanations of the replication crisis have been put forward, the fact that the crisis is more acute in these particular subfields (some areas of biomedicine but not others, social psychology but not animal-based experimental psychology) suggests that the cause might be some feature of the methodologies of these fields that is not shared with other scientific fields. One feature shared by these fields is null hypothesis significance testing with a significance level (α) of 5%, which is equivalent to 1.96 standard deviations from the mean. When particle physicists use significance testing, they regard a result as established only when a significance level of five standard deviations (5σ) is achieved. The fields in question, also in contrast to physics, lack strong basic theories to guide research and their experiments and trials often seek small effect sizes with small samples (Bird 2021).

Whether the features just mentioned are the cause of the replication crisis is not to be settled here. My point is that it is prima facie plausible that they might be. If so, that gives us a reason for thinking that while current methodological practices remain we should have a lower confidence in the truth of a successful theory in these fields than we would otherwise have had. These reasons, however, do not carry over to other fields. So a local Pessimistic Induction is in order, but not a global one.

Should methodological practices change (e.g. by employing a lower, more stringent value of α, or by concentrating resources on testing fewer, more plausible hypotheses) replication rates may improve. If that occurred, one might reason that the best explanation for the improved replication rate is that the methods employed by the field have become more reliable. And it would not, perhaps, be inappropriate to regard that reasoning as a species of the No Miracles Argument. Again, this is a local argument which might justify an increased confidence in the results of this field, without that raising our confidence in the results in other scientific fields.

9.6 Conclusion

Magnus and Callender (2004) characterize both the NMA and the PMI as 'wholesale' arguments for scientific realism and anti-realism respectively. They reject both as depending on the use of a base-rate fallacy. Instead, they say, discussions of realism and anti-realism should focus on retail arguments: arguments regarding specific scientific theories as to whether we should believe in the truth of those theories (and the entities they posit, etc.). The argument of this chapter is highly sympathetic to their conclusions, albeit for different reasons. The terminology I have used is also different from their 'retail' and 'wholesale'. I have used 'first-order' and 'metascientific' in order to signify that the objects of my contrast are the arguments of the scientists (about the specific scientific subject-matter) on the one hand and the arguments of the philosophers (about that scientific reasoning) on the other hand. I have also used 'global' and 'local' to subdivide metascientific arguments into those concerning the whole of science and those concerning some particular scientific sub-field or specific kind of scientific reasoning. Magnus and Callender's terminology of 'retail' and 'wholesale' correspond to my 'local' and 'global' since even retail arguments for realism or for anti-realism will be about the arguments of the scientists (rather than being identical to those arguments) and so will be metascientific.

Clearly there is some symmetry between the NMA and the PMI. The NMA is a combined philosophical and historical argument that seems to supplement the first-order arguments of the scientists whereas the PMI is a combined philosophical and historical argument argument that aims to undermine the first-order arguments of the scientists. This ought to strike us as odd. Are scientists underselling or overselling their theories? For example, a scientist considers with care the evidence and reasoning supporting a hypothesis h. She concludes that this warrants her having a moderately high degree of belief κ but no more. Assuming that the NMA were a good argument and she were to consider the NMA, what should her credence in h be now? Should it be greater than κ? If we say 'no', then the NMA is redundant—the epistemic work is all done by the first-order scientific arguments (as I have argued in Section 9.2). If, on the other hand, we say 'yes, her credence should increase', then we may ask why it should increase. If the NMA were to present new evidence relevant to h then one might see how her credence should change. But NMA says nothing about h in particular. An argument might rationally increase the scientist's credence in h if it were to increase the rational confidence in the methods and reasoning she uses. Again the NMA says nothing about her specific methodology. Perhaps the NMA increases her confidence in her methodology *qua* 'a method of science'—since the NMA tells us (in effect) to have confidence in the methods used in science. But why should it do that? First, her confidence in her methods might be the result of a careful assessment of the relevant evidence for their reliability (which will be local to her field). And the NMA cannot change that. Secondly, there is an implicit but mistaken assumption that the methods of science are unified, for example that they are all manifestations of a single scientific method. Without that assumption, there is no reasonable argument about the rest of science that has implications for how our scientist should assess the reliability of her methods.

It is an oddity of the NMA that it tells us that our first-order scientific arguments underestimate the true rational credibility of our theories. This does not *look* quite so odd if one thinks that there is a unifying method to science, such that reflecting on its reliability should boost our confidence in our first-order scientific reasoning and its results. This commitment is pretty much explicit in Boyd's and Psillos's talk of 'scientific methodology'. But in reality the oddity does not disappear—it simply shifts. For, as I have argued in Section 9.2, there is no general explanation of the success of science that is not equivalent to the conjunction of the particular explanations by specific theories of the data relevant to them. If there were, we would be able to give an epistemic boost to every successful theory, but without the addition of any new evidence. The NMA therefore assumes a conception of science as methodologically unified in order to avoid *immediate* accusations of illegitimate bootstrapping. But the accusations must return for science as a whole.

The PMI also relies on that conception as a central component of its argument, for its induction requires that there is something common to successful theories that is correlated with their future falsification. The most plausible version of the PMI takes the relevant feature to be the use of Inference to the Best/Only Explanation in the generation of theoretical beliefs and that it is underconsideration (our inability to generate all plausible potential explanations) that explains the past failure of successful theories and which permits the induction. However, even if IBE/IOE is a common feature of theoretical science, it does not follow that all such inferences are equally susceptible to the problem of underconsideration. On the contrary, there are good reasons for thinking that some areas of science may well be prone to the problem while others are not. Without an argument for the ubiquity of underconsideration, we have no PMI.

So I conclude that there is no global metascientific argument either for realism or for anti-realism. In Magnus and Callender's terminology, there are no wholesale arguments in either direction. However, that does not rule out *local* metascientific arguments. Here the NMA-1 and PMI are not quite on an equal footing. As just conceded, there can be a local argument for PMI as applied to a particular field or kind of question. On occasion we may see that an older successful theory is replaced by one that we did not conceive of (nor could reasonably have hoped to conceive of). Such failures of past (successful) theories may well give us reason to question our ability to generate all suitable alternatives in this kind of case. On the other hand, the arguments of Section 9.2 rule out local versions of NMA Mark 1: Truth is not explanatory either globally or locally. Nonetheless, something like a local version of NMA Mark 2 can have purchase for reasons parallel to a local version of the PMI. If the reliability of a method is initially uncertain then its track record can boost (as well as reduce) the credence given to conclusions reached using it.

In the light of the failure of global metascientific arguments, what should one think about the truth-claims of scientists? The answer is that there is no global answer to that question. One should always return to the first-order arguments: if a scientist says that one should believe that theory T is true and that the entities it posits exist, then one should examine the evidence for T (and its competitors) and look at the arguments that support T (relative to its competitors) on the basis of that evidence. One should ask how likely it is that there are unconceived alternatives out there—and a local PMI may inform one's thinking on that point. The core of this is precisely the first-order thinking that scientists engage in, possibly moderated by some degree of very local, second-order, metascientific reflection on the track record of the first-order reasoning of that type. But there is no metascientific shortcut that avoids having to engage with the first-order science itself.

This conclusion shares much in common with Penelope Maddy's (2007) Second Philosophy, which Wolff (2019) describes as 'quietistic naturalism'. The Second Philosopher starts from within the practice of science (hence naturalism), and does not step back from or outside of science in order to judge science from a non-scientific perspective (hence quietism). Thus the Second Philosopher rejects both anti-realism and realism insofar as these are extra-scientific theses. This restriction to first-order science leads the Second Philosopher also to emphasize the local in science. This quietistic determination to avoid advancing any theses outside of science leads Wolff to characterize Second Philosophy as a *stance*.[16]

One does not have to be quite so quietistic. After all, if one thinks that anti-realism and realism are mistaken, then one can reject the arguments put forward in favour of those views and point out their flaws (as this chapter attempts to do). Because those arguments are global, one's counterarguments can have a global character without that committing one to any global positive metascientific theses about science. In particular, rejecting global arguments for anti-realism does not commit one to global realism. Compare scepticism about perception. One can reject such global sceptical arguments with global replies (e.g. by rejecting the epistemologically internalist assumptions of scepticism) without thereby committing oneself to the positive global thesis that every ordinary perceptual belief delivers knowledge. Rather the global rejection of scepticism allows that perceptual beliefs *might* deliver knowledge; whether they actually so do on any particular occasion is a local matter to be determined in ordinary, local ways. Likewise one can reject a global argument for anti-realism without being committed to the view that all or most of our best theories are true or highly truthlike. What credence one gives to any particular theory will depend on the

[16] There are similarities to Fine's (1984) Natural Ontological Attitude (NOA).

local, first-order details (how well that theory is supported by the relevant scientific evidence and argument).[17]

But, it might be responded, will that not end up being a form of realism, because on taking the local, internal-to-science perspective, one will thereby endorse what the scientists say? Hence the suspicion that Second Philosophy is a sort of *sotto voce* realism.[18] Not quite, for the epistemic attitude that scientists take to the leading theories in their fields is not uniformly one of full-on belief, nor even a high degree of credence in a high degree of verisimilitude. As noted earlier, scientists often take a rather weaker view, especially of cutting-edge theories, holding them to be tentative or even likely to be superseded. Such attitudes depend on the field, and the current state of evidence, and so on. Endorsing the scientists' opinions does not lead to a blanket realism. Furthermore the possibility of local metascientific arguments allows for some adjustment of the scientists' credences, depending on the track record of the methods they use.[19] Responses to the replication crisis that suggest methodological changes have that character.

The conclusion that we must consider the local scientific arguments is of course frustrating for philosophers who cannot be experts in many fields of science—it is no surprise that it is tempting to reach for the NMA or PMI as a sword to cut through that Gordian knot. If we are not sufficiently expert to take a view on the first-order arguments then the next best thing we can do is to defer to those who are experts. If they are in agreement then it is highly likely that we would share in that agreement were we experts also. Again, this deference may be galling to philosophers, but we should not try to avoid it by appealing to metascientific philosophical arguments. The NMA in particular is a tempting argument for those who want to persuade others who are not scientifically expert of the truth of science's theories. But it is not possible to avoid the work of understanding and explaining the science by airily pointing to the 'success' of science and declaring that this success would be a miracle if science were not largely true. After all, what is this success? The uninitiated may be in a position to appreciate the success of science-based technology. But the successes of science are much more than this—they include such esoteric matters as the precise predictions made by fundamental theories in physics and the corresponding observations made using particle accelerators and radio telescopes. The non-expert is not in a position to appreciate for herself that such predictions and observations are successes. Nor

[17] In my view (Bird 2017) philosophical scepticism should be formulated as a modal thesis: knowledge of type X is not possible. The argument *Sceptical Empiricism* of Section 6.1 is just such an argument. A modest form of scientific realism aims to refute the sceptic's argument that knowledge is not possible. An ambitious (philosophical) realist aims to give a philosophical argument to show that this knowledge is not only possible but actual. This chapter endorses modest scientific realism but rejects ambitious scientific realism.

[18] Rather as NOA has been taken to be a form of realism 'without table-thumping' (Musgrave 1989).

[19] For reasons mentioned in the next paragraph, assessing the track record of a field or method is often not something a non-expert can do, and so will need to be carried out by a scientist with first-order expertise rather than by a non-expert philosopher, sociologist, or historian.

is the non-expert able to judge whether it would be a miracle if a false theory could generate such successes.[20]

In the light of the above, the best argument I can see that resembles the NMA and which delivers a general conclusion about science without delving into the first-order science goes something like this:

> Science-based technology is obviously very successful. So the scientists behind this technology are not charlatans or self-deceiving—they genuinely understand a lot of difficult science. So we can trust what those scientists say: in particular, if they say that the evidence they have for a certain theory is overwhelming evidence in its favour, then we should be willing to accept their testimony and believe those theories to be true. The various fields of science are inter-linked (e.g. through interdisciplinary collaboration). So the broad norms of science (respect for truth, tentativeness about cutting-edge theories, willingness to follow where the evidence leads and to question orthodoxy if need be, etc.) are common to all areas of science. Consequently we should extend the trust afforded to applied scientists (those behind successful technology) to pure scientists, and so we should also be willing to believe those theories that the latter agree to be overwhelmingly supported by the evidence.

This argument starts from the easily discernible success of technology and concludes that we should trust scientists' claims about their theories. The foundation in technological success and a normative unity of science means that this argument does not extend to experts outside of mainstream science (so excluding experts in, for example, theology or various 'parasciences'). While starting from something resembling the NMA, this argument diverges from it quite significantly, for it is not an argument for the truth of the theories of science but is an argument for the reliability of scientific experts. So it does not circumvent the first-order scientific arguments, but is an argument regarding the scientists' evaluation of those arguments. It is an argument that is relevant for those of us unable to assess the first-order arguments ourselves—if we could make those assessments reliably, we would not need this argument. And finally, this argument allows for agnosticism not only about non-science but also about some areas of science where additional evidence (such as failures of replicability or significant disagreement among experts) may call expert reliability into question.

[20] More generally, is appreciation of the success of a theory sufficient to ground an inference to its being true? The NMA is an instance of Inference to the Best Explanation (IBE). But IBE does not say simply 'the best (most successful) explanation of the evidence is likely to be true' *tout court*. There are riders and caveats. In particular, if T1 is the best explanation of the evidence but is only a marginally better explanation than its rival T2, then one should not infer that T1 is true (and T2 is false). So being told that X is a successful theory, even that it is *the* most successful theory, does not warrant believing X to be true or truthlike without further investigation of the alternative theories and the relative goodness of their explanations. Once again, we cannot get far with the metascientific argument—we are forced to consider the merits of the first-order scientific arguments.

Where does all this leave scientific realism and anti-realism? Since these are both global metascientific views, this chapter rejects both. Instead of looking for second-order arguments, we should look at first-order arguments. That is, we should ask of particular theories 'with what degree of confidence should I believe *this* theory?', and the answer is 'with the degree of confidence that is rationally mandated by the evidence and the reasoning from that evidence'. However, that answer is not in the current state of science—neutral between the optimistic and pessimistic attitudes towards science. For many fields of science are sufficiently methodologically sophisticated and well-evidenced that scientists do think that the evidence supports a high credence or even outright belief in certain theories of those fields. But that's consistent with there being other theories in those fields or theories in other fields where the evidence fails to give strong support to the theories. Only by looking at the fields, the theories, and the evidence can one say.

REFERENCES

Abbott, B. P. *et al.* (2016). Observation of gravitational waves from a binary black hole merger. *Physical Review Letters*, **116**, 061102. Authors: B. P. Abbott *et al.* of the LIGO Scientific Collaboration and Virgo Collaboration.

Achinstein, Peter (1983). *The Nature of Explanation.* Oxford: Oxford University Press.

Achinstein, Peter (2001). *The Book of Evidence.* New York, NY: Oxford University Press.

Adler, Jonathan (2015). Epistemological problems of testimony. In *The Stanford Encyclopedia of Philosophy* (Summer 2015 edn) (ed. E. N. Zalta). Metaphysics Research Lab, Stanford University. http://plato.stanford.edu/archives/sum2015/entries/testimony-episprob/.

Agassi, Joseph (2011). Verisimilitude. *Discusiones Filosóficas*, **12**, 61–86.

Aichhorn, Markus, Biermann, Silke, Miyake, Takashi, Georges, Antoine, and Imada, Masatoshi (2010). Theoretical evidence for strong correlations and incoherent metallic state in fese. *Physical Review B*, **82**.

Alai, Mario (2012). Levin and Ghins on the "no miracle" argument and naturalism. *European Journal for Philosophy of Science*, **2**.

Arnold, Alexander (2013). Some evidence is false. *Australasian Journal of Philosophy*, **91**, 165–72.

Assmus, Alexi (1995). Early history of X rays. *Beam Line*, **25**, 10–24.

ATLAS Collaboration (2013). Evidence for the spin-0 nature of the Higgs boson using ATLAS data. *Physics Letters B*, **726**, 120–44.

Bacon, Francis (1620). *Novum Organum.* Transl. and ed. Peter Urbach and John Gibson, 1994. La Salle, IL: Open Court.

Bacon, Francis (1627). New Atlantis. In *Francis Bacon The Major Works* (ed. B. Vickers). Oxford World's Classics edition 2008. Oxford: Oxford University Press.

Baker, Monya (2016). Is there a reproducibility crisis? *Nature*, **533**, 452–4.

Balfour, George W. (1855). English and German midwives (letter to the editor). *The Lancet*, **66**, 503.

Barnes, Eric (1991). Beyond verisimilitude: A linguistically invariant basis for scientific progress. *Synthese*, **88**, 309–39.

Baron, Jonathan (2014). Heuristics and biases. In *The Oxford Handbook of Behavioral Economics and the Law* (ed. E. Zamir and D. Teichman), pp. 3–27. Oxford: Oxford University Press.

Barrett, Jeffrey Alan (2008). Approximate truth and descriptive nesting. *Erkenntnis*, **68**, 213–24.

Bartlett, Frederic (1932). *Remembering.* Cambridge: Cambridge University Press.

Benjamin, Daniel J., Berger, James O., Johannesson, Magnus, Nosek, Brian A., Wagenmakers, E.-J., Berk, Richard, Bollen, Kenneth A., Brembs, Björn, Brown,

Lawrence, Camerer, Colin, Cesarini, David, Chambers, Christopher D., Clyde, Merlise, Cook, Thomas D., De Boeck, Paul, Dienes, Zoltan, Dreber, Anna, Easwaran, Kenny, Efferson, Charles, Fehr, Ernst, Fidler, Fiona, Field, Andy P., Forster, Malcolm, George, Edward I., Gonzalez, Richard, Goodman, Steven, Green, Edwin, Green, Donald P., Greenwald, Anthony, Hadfield, Jarrod D., Hedges, Larry V., Held, Leonhard, Ho, Teck Hua, Hoijtink, Herbert, Hruschka, Daniel J., Imai, Kosuke, Imbens, Guido, Ioannidis, John P. A., Jeon, Minjeong, Jones, James Holland, Kirchler, Michael, Laibson, David, List, John, Little, Roderick, Lupia, Arthur, Machery, Edouard, Maxwell, Scott E., McCarthy, Michael, Moore, Don, Morgan, Stephen L., Munafò, Marcus, Nakagawa, Shinichi, Nyhan, Brendan, Parker, Timothy H., Pericchi, Luis, Perugini, Marco, Rouder, Je, Rousseau, Judith, Savalei, Victoria, Schönbrodt, Felix D., Sellke, Thomas, Sinclair, Betsy, Tingley, Dustin, Van Zandt, Trisha, Vazire, Simine, Watts, Duncan J., Winship, Christopher, Wolpert, Robert L., Xie, Yu, Young, Cristobal, Zinman, Jonathan, and Johnson, Valen E. (2018). Redefine statistical significance. *Nature Human Behaviour*, **2**, 6–10.

Bergmann, Michael (2006). *Justification Without Awareness: A defense of epistemic externalism.* Oxford: Oxford University Press.

Berkeley, George (1710). *A Treatise Concerning the Principles of Human Knowledge.* Dublin: Jeremy Pepyat.

Bird, Alexander (2000). *Thomas Kuhn.* Chesham: Acumen.

Bird, Alexander (2004). Is evidence non-inferential? *Philosophical Quarterly*, **54**, 252–65.

Bird, Alexander (2005*a*). Abductive knowledge and Holmesian inference. In *Oxford Studies in Epistemology Vol.1* (ed. T. S. Gendler and J. Hawthorne), pp. 1–31. Oxford: Oxford University Press.

Bird, Alexander (2005*b*). Naturalizing Kuhn. *Proceedings of the Aristotelian Society*, **105**, 109–27.

Bird, Alexander (2007*a*). Incommensurability naturalized. In *Rethinking Scientific Change and Theory Comparison* (ed. L. Soler, H. Sankey, and P. Hoyningen-Huene), Volume 255 of *Boston Studies in the Philosophy of Science*, pp. 21–39. Dordrecht: Springer.

Bird, Alexander (2007*b*). Justified judging. *Philosophy and Phenomenological Research*, **74**, 81–110.

Bird, Alexander (2007*c*). *Nature's Metaphysics: Laws and properties.* Oxford: Oxford University Press.

Bird, Alexander (2007*d*). What is scientific progress? *Noûs*, **41**, 64–89.

Bird, Alexander (2008). Scientific progress as accumulation of knowledge: a reply to Rowbottom. *Studies in History and Philosophy of Science*, **39**, 279–81.

Bird, Alexander (2010*a*). Eliminative abduction—examples from medicine. *Studies in History and Philosophy of Science*, 345–52.

Bird, Alexander (2010*b*). The epistemology of science—a bird's-eye view. *Synthese*, **175**, 5–16.

Bird, Alexander (2010*c*). Social knowing. *Philosophical Perspectives*, **24**, 23–56.

Bird, Alexander (2012). *The Structure of Scientific Revolutions* and its significance. *British Journal for the Philosophy of Science*, **63**, 859–83.

Bird, Alexander (2014). When is there a group that knows? Distributed cognition, scientific knowledge, and the social epistemic subject. In *Essays in Collective Epistemology* (ed. J. Lackey), pp. 42–63. Oxford: Oxford University Press.

Bird, Alexander (2016). Scientific progress. In *The Oxford Handbook of Philosophy of Science* (ed. P. Humphreys), pp. 544–563. Oxford: Oxford University Press.

Bird, Alexander (2017). Scientific realism and epistemology. In *Routledge Handbook of Scientific Realism* (ed. J. Saatsi), pp. 419–33. Abingdon: Routledge.

Bird, Alexander (2018). Evidence and inference. *Philosophy and Phenomenological Research*, **96**, 299–317.

Bird, Alexander (2020*a*). How can loveliness be a guide to truth? Inference to the Best Explanation and exemplars. In *The Aesthetics of Science: Beauty, imagination and understanding* (ed. M. Ivanova and S. French), pp. 125–45. Abingdon: Routledge.

Bird, Alexander (2020*b*). Scientific realism and three problems for inference to the best explanation. In *New Approaches to Scientific Realism* (ed. W. J. Gonzalez), pp. 48–67. Berlin: De Gruyter.

Bird, Alexander (2021). Understanding the replication crisis as a base-rate fallacy. *British Journal for the Philosophy of Science*, **72**, 965–93.

Bird, Alexander and Pettigrew, Richard (2021). Internalism, externalism, and the KK principle. *Erkenntnis*, **86**, 1713–32.

Bishop, Michael and Trout, J. D. (2004). *Epistemology and the Psychology of Human Judgment*. Oxford: Oxford University Press.

Blachowicz, James (2009). How science textbooks treat scientific method: A philosopher's perspective. *British Journal for the Philosophy of Science*, **60**, 303–44.

Bogen, James (2017). Theory and observation in science. In *The Stanford Encyclopedia of Philosophy* (Summer 2017 edn) (ed. E. N. Zalta). Metaphysics Research Lab, Stanford University. https://plato.stanford.edu/archives/sum2017/entries/science-theory-observation/.

Bogen, James and Woodward, James (1988). Saving the phenomena. *Philosophical Review*, **97**, 302–52.

BonJour, Laurence (1980). Externalist theories of empirical knowledge. In *Midwest Studies in Philosophy V: Studies in Epistemology* (ed. P. French, T. Uehling Jr., and H. Wettstein), pp. 53–74. Minneapolis: University of Minnesota Press.

BonJour, Laurence (1985). *The Structure of Empirical Knowledge*. Cambridge, MA: Harvard University Press.

Boyce, Kenneth (2018). Proper functionalism. *Internet Encyclopedia of Philosophy*. http://www.iep.utm.edu/prop-fun/ accessed 24 February 2018.

Boyce, Kenneth and Plantinga, Alvin (2012). Proper functionalism. In *The Continuum Companion to Epistemology* (ed. A. Cullison). London: Continuum.

Boyd, Richard N. (1981). Scientific realism and naturalistic epistemology. In *PSA 1980* (ed. P. D. Asquith and T. Nickles), Volume 1980, v.2 (Symposia and Invited Papers), pp. 613–62. East Lansing, MI: Philosophy of Science Association.

Boyd, Richard N. (1983). On the current status of the issue of scientific realism. *Erkenntnis*, **19**, 45–90.

Boyd, Richard N. (1985). Lex orandi est lex credendi. In *Images of Science: Essays on Realism and Empiricism* (ed. P. M. Churchland and C. A. Hooker). Chicago, IL: University of Chicago Press.

Boyd, Richard N. (1989). What realism implies and what it does not. *Dialectica*, **43**, 5–29.

Boyd, Richard N. (1990*a*). Realism, approximate truth and philosophical method. In *Scientific Theories* (ed. C. W. Savage), Volume 14. Minneapolis, MN: University of Minnesota Press.

Boyd, Richard N. (1990*b*). Realism, conventionality, and 'realism about'. In *Meaning and Method: Essays in Honor of Hilary Putnam* (ed. G. Boolos), pp. 171–95. Cambridge: Cambridge University Press.

Boyle, Robert (1661). *The Sceptical Chymist.* London: Crooke. Page references to the 1911 edition published by Dent, London.

Bradley, Francis Herbert (1914). *Essays on Truth and Reality.* Oxford: Clarendon Press.

Bragg, William (1936). The progress of physical science. In *Scientific Progress* (ed. J. Jeans, W. Bragg, E. V. Appleton, E. Mellanby, J. B. S. Haldane, and J. S. Huxley). London: George Allen and Unwin.

Brewer, Bill (2005). Perceptual experience has conceptual content. In *Contemporary Debates in Epistemology* (ed. M. Steup and E. Sosa), pp. 2017–30. Oxford: Blackwell.

Brewer, William F. and Lambert, Bruce L. (2001). The theory-ladenness of observation and the theory-ladenness of the rest of the scientific process. *Philosophy of Science*, **68**, S176–86.

Briggs, Rachel (2009). Distorted reflection. *Philosophical Review*, **118**, 59–85.

Brown, Matthew (2009). Science as socially distributed cognition: Bridging philosophy and sociology of science. In *Foundations of the Formal Sciences VII* (ed. K. Francois, B. Löwe, T. Müller, and B. V. Kerkhove).

Browne, Thomas (1646). *Pseudodoxia Epidemica, or, Enquiries into very many received tenents, and commonly presumed truths* (6th (1672) edn). London: Printed by T.H. for E. Dod.

Budd, Graham and Mann, Richard (2020). Survival and selection biases in easrly animal evolution and a source of systematic overestmiation in molecular clocks. *Interfase Focis*, **10**.

Burge, Tyler (1993). Content preservation. *Philosophical Review*, **102**, 457–89.

Burgin, Mark (2009). *Theory of Information: Fundamentality, diversity and unification.* Singapore: World Scientific.

Burnyeat, Myles (1984). Aristotle on understanding knowledge. In *Aristotle on Science: the Posterior Analytics* (ed. E. Berti), Proceedings of the Eighth Symposium Aristotelicum, pp. 97–139. Padua: Editrice Antenore.

Bykvist, Krister and Hattiangadi, Anandi (2013). Belief, truth, and blindspots. In *The Aim of Belief* (ed. T. Chan), pp. 100–22. Oxford: Oxford University Press.
Canciani, Mark, Arnellos, Argyris, and Moreno, Alvaro (2019). Revising the superorganism: An organizational approach to complex eusociality. *Frontiers in Psychology*, **10**, 1–12.
Carter, K. Codell and Carter, Barbara R. (2005). *Childbed Fever: A Scientific Biography of Ignaz Semmelweis.* Edison, NJ: Aldine Transaction.
Cevolani, Gustavo and Tambolo, Luca (2013). Progress as approximation to the truth: A defence of the verisimilitudinarian approach. *Erkenntnis*, **78**, 921–35.
Chakrabarti, Arindam (1994). Review essay: *Testimony: A Philosophical Study. Philosophy and Phenomenological Research*, **54**, 965–72.
Chakravartty, Anjan (2007). *A Metaphysics for Scientific Realism: Knowing the Unobservable.* Cambridge: Cambridge University Press.
Chang, Hasok (2005). A case for old-fashioned observability, and a reconstructed constructive empiricism. *Philosophy of Science*, **72**, 876–87.
Clark, Andy (1997). *Being There: Putting brain, body, and world together again.* Cambridge, MA: MIT Press.
Clark, Andy and Chalmers, David (1998). The extended mind. *Analysis*, **58**, 7–19.
Coady, C. A. J. (1992). *Testimony.* Oxford: Clarendon Press.
Coffman, E. J. (2008). Warrant without truth? *Synthese*, **162**, 173–94.
Cohen, L. Jonathan (1980). What has science to do with truth? *Synthese*, **45**, 489–510.
Cohen, Stewart (1998). Contextualist solutions to epistemological problems: Scepticism, Gettier, and the lottery. *Australasian Journal of Philosophy*, **76**, 289–306.
Cohen, Stewart (2000). Contextualism and skepticism. *Philosophical Issues*, **10**, 94–107.
Collingwood, Robin George (1956). *The Idea of History.* New York: Oxford University Press.
Comte, Auguste (1892/1970). *Cours de philosophie positive* (5th edn). Paris: Société Positiviste. Translated by Frederick Ferré as *Introduction to Positive Philosophy* Indianapolis: Bobbs-Merrill.
Corlett, J. Angelo (1996). *Analyzing Social Knowledge.* Lanham, MD: Rowman and Littlefield.
Davis, Christopher A., Ahijevych, David A., Haggerty, Julie A., and Mahoney, Michael J. (2014). Observations of temperature in the upper troposphere and lower stratosphere of tropical weather disturbances. *Journal of the Atmospheric Sciences*, **71**, 1593–608.
de Finetti, Bruno (1974). *Theory of Probability.* New York: John Wiley.
Dellsén, Finnur (2016). Scientific progress: Knowledge versus understanding. *Studies in History and Philosophy of Science*, **56**, 72–83.
Dennett, Daniel (1971). Intentional systems. *Journal of Philosophy*, **68**, 87–106.
Dennett, Daniel (1987). *The Intentional Stance.* Cambridge, MA: MIT Press.
Descartes, René (1637/1960). *Discourse on Method.* Harmondsworth: Penguin. Transl. A. Wollaston.

Doll, Richard and Hill, Austin Bradford (1954). The mortality of doctors in relation to their smoking habits. *British Medical Journal*, **4877**, 1451–5.

Douglas, Heather (2014). Pure science and the problem of progress. *Studies In History and Philosophy of Science Part A*, **46**, 55–63.

Douven, Igor (2017*a*). Inference to the Best Explanation. What is it? And why should we care? In *Best Explanations: New Essays on Inference to the Best Explanation*, pp. 7–24. Oxford: Oxford University Press.

Douven, Igor (2017*b*). Peirce on abduction (Supplement to 'Abduction'). In *The Stanford Encyclopedia of Philosophy* (Summer 2017 edn) (ed. E. N. Zalta). Metaphysics Research Lab, Stanford University.

Doyle, Arthur Conan (1953*a*). The adventure of the Bruce-Partington plans. In *The Complete Sherlock Holmes*, Volume II. New York: Doubleday.

Doyle, Arthur Conan (1953*b*). The sign of four. In *The Complete Sherlock Holmes*, Volume I. New York: Doubleday.

Dretske, Fred (1981). *Knowledge and the Flow of Information*. Cambridge, MA: MIT Press.

Duhem, Pierre (1914/1954). *La théorie physique, son objet et sa structure*. Paris: Chevalier et Rivière. Translated as *The Aim and Structure of Physical Theory* by P. Wiener, Princeton NJ: Princeton University Press, 1954.

Dunn, Jeffrey (2014). Inferential evidence. *American Philosophical Quarterly*, **51**, 201–13.

Durkheim, Émile (1893). *De la division du travail social*. Paris: Alcan. Translated as *The Division of Labor in Society*, by W. D. Halls, New York, NY: The Free Press, 1984.

Durkheim, Émile (1894). *Les règles de la méthode sociologique*. Paris: Alcan. Translated as *The Rules of Sociological Method*, by W. D. Halls, and edited with an introduction by S. Lukes, New York: The Free Press, 1982.

Earman, John (1992). *Bayes or Bust?* Cambridge, MA: Bradford.

Edlén, Bengt (1943). Die Deutung der Emissionslinien im Spektrum der Sonnenkorona. *Zeitschrift für Astrophysik*, **20**, 30–64.

Egan, Andy and Elga, Adam (2005). I can't believe I'm stupid. *Philosophical Perspectives*, **19**, 77–93.

Einstein, Albert (1905/1956). *Investigations on the Theory of the Brownian Movement*. New York: Dover. R. Fürth (ed.), A. D. Cowper (trans.).

Einstein, Albert (1991). *Autobiographical Notes*. La Salle, IL: Open Court.

Elgin, Catherine Z. (1996). *Considered Judgment*. Princeton: Princeton University Press.

Elgin, Catherine Z. (2004). True enough. *Philosophical Issues*, **14**, 113–31.

Elgin, Catherine Z. (2007). Understanding and the facts. *Philosophical Studies*, **132**, 33–42.

Epstein, Brian (2015). *The Ant Trap*. Oxford: Oxford University Press.

Fahrbach, Ludwig (2009). Pessimistic meta-induction and the exponential growth of science. In *Reduction–Abstraction–Analysis* (ed. A. Hieke and H. Leitgeb), pp. 95–111. Frankfurt: Ontos.

Fahrbach, Ludwig (2011*a*). How the growth of science ends theory change. *Synthese*, **180**, 139–55.

Fahrbach, Ludwig (2011*b*). Theory change and degrees of success. *Philosophy of Science*, **78**, 1283–92.

Falconer, Isobel (2004). Charles Augustin Coulomb and the fundamental law of electrostatics. *Metrologia*, **41**, S107–S114.

Fassio, Davide (2018). The aim of belief. *Internet Encyclopedia of Philosophy*. http://www.iep.utm.edu/beli-aim/ accessed 25 February 2018.

Faulkner, Paul (1998). David Hume's reductionist epistemology of testimony. *Pacific Philosophical Quarterly*, **79**, 302–13.

Fine, Arthur I. (1984). The natural ontological attitude. In *Scientific Realism* (ed. J. Leplin), pp. 261–77. Berkeley, CA: University of California Press.

Fodor, Jerry A. (1984). Observation reconsidered. *Philosophy of Science*, **51**, 23–43.

Foley, Richard (1992). The epistemology of belief and the epistemology of degrees of belief. *American Philosophical Quarterly*, **29**, 111–21.

Franklin, James (2009). *What Science Knows and How It Knows It*. New York, NY: Encounter Books.

Friis-Hansen, Bent J., Holiday, Malcolm, Stapleton, Thomas, and Wallace, William M. (1951). Total body water in children. *Pediatrics*, **7**, 321–7.

Gardner, A. and Grafen, A. (2009). Capturing the superorganism: a formal theory of group adaptation. *Journal of Evolutionary Biology*, **22**, 659–71.

Gardner, Ian, Willeberg, Preben, and Mousing, Jan (2002). Empirical and theoretical evidence for herd size as a risk factor for swine diseases. *Animal Health Research Reviews (1):*, **3**, 43–55.

Germano, Giuseppe (2013). New editorial perspectives on Fibonacci's *Liber Abaci*. *Reti Medievali Rivista*, **14**, 157–73.

Ghins, Michel (2002). Putnam's No-Miracle Argument: A critique. In *Recent Themes in the Philosophy of Science* (ed. S. P. Clarke and T. D. Lyons), pp. 121–38. Dordrecht: Kluwer.

Gibbard, Alan (2005). Truth and correct belief. *Philosophical Issues*, **15**, 338–50.

Giere, Ronald N. (2002*a*). Distributed cognition in epistemic cultures. *Philosophy of Science*, **69**, 637–44.

Giere, Ronald N. (2002*b*). Scientific cognition as distributed cognition. In *The Cognitive Basis of Science* (ed. P. Carruthers, S. Stich, and M. Siegal), pp. 285–99. Cambridge: Cambridge University Press.

Gigerenzer, Gerd (1991). How to make cognitive illusions disappear: Beyond "heuristics and biases". *European Review of Social Psychology*, **2**, 83–115.

Gigerenzer, Gerd (2007). *Gut Feelings. Short cuts to better decision making*. New York, NY: Penguin.

Gilbert, Margaret (1987). Modelling collective belief. *Synthese*, **73**, 185–204.

Gilbert, Margaret (1989). *On Social Facts*. London: Routledge.

Gilbert, Margaret (1994). Remarks on collective belief. In *Socializing Epistemology: The Social Dimensions of Knowledge* (ed. F. F. Schmitt), pp. 235–56. Lanham, MD: Rowman and Littlefield.

Gilbert, Margaret (2000). *Sociality and Responsibility: New essays in plural subject theory.* Lanham, MD: Rowman and Littlefield.

Gilbert, Margaret (2004). Collective epistemology. *Episteme*, **1**, 95–107.

Gillies, Donald (2005). Hempelian and Kuhnian approaches in the philosophy of medicine: the Semmelweis Case *Studies in History and Philosophy of Biological and Biomedical Sciences*, **36**, 159–81.

Ginet, Carl (1975). *Knowledge, Perception, and Memory.* Dordrecht: Reidel.

Glanzberg, Michael (2016). Truth. In *The Stanford Encyclopedia of Philosophy* (Winter 2016 edn) (ed. E. N. Zalta). Metaphysics Research Lab, Stanford University. https://plato.stanford.edu/archives/win2016/entries/truth/.

Glymour, Clark N. (1980). *Theory and Evidence.* Princeton, NJ: Princeton University Press.

Glynn, Ian (2010). *Elegance in Science.* Oxford: Oxford University Press.

Goldberg, Sanford (2007). *Anti-Individualism: Mind and language, knowledge and justification.* Cambridge: Cambridge University Press.

Goldman, Alvin I. (1999). *Knowledge in a Social World.* Oxford: Clarendon Press.

Goldman, Alvin I. (2009). Williamson on knowledge and evidence. In *Williamson on Knowledge* (ed. P. Greenough and D. Pritchard), pp. 73–91. Oxford: Oxford University Press.

Goldstein, Daniel G. and Gigerenzer, Gerd (2002). Models of ecological rationality: The recognition heuristic. *Psychological Review*, **109**, 75–90.

Golgi, Camillo (1898). Intorno alla struttura delle cellule nervose. *Bollettino della Società Medico-Chirurgica di Pavia*, **13**, 316.

Green, Charles and Cook, James (1771). Observations made, by appointment of the Royal Society, at King George's Island in the South Sea. *Philosophical Transactions of the Royal Society of London*, **61**, 397–421.

Grimm, Stephen R. (2006). Is understanding a species of knowledge? *British Journal for the Philosophy of Science*, **57**, 515–35.

Grimm, Stephen R. (2014). Understanding as knowledge of causes. In *Virtue Epistemology Naturalized* (ed. A. Fairweather), Synthese Library (Studies in Epistemology, Logic, Methodology, and Philosophy of Science), pp. 329–45. Cham: Springer.

Grotrian, Walter (1939). Zur Frage der Deutung der Linien im Spektrum der Sonnenkorona. *Naturwissenschaften*, **27**, 214.

Hacking, Ian (1983). *Representing and Intervening.* Cambridge: Cambridge University Press.

Hao, Yajiang, Inhester, Ludger, Son, Sang-Kil, and Santra, Robin (2019). Theoretical evidence for the sensitivity of charge-rearrangement-enhanced x-ray ionization to molecular size. *Physical Review A*, **100**.

Hardwig, John (1985). Epistemic dependence. *Journal of Philosophy*, **82**, 335–49.

Hardwig, John (1991). The role of trust in knowledge. *Journal of Philosophy*, **88**, 693–708.

Harker, David (2008). On the predilections for predictions. *British Journal for the Philosophy of Science*, **59**, 429–53.

Harman, Gilbert (1965). The inference to the best explanation. *Philosophical Review*, **74**, 88–95.

Haukioja, Jussi (2013). Different notions of response-dependence. In *Varieties of Dependence* (ed. M. Hoeltje, B. Schnieder, and A. Steinberg), pp. 167–90. Munich: Philosophia Verlag.

Heering, Peter (1992). On Coulomb's inverse square law. *American Journal of Physics*, **60**, 988–94.

Heering, Peter (1994). The replication of the torsion balance experiment: The inverse square law and its refutation by early 19th-century German physicists. In *Restaging Coulomb. Usages, Controverses et Réplications autour de la Balance de Torsion* (ed. C. Blondel and M. Dörries), Volume IV of *Biblioteca di Nuncius*, pp. 47–66. Florence: Leo S. Olschki.

Hempel, Carl G. (1952). *Fundamentals of Concept Formation in Empirical Science*. Chicago, IL: Unvieristy of Chicago Press.

Henderson, Leah (2014). Bayesianism and Inference to the Best Explanation. *British Journal for the Philosophy of Science*, **65**, 687–715.

Hendry, Robin Findlay (2018). Scientific realism and the history of chemistry. *Spontaneous Generations, Vol. 9, No. 1 (2018)*, **9**, 108–117.

Herschel, William (1795). On the nature and construction of the sun and fixed stars. *Philosophical Transactions of the Royal Society of London*, **85**, 46–72.

Heuer, Richards J. (1999). *Psychology of Intelligence Analysis*. Washington, D.C.: Center for the Study of Intelligence, Central Intelligence Agency.

Hills, Alison (2016). Understanding why. *Noûs*, **50**, 661–88.

Hilpinen, Risto (1976). Approximate truth and truthlikeness. In *Formal Methods in the Methodology of the Empirical Sciences* (ed. M. Przelecki, A. Szaniawski, and R. Wójcicki), pp. 19–42. Dordrecht: Reidel.

Hilpinen, Risto (1988). Knowledge and conditionals. *Philosophical Perspectives*, **2**, 157–82.

Hirschhorn, Joel N., Lohmueller, Kirk, Byrne, Edward, and Hirschhorn, Kurt (2002). A comprehensive review of genetic association studies. *Genetics in Medicine*, **4**, 45–61.

Hoefer, Carl (2020). Scientific realism without the quantum. In *Scientific Realism and the Quantum* (ed. S. French and J. Saatsi), pp. 19–34. Oxford: Oxford University Press.

Hoefer, Carl and Martí, Genoveva (2020). Realism, reference and perspective. *European Journal for Philosophy of Science*, **10**.

Hölldobler, Bert and Wilson, Edward O. (2009). *The Superorganism. The Beauty, Elegance, and Strangeness of Insect Societies*. New York, NY: W. W. Norton.

Horton, Richard (1998). Editor's reply. *The Lancet*, **351**, 908–9.

Horwich, Paul (1982). *Probability and Evidence*. Cambridge: Cambridge University Press.

Horwich, Paul (1990). *Truth*. Oxford: Basil Blackwell.

Hoyningen-Huene, Paul (2013). *Systematicity: The Nature of Science*. Oxford: Oxford University Press.

Hróbjartsson, Asbørn, Gøtzsche, Peter C., and Gluud, Christian (1998). The controlled clinical trial turns 100 years: Fibiger's trial of serum treatment of diphtheria. *British Medical Journal*, **317**, 1243.

Huang, Wei, Chien, Po-Hsiu, McMillen, Kyle, Patel, Sawankumar, Tedesco, Joshua, Zeng, Li, Mukherjee, Subhrangsu, Wang, Binghao, Chen, Yao, Wang, Gang, Wang, Yang, Gao, Yanshan, Bedzyk, Michael J., DeLongchamp, Dean M., Hu, Yan-Yan, Medvedeva, Julia E., Marks, Tobin J., and Facchetti, Antonio (2020). Experimental and theoretical evidence for hydrogen doping in polymer solution-processed indium gallium oxide. *Proceedings of the National Academy of Sciences*, **117**, 18231–39.

Huemer, Michael (2009). Explanationist aid for the theory of inductive logic. *British Journal for the Philosophy of Science*, **60**, 345–75.

Hughes, Nick (2014). Consistency and evidence. *Philosophical Studies*, **169**, 333–38.

Hume, David (1748). *Philosophical Essays Concerning Human Understanding.* London: Millar. References to the second edition of *Enquiries Concerning the Human Understanding and the Principals of Morals of David Hume* edited by L. A. Selby-Bigge, 1902. Oxford: Clarendon Press.

Hume, David (1757). Of the standard of taste. In *Essays Moral, Political, Literary.* References to the edition edited by Eugene F. Miller. Indianapolis, IN: Liberty Fund, 1987.

Hunter, Michael (1989). *Establishing the New Science.* Woodbridge, Suffolk: The Boydell Press.

Hutchins, Edwin (1995*a*). *Cognition in the Wild.* Cambridge, MA: MIT Press.

Hutchins, Edwin (1995*b*). How a cockpit remembers its speeds. *Cognitive Science*, **19**, 265–88.

Hyman, John (1999). How knowledge works. *Philosophical Quarterly*, **49**, 433–51.

Ierodiakonou, Katerina (2016). Theophrastus. In *The Stanford Encyclopedia of Philosophy* (Summer 2016 edn) (ed. E. N. Zalta). Metaphysics Research Lab, Stanford University.

Ivanova, Milena (2010). Pierre Duhem's good sense as a guide to theory choice good sense as a guide to theory choice. *Studies in History and Philosophy of Science*, **41**, 58–64.

Ivanova, Milena (2021). *Duham and Holism.* Cambridge: Cambridge University Press.

Jaynes, Edwin (1957). Information theory and statistical mechanics. *Physical Review*, **106**, 620–30.

Jaynes, Edwin (1968). Prior probabilities. *Institute of Electrical and Electronic Engineers Transactions on Systems Science and Cybernetics*, **SSC-4**, 227–41.

Jeffrey, Richard (1983). *The Logic of Decision* (2nd edn). Chicago: University of Chicago Press.

Jones, Jean (1996). James Hutton. In *The Scottish Enlightenment 1730-1790 A Hotbed of Genius* (ed. D. Daiches, P. Jones, and J. Jones). The Saltire Society.

Joyce, James M. (1998). A nonpragmatic vindication of probabilism. *Philosophy of Science*, **65**, 575–603.

Kanai, Ryota and Banissy, Michael (2010). Are two heads better than one? It depends. *Scientific American.* August 2012. URL: <http://www.scientificamerican.com/article.cfm?id=are-two-heads-better-than> accessed 25 Sept 2012.

Kemeny, John (1955). Fair bets and inductive probabilities. *Journal of Symbolic Logic*, **20**, 263–73.

Keynes, John Maynard (1921). *A Treatise on Probability.* London: Macmillan.

Kiesewetter, Benjamin (2017). *The Normativity of Rationality.* Oxford: Oxford University Press.

King, Ross D., Rowland, Jem, Oliver, Stephen G., Young, Michael, Aubrey, Wayne, Byrne, Emma, Liakata, Maria, Markham, Magdalena, Pir, Pinar, Soldatova, Larisa N., Sparkes, Andrew, Whelan, Kenneth E., and Clare, Amanda (2009). The automation of science. *Science*, **324**, 85–9.

King, Ross D., Whelan, Kenneth E., Jones, Ffion M., Reiser, Philip G. K., Bryant, Christopher H., Muggleton, Stephen H., Kell, Douglas B., and Oliver, Stephen G. (2004). Functional genomic hypothesis generation and experimentation by a robot scientist. *Nature*, **427**, 247–52.

Kitcher, Philip (1993). *The Advancement of Science.* New York: Oxford University Press.

Knorr Cetina, Karin (1999). *Epistemic Cultures: How the sciences make knowledge.* Cambridge, MA: Harvard University Press.

Kolmogorov, Andrey Nikolaevich (1950). *The Foundations of the Theory of Probability.* New York: Chelsea Publishing Company.

Kornblith, Hilary (2002). *Knowledge and its Place in Nature.* New York: Oxford University Press.

Kosso, Peter (1988). Dimensions of observability. *British Journal for the Philosophy of Science*, **39**, 449–67.

Kragh, Helge (1990). *Dirac: A scientific biography.* Cambridge: Cambridge University Press.

Kuhn, Thomas S. (1962). *The Structure of Scientific Revolutions.* Chicago, IL: University of Chicago Press.

Kuhn, Thomas S. (1970). *The Structure of Scientific Revolutions* (2nd edn). Chicago, IL: University of Chicago Press.

Kuhn, Thomas S. (1977). Objectivity, value judgment, and theory choice. In *The Essential Tension*, pp. 320–39. Chicago, IL: University of Chicago Press.

Kuipers, Theo (2000). *From Instrumentalism to Constructive Realism.* Dordrecht: Kluwer.

Kümmerli, Rolf (2009). Social insects—superorganisms or just superb organisms? *Current Biology*, **19**.

Kusch, Martin (2002). *Knowledge by Agreement: The Programme of Communitarian Epistemology.* Oxford: Oxford University Press.

Kvanvig, Jonathan (2003). *The Value of Knowledge and the Pursuit of Understanding.* New York, NY: Cambridge University Press.

Lackey, Jennifer (2014). Socially extended knowledge. *Philosophical Issues*, **24**, 282–98.

Ladyman, James (2000). What's really wrong with Constructive Empiricism? Van Fraassen and the metaphysics of modality. *British Journal for the Philosophy of Science*, **51**, 837–56.

Ladyman, James (2014). Structural realism. In *The Stanford Encyclopedia of Philosophy* (Winter 2016 edn) (ed. E. N. Zalta). Metaphysics Research Lab, Stanford University.

Ladyman, James (2018). Scientifc realism again. *Spontaneous Generations*, **9**, 99–107.

Lagemann, Robert T. (1977). New light on old rays. *American Journal of Physics*, **45**, 281–4.

Lakatos, Imre (1970). Falsification and the methodology of scientific research programmes. In *Criticism and the Growth of Knowledge* (ed. I. Lakatos and A. Musgrave), pp. 91–195. Cambridge: Cambridge University Press.

Laudan, Larry (1977). *Progress and its Problems: Toward a theory of scientific growth.* Berkeley: University of California Press.

Laudan, Larry (1981*a*). A confutation of convergent realism. *Philosophy of Science*, **48**, 19–48.

Laudan, Larry (1981*b*). A problem-solving approach to scientific progress. In *Scientific Revolutions* (ed. I. Hacking), pp. 144–55. Oxford: Oxford University Press.

Leakey, Louis S. B. (1960). Finding the world's earliest man. *National Geographic*, **118**, 420–35.

Lear, Jonathan (1988). *Aristotle: the desire to understand.* Cambridge: Cambridge University Press.

Leitgeb, Hannes and Pettigrew, Richard (2010*a*). An objective justification of Bayesianism I: Measuring inaccuracy. *Philosophy of Science*, **77**, 201–35.

Leitgeb, Hannes and Pettigrew, Richard (2010*b*). An objective justification of Bayesianism II: The consequences of minimizing inaccuracy. *Philosophy of Science*, **77**, 236–72.

Lesher, James H. (2001). On Aristotelian ἐπιστήμη as 'understanding'. *Ancient Philosophy*, **21**, 45–55.

Levin, Michael (1984). What kind of explanation is truth? In *Scientific Realism* (ed. J. Leplin), pp. 124–39. Berkeley, CA: University of California Press.

Lewis, David K. (1969). *Convention: A Philosophical Study.* Cambridge, MA: Harvard University Press.

Lewis, David K. (1980). A subjectivist's guide to objective chance. In *Studies in Inductive Logic and Probability* (ed. R. C. Jeffrey), Volume II, pp. 263–93. Berkeley: University of California Press. Reprinted in Lewis (1986).

Lipton, Peter (1993). Is the best good enough? *Proceedings of the Aristotelian Society*, **93**, 89–104.

Lipton, Peter (1998). The epistemology of testimony. *Studies in History and Philosophy of Science*, **29**, 1–31.

Lipton, Peter (2000). Tracking track records. *Proceedings of the Aristotelian Society Supplementary Volume*, **74**, 179–205.

Lipton, Peter (2001). Is explanation a guide to inference? A reply to Wesley Salmon. In *Explanation: Theoretical Approaches and Applications* (ed. G. Hon and S. Rakover), pp. 93–120. Dordrecht: Kluwer.

Lipton, Peter (2004). *Inference to the Best Explanation* (2nd edn). London: Routledge.

List, Christian and Pettit, Philip (2006). Group agency and supervenience. *The Southern Journal of Philosophy*, **54**, 85–105.

List, Christian and Pettit, Philip (2011). *Group Agency*. Oxford: Oxford University Press.

Littlejohn, Clayton (2011). Evidence and knowledge. *Erkenntnis*, **74**, 241–62.

Littlejohn, Clayton (2013). No evidence is false. *Acta Analytica*, **28**, 145–59.

Locke, John (1690). *An Essay Concerning Human Understanding* (1964, A. D. Woozley edn). London: Fontana.

Luzzi, Federico (2010). Counter-closure. *Australasian Journal of Philosophy*, **88**, 673–83.

Maddy, Penelope (2007). *Second Philosophy*. Oxford: Oxford University Press.

Magnus, P. D. (2007). Distributed cognition and the task of science. *Social Studies of Science*, **37**, 297–310.

Magnus, P. D. and Callender, Craig (2004). Realist ennui and the base rate fallacy. *Philosophy of Science*, **71**, 320–38.

Maher, Patrick (1996). Subjective and objective confirmation. *Philosophy of Science*, **65**, 149–74.

Mahtani, Anna (2012). Diachronic Dutch book arguments. *Philosophical Review*, **121**, 443–50.

Martínez, Alberto A. (2006). Replication of Coulomb's torsion balance experiment. *Archive for History of Exact Sciences*, **60**, 517–63.

McAllister, James W. (1999). *Beauty and Revolution in Science*. Ithaca, NY: Cornell University Press.

Mcauliffe, William H. B. (2015). How did abduction get confused with inference to the best explanation? *Transactions of the Charles S. Peirce Society*, **51**, 300–19.

McHugh, Conor (2011). What do we aim at when we believe? *dialectica*, **65**, 369–92.

McHugh, Conor (2014). Fitting belief. *Proceedings of the Aristotelian Society*, **114**, 167–87.

Mellor, D. H. (1982). The reduction of society. *Philosophy*, **57**, 51–75.

Mill, John Stuart (1843). *A System of Logic, Ratiocinative and Inductive*. John W. Parker. Reference to the 1974 edition, J. M. Robson ed. Toronto: University of Toronto Press.

Miller, David (1974). Popper's qualitative theory of verisimilitude. *British Journal for the Philosophy of Science*, **25**, 166–77.

Millikan, Ruth G. (1989). In defense of proper functions. *Philosophy of Science*, **56**, 288–302.

Millikan, Ruth G. (1993). Naturalist reflections on knowledge. In *White Queen Psychology and Other Essays for Alice* (ed. R. Millikan), pp. 241–64. Cambridge, MA: MIT Press.

Milne, Peter (1996). log [P(h/eb)/P(h/b)] is the one true measure of confirmation. *Philosophy of Science*, **63**, 21–6.

Mizrahi, Moti (2013). What is scientific progress? Lessons from scientific practice. *Journal for General Philosophy of Science*, **44**, 375–90.

Moore, George Edward (1899). The nature of judgment. *Mind*, **8**, 176–93.

Moore, George Edward (1902). Truth and falsity and error. In *Dictionary of Philosophy and Psychology* (ed. J. M. Baldwin), Volume 2, pp. 716–18. London: Macmillan.

Musgrave, Alan (1989). Noa's ark—fine for realism. *Philosophical Quarterly*, **39**, 383–98.

Nagel, Ernest (1961). *The Structure of Science: Problems in the Logic of Scientific Explanation*. New York: Harcourt, Brace & World.

Nagel, Jennifer (2015). Sensitive knowledge: Locke on skepticism and sensation. In *The Blackwell Companion to Locke* (ed. M. Stuart), pp. 313–33. Malden, MA: Wiley-Blackwell.

Neander, Karen (1991). Function as selected effects: A conceptual analyst's defense. *Philosophy of Science*, **58**, 168–84.

Nelkin, Dana K. (2000). The lottery paradox, knowledge, and rationality. *Philosophical Review*, **109**, 373–409.

Neta, Ram (2007). How to naturalize epistemology. In *New Waves in Epistemology* (ed. V. F. Hendricks and D. Pritchard), pp. 324–53. Basingstoke: Palgrave Macmillan.

Neurath, Otto (1932). Sozialbehaviorismus. *Sociologicus*, **8**, 281–8.

Newton, Hannah (2012). *The Sick Child in Early Modern England, 1580–1720*. Oxford: Oxford University Press.

Niiniluoto, Ilkka (1977). On the truthlikeness of generalizations. In *Basic Problems in Methodology and Linguistics* (ed. R. E. Butts and J. Hintikka), pp. 121–47. Dordrecht: Reidel.

Niiniluoto, Ilkka (1980). Scientific progress. *Synthese*, **45**, 427–62. (Reprinted as c.5 of Niiniluoto 1984.).

Niiniluoto, Ilkka (1984). *Is Science Progressive?* Dordrecht: Reidel.

Niiniluoto, Ilkka (1987). *Truthlikeness*. Dordrecht: Reidel.

Niiniluoto, Ilkka (1999). *Critical Scientific Realism*. Oxford: Oxford University Press.

Niiniluoto, Ilkka (2011). Scientific progress. In *The Stanford Encyclopedia of Philosophy* (Summer 2011 edn) (ed. E. N. Zalta). The Stanford Encyclopedia of Philosophy.

Niiniluoto, Ilkka (2014). Scientific progress as increasing verisimilitude. *Studies in History and Philosophy of Science Part A*, **46**, 73–7.

Norton, David Fate (1981). The myth of 'British Empiricism'. *History of European Ideas*, **1**, 331–344.

Norton, John (2003). A material theory of induction. *Philosophy of Science*, **70**, 647–70.

Nozick, Robert (1981). *Philosophical Explanations.* Cambridge, MA: Harvard University Press.

Nuland, Sherwin B. (2003). *The Doctors' Plague.* New York: Norton.

O'Connell, Lindsey (2018). The Meckel–Serres conception of recapitulation. *Embryo Project Encyclopedia.* accessed 1 March 2021.

Oddie, Graham (1986). *Likeness to Truth.* Western Ontario Series in Philosophy of Science. Dordrecht: Reidel.

Oddie, Graham (2014). Truthlikeness. In *The Stanford Encyclopedia of Philosophy* (Summer 2014 edn) (ed. E. N. Zalta). Metaphysics Research Lab, Stanford University.

Okasha, Samir (2013). On a flawed argument against the KK principle. *Analysis,* **73**, 80–6.

Owens, David John (2003). Does belief have an aim? *Philosophical Studies,* **115**, 283–305.

Palermos, Spyridon Orestis (2020). Epistemic collaborations: Distributed cognition and virtue reliabilism. *Erkenntnis.*

Papineau, David (1978). *For Science in the Social Sciences.* Basingstoke: Macmillan.

Papineau, David (1988). Does the sociology of science discredit science? In *Relativism and Realism in Science* (ed. R. Nola), pp. 37–57. Dordrecht: Kluwer.

Papineau, David (1993). *Philosophical Naturalism.* Oxford: Blackwell.

Papineau, David (2013). There are no norms of belief. In *The Aim of Belief* (ed. T. Chan), pp. 64–79. New York, NY: Oxford University Press.

Pappas, George (2017). Internalist vs. externalist conceptions of epistemic justification. In *The Stanford Encyclopedia of Philosophy* (Fall 2017 edn) (ed. E. N. Zalta). Metaphysics Research Lab, Stanford University. https://plato.stanford.edu/archives/fall2017/entries/justep-intext/.

Parnell, Clare E. and Moortel, Ineke De (2012). A contemporary view of coronal heating. *Philosophical Transactions of the Royal Society A,* **370**, 3217–40.

Parsons, Talcott (1961). *Theories of Society: foundations of modern sociological theory.* New York, NY: Free Press.

Peacocke, Christopher (1983). *Sense and Content.* Oxford: Oxford University Press.

Peacocke, Christopher (2001). Does perception have a nonconceptual content? *Journal of Philosophy,* **98**, 23–64.

Peirce, Charles S. (1902). Truth and falsity and error (logical). In *Dictionary of Philosophy and Psychology* (ed. J. M. Baldwin), Volume 2, pp. 718–20. London: Macmillan.

Pettit, Philip (2003). Groups with minds of their own. In *Socializing Metaphysics* (ed. F. F. Schmitt), pp. 167–93. Lanham, MD: Rowman and Littlefield.

Plantinga, Alvin (1993). *Warrant and Proper Function.* New York, NY: Oxford University Press.

Playfair, John (1805). Biographical account of the life of Dr James Hutton F.R.S. Edin. *Transactions of the Royal Society of Edinburgh,* **5**.

Poincaré, Henri (1943). *La Science et l'Hypothèse.* Paris: Edition Flammarion. First published 1902; references to English language edition Dover, New York 1952.

Popper, Karl (1963). *Conjectures and Refutations: The Growth of Scientific Knowledge.* London: Routledge and Kegan Paul.

Popper, Karl (1979). *Objective Knowledge* (Revised edn). Oxford: Clarendon Press.

Popper, Karl and Wächterschäuser, Günter (1990). Progenote or protogenote? *Science*, **250**, 1070.

Poston, Ted (2014). *Reason and Explanation.* Basingstoke: Palgrave Macmillan.

Prichard, Harold Arthur (1950). *Knowledge and Perception.* Oxford: Clarendon Press.

Psillos, Stathis (1999). *Scientific Realism: How Science Tracks Truth.* London: Routledge.

Psillos, Stathis (2004). Inference to the best explanation and Bayesianism. In *Induction and Deduction in the Sciences* (ed. F. Stadler), pp. 83–91. Kluwer.

Putnam, Hilary (1975*a*). What is mathematical truth? In *Mathematics, Matter, and Method: Philosophical Papers, Volume 1*, pp. 60–78. Cambridge: Cambridge University Press.

Putnam, Hilary (1975*b*). What is "realism"? *Proceedings of the Aristotelian Society*, **76**, 177–94.

Quine, Willard V. (1951). Two dogmas of empiricism. *Philosophical Review*, **60**, 20–43.

Quine, Willard V. (1990). *Pursuit of Truth.* Cambridge, MA: Harvard University Press.

Quinton, Anthony (1976). Social objects. *Proceedings of the Aristotelian Society*, **76**, 1–27.

Raffman, Diana (1995). On the persistence of phenomenology. In *Conscious Experience* (ed. T. Metzinger), pp. 29–302. Paderborn: Schoningh Verlag.

Ramsey, Frank P. (1927). Facts and propositions. *Aristotelian Society Supp. Vol.*, **7**, 153–70.

Robinson, Andrew (2006). *The Last Man Who Knew Everything.* Oxford: Oneworld.

Rosen, Gideon (1994). Objectivity and modern idealism: What is the question? In *Philosophy in Mind* (ed. M. Michael and J. O'Leary-Hawthorne), pp. 277–319. Dordrecht: Kluwer Academic Publishers.

Rowbottom, Darrell (2008). N-rays and the semantic view of scientific progress. *Studies in History and Philosophy of Science*, **39**, 277–8.

Rowbottom, Darrell (2010). What scientific progress is not: Against Bird's epistemic view. *International Studies in the Philosophy of Science*, **24**, 241–55.

Rowbottom, Darrell (2019). *The Instrument of Science.* Abingdon: Routledge.

Russell, Bertrand (1903). *Principles of Mathematics.* Cambridge: Cambridge University Press.

Russell, Bertrand (1919). The philosophy of logical atomism. *The Monist*, **29**, 345–80.

Sakamoto, Manabu, Venditti, Chris, and Benton, Michael J. (2017). 'Residual diversity estimates' do not correct for sampling bias in palaeodiversity data. *Methods in Ecology and Evolution*, **8**, 453–9.

Salmon, Wesley C. (1971). Statistical explanation. In *Statistical Explanation and Statistical Relevance* (ed. W. C. Salmon), Pittsburgh, pp. 29–87. University of Pittsburgh Press.

Salmon, Wesley C. (1990). Rationality and objectivity in science or Tom Kuhn meets Tom Bayes. In *Scientific Theories* (ed. C. W. Savage), Volume 14 of *Minnesota Studies in the Philosophy of Science*, pp. 175–204. Minneapolis, MN: University of Minnesota Press.

Salmon, Wesley C. (2001). Explanation and confirmation: A Bayesian critique of inference to the best explanation. In *Explanation: Theoretical Approaches and Applications* (ed. G. Hon and S. S. Rakover), pp. 61–91. Dordrecht: Kluwer.

Salomon, Albert (1945). German sociology. In *Twentieth Century Sociology* (ed. G. Gurvitch and W. E. Moore). New York, NY: Philosophical LIbrary.

Savage, Leonard J. (1954). *The Foundations of Statistics*. New York, NY: Dover.

Schlick, Moritz (1934). Über das Fundament der Erkenntnis. *Erkenntnis*, **4**, 79–99.

Schmitt, Frederick F. (1994). The justification of group beliefs. In *Socializing Epistemology: The Social Dimensions of Knowledge* (ed. F. F. Schmitt), pp. 257–87. Lanham, MD: Rowman and Littlefield.

Schmitt, Frederick F. (2003). Socializing metaphysics: An introduction. In *Socializing Metaphysics* (ed. F. F. Schmitt), pp. 1–37. Lanham, MD: Rowman and Littlefield.

Schoeller, Dale A. (1989). Changes in total body water with age. *The American Journal of Clinical Nutrition*, **50**, 1176–81.

Schupbach, Jonah N. (2016). Inference to the best explanation, cleaned up and made respectable. In *Best Explanations: New Essays on Inference to the Best Explanation* (ed. K. McCain and T. Poston), pp. 39–64. Oxford: Oxford University Press.

Semmelweis, Ignaz (1861). *Die Aetiologie, der Begriff und die Prophylaxis des Kindbettfiebers*. Pest: Hartleben. References to the translation by K. Codell Carter 1983 *The Etiology, Concept, and Prophylaxis of Childbed Fever* Madison, WI: University of Wisconsin Press.

Shah, Nishi (2003). How truth governs belief. *Philosophical Review*, **112**, 447–82.

Shah, Nishi and Velleman, J. David (2005). Doxastic deliberation. *Philosophical Review*, **114**, 497–534.

Shan, Yafeng (2019). A new functional approach to scientific progress. *Philosophy of Science*, **86**, 739–58.

Shapere, Dudley (1982). The concept of observation in science and philosophy. *Philosophy of Science*, **49**, 485–525.

Shermer, Michael (2002). *Why People Believe Weird Things: Pseudoscience, Superstition, and Other Confusions of Our Time* (Revised edn). New York, NY: Henry Holt.

Simon, Michael A. (1969). When is a resemblance a family resemblance? *Mind*, **78**, 408–16.

Sinnott-Armstrong, Walter (2018). *Think Again*. London: Pelican.

Sliwa, Paulina (2015). Understanding and knowing. *Proceedings of the Aristotelian Society*, **115**, 57–74.

Sosa, Ernest (1993). Proper functionalism and virtue epistemology. *Noûs*, **27**, 51–65.

Sozou, Peter D., Lane, Philip R., Addis, Mark, and Gobet, Fernand. (2017) Computational scientific discovery. In *Springer Handbook of Model-Based Science* (ed. L. Magnani and T. Bertolotti), pp. 719–34. Cham: Springer.

Spencer, Herbert (1874–96). *The Principles of Sociology.* Volumes I-III. London: Williams and Norgate.

Spencer, Herbert (1911). *Essays on Education and Kindred Subjects.* London: Dent.

Stalnaker, Robert (1970). Probabilities and conditionals. *Philosophy of Science*, **37**, 64–80.

Stanford, P. Kyle (2006). *Exceeding Our Grasp: Science, History, and the Problem of Unconceived Alternatives.* New York: Oxford University Press.

Stanley, Jason and Williamson, Timothy (2001). Knowing how. *Journal of Philosophy*, **98**, 411–44.

Starkie, Thomas (1834). *A Practical Treatise on the Law of Evidence* (5th edn). Philadelphia PA: Nicklin and Johnson.

Steglich-Petersen, Asbørn (2006). No norm needed: On the aim of belief. *Philosophical Quarterly*, **56**, 499–516.

Stimson, Dorothy (1949). *Scientists and Amateurs.* London: Sigma Books.

Stolt, Carl-Magnus, Klein, George, and Jansson, Alfred T. R. (2004). An analysis of a wrong Nobel Prize—Johannes Fibiger, 1926: A study in the Nobel archives. In *Advances in Cancer Resarch* (ed. G. F. Vande Woude and G. Klein), Volume 92, pp. 1–12. San Diego, CA: Elsevier.

Sullivan, Patrick F. (2007). Spurious genetic associations. *Biological Psychiatry*, **61**, 1121–6.

Sullivan-Bisset, Ema (2017). Biological function and epistemic normativity. *Philosophical Explorations*, **20**(sup 1), 94–110.

Suppes, Patrick (1966). A bayesian approach to the paradoxes of confirmation. In *Aspects of Inductive Logic* (ed. J. Hintikka and P. Suppes), Studies in Logic and the Foundations of Mathematics, pp. 198–207. Amsterdam: North-Holland.

Sutton, Clive (1994). 'Nullius in verba' and 'nihil in verbis': public understanding of the role of language in science. *British Journal for the History of Science*, **27**, 55–64.

Sutton, Jonathan (2007). *Without Justification.* Cambridge, MA: MIT Press.

Tambolo, Luca (2011). Scientific progress, verisimilitude, and evidence. *Logic and Philosophy of Science*, **9**, 561–7.

Teller, Paul (2001). Whither constructive empiricism? *Philosophical Studies*, **106**, 123–50.

Thorndike, Lynn (1914). Roger Bacon and experimental method in the Middle Ages. *The Philosophical Review*, **23**, 271–98.

Tichý, Pavel (1974). On Popper's definitions of verisimilitude. *British Journal for the Philosophy of Science*, **25**, 155–60.

Tollefsen, Deborah Perron (2002). Collective intentionality and the social sciences. *Philosophy of the Social Sciences*, **32**, 25–50.

Tollefsen, Deborah Perron (2006). From extended mind to collective mind. *Cognitive Systems Research*, **7**, 140–50.

Toon, Adam (2014). Empiricism for cyborgs. *Philosophical Issues*, **24**, 409–25.

Tuomela, Raimo (1992). Group beliefs. *Synthese*, **91**, 285–318.

Tuomela, Raimo (2004). Group knowledge analyzed. *Episteme*, **1**, 109–27.

Tversky, Amos and Kahneman, Daniel (1974). Judgment under uncertainty: Heuristics and biases. *Science*, **185**, 1124–31.

Vallery-Radot, René (1926). *The Life of Pasteur.* Garden City, NY: Garden City Publishing Company. Translated by Mrs R. L. Devonshire.

van Fraassen, Bas (1980). *The Scientific Image.* Oxford: Oxford University Press.

van Fraassen, Bas (1989). *Laws and Symmetry.* Oxford: Oxford University Press.

van Fraassen, Bas (2001). Constructive empiricism now. *Philosophical Studies*, **106**, 151–70.

Vanderschraaf, Peter and Sillari, Giacomo (2005). Common knowledge. In *The Stanford Encyclopedia of Philosophy* (Winter 2005 edn) (ed. E. N. Zalta). Metaphysics Research Lab, Stanford University.

Velleman, J. David (2000). On the aim of belief. In *The Possibility of Practical Reason*, pp. 244–81. Oxford: Oxford University Press.

Vineberg, Susan (1996). Eliminative induction and Bayesian confirmation theory. *Canadian Journal of Philosophy*, **26**, 257–66.

von Wright, Georg Henrik (1951). *A Treatise on Induction and Probability.* London: Routledge and Kegan Paul.

Walker, David (2009). *A Kuhnian Defence of Inference to the Best Explanation.* Ph. D. thesis, University of Bristol.

Walker, David (2012). A Kuhnian defence of inference to the best explanation. *Studies in History and Philosophy of Science*, **43**, 64–73.

Walsh, Denis (1996). Fitness and function. *British Journal for the Philosophy of Science*, **47**, 553–74.

Warfield, Ted A. (2005). Knowledge from falsehood. *Philosophical Perspectives*, **19**, 405–16.

Weber, Marcel (2009). The crux of crucial experiments: Duhem's problems and inference to the best explanation. *British Journal for the Philosophy of Science*, **60**, 19–49.

Wedgwood, Ralph (1997). The essence of response-dependence. *European Review of Philosophy*, **3**, 31–54.

Wedgwood, Ralph (2002*a*). The aim of belief. *Philosophical Perspectives*, **16**, 267–97.

Wedgwood, Ralph (2002*b*). Internalism explained. *Philosophy and Phenomenological Research*, **65**, 349–69.

Weisberg, Jonathan (2009). Locating IBE in the Bayesian framework. *Synthese*, **167**, 125–44.

Whitcomb, Dennis (2010). Curiosity was framed. *Philosophy and Phenomenological Research*, **81**, 664–87.

Whiting, Daniel (2013). Nothing but the truth. In *The Aim of Belief* (ed. T. Chan), pp. 184–203. Oxford: Oxford University Press.

Wilkins, Emma (2014). Margaret Cavendish and the Royal Society. *Notes and Records of the Royal Society of London*, **68**, 245–60.

Williams, Bernard (1973). Deciding to believe. In *Problems of the Self*, pp. 136–51. Cambridge: Cambridge University Press.

Williamson, Jon (2010). *In Defence of Objective Bayesianism*. Oxford: Oxford University Press.

Williamson, Jon (2015). Deliberation, judgement and the nature of evidence. *Economics and Philosophy*, **31**, 27–65.

Williamson, Timothy (1996*a*). Cognitive homelessness. *Journal of Philosophy*, **93**, 554–73.

Williamson, Timothy (1996*b*). Knowing and asserting. *Philosophical Review*, **105**, 489–523.

Williamson, Timothy (1997). Knowledge as evidence. *Mind*, **106**, 1–25.

Williamson, TImothy (1998). Conditionalizing on knowledge. *British Journal for the Philosophy of Science*, **49**, 89–121.

Williamson, Timothy (2000). *Knowledge and its Limits*. Oxford: Oxford University Press.

Wolff, Johanna (2019). Naturalistic quietism or scientific realism. *Synthese*, **196**, 485–98.

Worrall, John (1989). Structural rrealism: The best of both worlds? *dialectica*, **43**, 99–124.

Wray, K. Brad (2001). Collective belief and acceptance. *Synthese*, **129**, 319–33.

Wray, K. Brad (2007). A selectionist explanation for the success and failures of science. *Erkenntnis*, **67**, 81–9.

Wray, K. Brad (2018). *Resisting Scientific Realism*. Cambridge: Cambridge University Press.

Wright, John P. (1983). *The Sceptical Realism of David Hume*. Manchester: Manchester University Press.

Wright, Larry (1973). Functions. *Philosophical Review*, **82**, 139–68.

Zagzebski, Linda (1994). The inescapability of gettier problems. *The Philosophical Quarterly*, **44**, 65–73.

Zilsel, Edgar (1945). The genesis of the concept of scientific progress. *Journal of the History of Ideas*, **6**, 325–49.

INDEX